Tourism Social Science Series
Volume 27

Cognitive Psychology and Tourism

Tourism Social Science Series

Series Editor: **Jafar Jafari**

University of Wisconsin-Stout, USA
Email: jafari@uwstout.edu

Volume Authors: **Noel Scott, Brent Moyle, Ana Cláudia Campos, Liubov Skavronskaya and Biqiang Liu**, with contributions from other authors

The books in *Tourism Social Science Series* (TSSSeries) are intended to systematically and cumulatively contribute to the formation, embodiment and advancement of knowledge in the field of tourism.

The TSSSeries' multidisciplinary framework and treatment of tourism includes application of theoretical, methodological and substantive contributions from such fields as anthropology, business administration, ecology, economics, geography, history, hospitality, leisure, planning, political science, psychology, recreation, religion, sociology, transportation, etc., but it significantly favours state-of-the-art presentations, works featuring new directions and especially the cross-fertilisation of perspectives beyond each of these singular fields. While the development and production of this book series is fashioned after the successful model of *Annals of Tourism Research*, the TSSSeries further aspires to assure each theme a comprehensiveness possible only in book-length academic treatment. Each volume in the series is intended to deal with a particular aspect of this increasingly important subject, thus to play a definitive role in the enlarging and strengthening of the foundation of knowledge in the field of tourism, and consequently to expand the frontiers of knowledge into the new research and scholarship horizons ahead.

Published TSSSeries Titles

Volume 26:	Fashion and Tourism: Parallel Stories, *Maria Gravari-Barbas and Nadzeya Sabatini*
Volume 25:	Space Tourism: The Elusive Dream, *Erik Cohen and Sam Spector*
Volume 24:	Authenticity & Tourism: Materialities, Perceptions, Experiences, *Jillian M. Rickly and Elizabeth S. Vidon*
Volume 23:	Social Conflict and Harmony: Tourism in China's Multi-ethnic Communities, *Liping Cai*
Volume 22:	Tourism Research Paradigms: Critical and Emergent Knowledges, *Ana María Munar and Tazim Jamal*
Volume 21:	Tourism Education: Global Issues and Trends, *Pauline J. Sheldon and Cathy H. C. Hsu*
Volume 20:	Tourism Research Frontiers: Beyond the Boundaries of Knowledge, *Donna Chambers and Tijana Rakić*
Volume 19:	Geographies of Tourism: European Research Perspectives, *Julie Wilson and Salvador Anton Clavé*
Volume 18:	Tourism Social Media: Transformations in Identity, Community and Culture, *Ana María Munar, Szilvia Gyimóthy and Liping Cai*
Volume 17:	Culture and Society in Tourism Contexts, *A-M. Nogués-Pedregal*
Volume 16:	The Discovery of Tourism Economics, *Larry Dwyer*
Volume 15:	The Study of Tourism: Foundations from Psychology, *Philip Pearce*
Volume 14:	Modern Mass Tourism, *Julio R. Aramberri*
Volume 13:	The Discovery of Tourism, *Stephen L. J. Smith*
Volume 12:	The Sociology of Tourism: European Origins and Developments, *Graham M. S. Dann and Giuli Liebman Par*

Tourism Social Science Series
Volume 27

COGNITIVE PSYCHOLOGY AND TOURISM

NOEL SCOTT
Edith Cowan University, Australia

BRENT MOYLE
Griffith University, Australia

ANA CLÁUDIA CAMPOS
University of Algarve, Portugal

LIUBOV SKAVRONSKAYA
Griffith University, Australia

AND

BIQIANG LIU
Griffith University, Australia

– With contributions from other authors

emerald
PUBLISHING

United Kingdom – North America – Japan – India – Malaysia – China

Emerald Publishing Limited
Emerald Publishing, Floor 5, Northspring, 21-23 Wellington Street, Leeds LS1 4DL

First edition 2024

British Library Cataloguing in Publication Data
A catalogue record for this book is available from the British Library

ISBN: 978-1-80262-580-6 (Print)
ISBN: 978-1-80262-579-0 (Online)
ISBN: 978-1-80262-581-3 (Epub)

ISSN: 1571-5043 (Series)

Contents

List of Figures and Tables

About the Contributors

Marcel Bastiaansen, PhD, is a Professor of Leisure and Tourism Experience at Tilburg University, Netherlands. He obtained an MSc in Theoretical and Experimental Psychology (1996) and a PhD in Cognitive Neuroscience (2000). In 2013, Marcel joined Breda University of Applied Sciences (BUas), refocused his research agenda on Leisure and Tourism, and co-founded and currently directs BUas' Experience Lab. In 2021, Marcel was appointed full professor in Leisure and Tourism Experience at BUas and at Tilburg University. He has published more than 100 papers in international peer-reviewed journals, obtained more than USD 1 million in research funding and delivered 30 invited lectures in 15 different countries.

Ana Cláudia Campos, PhD, is currently a Associate Lecturer at University of Algarve, Portugal and full member of the CinTurs, Research Centre for Tourism, Sustainability and Well-being at the University of Algarve. Her interests in tourism research are currently focused on topics such as service marketing, tourism and destination marketing, tourism experience, cognitive psychology and tourism and creative tourism and tourism co-creation.

Lynn I-Ling Chen, PhD, is an Assistant Professor of the Department of Leisure Management at National Pingtung University in Taiwan. She is also a Visiting Fellow at the Centre for Tourism Research, Wakayama University in Japan. She has been leading government projects on mindfulness in tourism for indigenous communities, lifestyle retreats and mountain resort business, after the completion of her PhD at the University of Queensland in Australia in 2015. She researches mindful encounters with people and environment from an Eastern philosophical perspective as she pursues knowing and being. She has dedicated 17 years to the public sector in Taiwan. Creative tourism design empowers her to be more reflexive.

Zhiming Deng, PhD, received her PhD degree in tourism management in The University of Queensland in Australia. Her research focuses on the technological affordances of travel live streaming in facilitating parasocial interaction and relationship. In 2022, she obtained The Hong Kong Polytechnic University Distinguished Postdoctoral Fellowship Scheme for her proposal on designing interventions in travel live streaming for promoting pro-environmental decisions for hotel products. In the same year, she supported a project as a Co-I on developing hotel environmental sustainability index. Her research interests include information technology on tourist behaviours, travel experience design, emotion coping strategies, eye-tracking and pro-environmental behaviours.

Lihua Gao (Grace), PhD, received her PhD degree in tourism from the University of Queensland (Australia) in 2013. She works now as an Associate Professor in the Department of Tourism Management, School of Social Science, Soochow University (China). She is also a researcher at the Academy of Culture and Tourism Research. Her research interests focus on tourism experiences, consumer behaviour and tourism destination management.

Hana Hadinejad (Arghavan), PhD, is a Lecturer at the University of South Australia. She has both industry and academic experience in tourism and destination marketing. Her research focus is on emotion, attitude, physiological technologies, cognitive psychology, social psychology and destination marketing. Hana's research has been published in quality journals such as *Tourism Management* (A*), *Journal of Travel Research* (A*) and *Journal of Sustainable Tourism* (A*). Hana won the Emerald Literati Award for the Highly Commended Paper in 2019 and 2020.

Yawei Jiang, PhD, is a Lecturer in the Department of Tourism, Sport and Hotel Management, Griffith University. She received her PhD degree in tourism crisis management from the University of Queensland in 2021. She is an experienced qualitative researcher with a demonstrated history of working in the higher education industry. She is also experienced in using psycho-physiological measures in researching human behaviours (e.g. eye-tracking, skin conductance, facial expressions). Her research interests include tourism crisis and disaster management, organisational/employee resilience, dynamic capabilities, stakeholder collaboration and strategic management. She has been involved with a variety of projects working with local and regional destinations.

Shan Jiang (Jaky), PhD, is an Assistant Professor at Capital Normal University and holds a PhD in Tourism from the University of Queensland, Australia. Her research interests include tourist psychology, AR experience, destination marketing and means-end chain theory. She has published papers in various international academic journals, such as *Current Issues in Tourism* and *Asia Pacific Journal of Tourism Research*. She wrote a book on travel motivation and cultural value, as well as numbers of book chapters in tourism design and destination marketing. She currently is an editorial board member of the *Journal of Destination Marketing and Management*.

Lee Kannis-Dymand, PhD, is a Senior Lecturer in Psychology, University of the Sunshine Coast, Australia. Lee obtained a PhD (Psychology) and Postgraduate Diploma in Clinical Psychology from the University of Canterbury, New Zealand, and a Postgraduate Diploma in Cognitive Behaviour Therapy from the University of Oxford, England. His research interests are in awe, tourism psychology, visitor and user experience psychology, nature and well-being, gaming, psychedelics and environmental psychology.

Anna Kralj, PhD, is a Senior Lecturer at the Department of Tourism Sport and Hotel Management in the Griffith Business School at Griffith University. Anna's research interest is the tourism and hospitality workforce, with a particular interest in service management, organisational behaviour and human resources management. Before joining academia, Anna had an extensive career in the hotel industry at a management level. Anna's professional experience is primarily in food and beverage operations and human resource management. Anna has widely published in leading hospitality and tourism academic journals and has completed various industry-sponsored research projects.

Dung Le (Jenny), PhD, is an Assistant Professor in the College of Business and Management, VinUniversity. She has more than 10 years of teaching experiences at various universities in Vietnam, Australia and the United States. Dr Le completed her doctoral programme at Griffith University, Australia, in 2020 and has published many papers in the top-ranking tourism and hospitality journals, including *International Journal of Hospitality Management*, *Journal of Sustainable Tourism*, *Current Issues in Tourism* and *Tourism Review*. Her teaching and research interests include eye-tracking technology, digital marketing, emotions, experience design, crisis management and sustainability. Also, she participated in several consulting projects for industry partners (e.g. Mantra hotel group, Intrepid Travel group, Vietnam Electricity group) and provided practical implications based on customer research.

Truc H. Le, PhD, is a Lecturer in the Department of Tourism, Sport and Hotel Management at Griffith University, Australia. She graduated with a PhD at Griffith University and has worked as a research consultant at EarthCheck Australia specialising in risk and resilience and sustainable tourism. Her research focus includes authenticity, risks and resilience, risk perceptions, cruise tourism, food tourism, user-generated content analytics and machine learning. She has published in various top-tier international tourism and hospitality journals.

Yaoqi Li is a Professor and PhD Supervisor at the School of Tourism Management, Sun Yat-sen University. He obtained his PhD in School of Business, Sun Yat-sen University. His research interests include consumer behaviour, service marketing and tourism management. He also serves as the Head of Department of Hotel Management at Sun Yat-sen University. He has published more than 50 journal articles including Tourism Management and Annals of Tourism Research and been funded with more than nine grants or projects.

Shanshi Li, PhD, is an Associate Professor of Tourism in the School of Management, Xiamen University, China. He obtained his PhD from The University of Queensland, Australia in 2017. His research interests are in destination marketing, tourism crisis management and research methods. Dr Li's academic research work has been published in leading academic journals including *Journal of Travel Research*, *International Journal of Hospitality Management*, *International Journal of Contemporary Hospitality Management*, *Current Issues in Tourism* and *Journal of Hospitality and Tourism Research*.

Biqiang Liu is a PhD candidate at the Department of Tourism, Sport and Hotel Management, Griffith University, Australia. He graduated from the School of Tourism Management, Sun Yat-sen University, China, as a Master in 2020. For his master's thesis, he dealt with the effectiveness of celebrity endorsement in destination marketing by adopting eye-tracking techniques. He has published over 15 research articles in top-tier journals, including *Tourism Management and Annals of Tourism Research.* His current research explores the intersection between tourist experiences and tourism marketing, with an interest in virtual reality and demonstrated expertise in eye tracking.

Siqi Lu (Emily) is a PhD candidate in the Department of Tourism, Sport and Hotel Management, Griffith University. She holds a master's degree in Tourism, Hotel, and Event Management at the University of Queensland, Australia. Her research interest areas are accessible and inclusive tourism, employees with disabilities in the hospitality industry, technological innovations in museum visitor experiences.

Jianyu Ma, PhD, is an Associate Professor of Tourism Management in the Institute of Tourism, Shanghai Normal University, PR China. She received her PhD from the School of Tourism, University of Queensland. Her research interests include tourism experience, visitor experience design, destination marketing and management. She is working on experimental studies collecting psychophysiological data to better understand visitors' simultaneous responses of designed stimuli in tourism contexts.

Ondrej Mitas, PhD, born 1982 (Bratislava) and citizen of the United States and European Union. He researches tourists' emotions and quality of life from a positive psychological perspective. His ultimate goal is to guide the tourism industry, governments and tourists themselves to make choices that will maximise their enjoyment of life.

Brent Moyle, PhD, is a Professor of Tourism in the Department of Tourism, Sport and Hotel Management at Griffith University, Australia and a Visiting Professor at Hong Kong Polytechnic University. His research interests include sustainable tourism, tourism experiences and regional development. He takes an interdisciplinary approach to tourism research, partnering with researchers and communities to maximise outcomes. Brent takes pride in conducting research at the interface between theory and practice, engaging extensively with industry partners to complete research with impact. His passion for sustainable regional development laid the foundation for a number of long-term collaborations with local government and parks agencies. He has co-authored over 100 academic publications and supervised 7 doctoral students to completion.

Hans Revers, BEng, MSc, is a PhD candidate at Tilburg University (Netherlands), Department of Cognitive Neuropsychology. He studies the neural basis of emotions and uses virtual reality to elicit highly emotional and immersive experiences. Hans has developed in-depth expertise in electro-encephalography

(EEG) and machine learning–based classification of EEG patterns. He has published several papers and has presented his work at various scientific conferences.

Vikki Schaffer, PhD, with a keen interest in cognitive appraisal and immersive tourism experience design, Dr Vikki Schaffer's research portfolio covers diverse topics including user experience design (virtual and real), human–wildlife interactions, emotion elicitation and influences with a focus on awe, marine tourism, citizen science, inclusivity and diversity and conservation and environmental psychology, communication and stakeholder engagement.

Noel Scott, PhD, is an Adjunct Professor of Tourism Management in Edith Cowan University, Mataram University, Taylor's University and the University of Sunshine Coast, Australia. His research interests include tourism experiences, destination management and marketing. He is a frequent speaker at international academic and industry conferences. He has over 300 academic articles published including 19 books. He has supervised 23 doctoral students to the successful completion of their theses. He is on the Editorial Board of 10 journals, a member of the International Association of China Tourism Scholars and a Fellow of the Council for Australasian Tourism and Hospitality Education. Prior to starting his academic career in 2001, Dr Scott worked as a Senior Manager in a variety of businesses including Tourism and Events Queensland.

Liubov Skavronskaya, PhD, is a Lecturer at Griffith College, Australia. She enjoys working with both domestic and international students and priorities the creation of a supportive learning environment while focusing on different aspects of students' well-being. In addition to her teaching duties, Liubov is also working as a Research Assistant at the Creative Arts Research Institute at Griffith University. She enjoys being a supervisor/mentor and contributing to the success of other students, which is her current priority. Liubov conducts research in English, Mandarin and Russian languages, with a focus on visitor experience, immersive technologies, tourist emotions and China tourism.

Wim Strijbosch, PhD, is a Researcher and Lecturer at Breda University of Applied Sciences and is a senior lab member of the university's Experience Lab. Wim has obtained both his BBA in Leisure Management (2014) and his MSc in Leisure Studies (2016) at Breda University of Applied Sciences. He obtained his PhD with distinction at Tilburg University. In his dissertation, Wim has studied the temporal dynamics of emotions in tourism and leisure experiences through the use of neuro-imaging and psychophysiological methods. Wim has particular interest in the field of themed entertainment and serves as a board member for the Europe-Middle East board of the Themed Entertainment Association.

Billy Sung, PhD, is an Associate Professor at the School of Management and Marketing, Curtin University, Australia. He specialises in digital marketing and consumer psychology. His research to date has been based on the study of emotion and the application of psychophysiological methodology in multiple disciplines including psychology, marketing, health, nursing and robotics. He also leads the Consumer Research Lab at Curtin University, which specialises in the use of biometric techniques such as eye tracking, facial expression and brainwave analyses to conduct consumer research.

Xinyan Wei is a graduate student in School of Management, Xiamen University, China. She obtained her bachelor's degree in Management from Shandong University, China in 2021, and her research interests are destination marketing. Ms Wei is studying with Dr Shanshi Li, who completed his doctorate at The University of Queensland. Ms Wei hopes to continue her doctoral studies in the area of cognitive psychology applied to tourism.

Juan Yang is a PhD candidate at the Department of Tourism, Sport and Hotel Management, Griffith University, Australia. Juan Yang earned her Master of International Tourism Hospitality Management and Graduate Diploma of Research Studies in Business, both from Griffith University in 2019 and 2020. Her publications are predominantly in risk perception, study destination choice and co-creation. She earned the award of the Hospitality and Society Best Paper at CAUTHE 2021 with her team. She is currently combining her interests to explore the community-based co-design impacts on meaningful tourism experiences in the early development stage destination.

Rui Zhang is the Director of Leisure and Tourism Programme (Australia-China), at the International School of Hainan Tropical Ocean University. She obtained her PhD from The Griffith University, Australia in 2019. Her research interests are in destination management, consumer behaviour in tourism and cognitive psychology in tourism. Dr Zhang's academic research work has been published in Current Issues in Tourism, Tourism Review, and Tourism Tribune. Dr Zhang completed her thesis on the structuring of beach holiday choice using schema and the difficulty of changing such schema.

Acknowledgments

This book is partly financed by National Funds provided by FCT, Foundation for Science and Technology through project UIDB/04020/2020.

Acknowledgements

SECTION 1
INTRODUCTION TO THE COGNITIVE APPROACH

Chapter 1

INTRODUCTION

Noel Scott
Edith Cowan University, Australia

Brent Moyle
Griffith University, Australia

Ana Cláudia Campos
University of Algarve, Portugal

Liubov Skavronskaya
Griffith College, Australia

Biqiang Liu
Griffith University, Australia

The purpose of this volume is to provide a foundation for the application of cognitive science theories and concepts to fields such as tourism, hospitality, events, leisure and services. The contents have been compiled from a 10-year programme of doctoral research, as well as chapters contributed by experts in the field. The primary intended audience of the volume are tourism researchers, doctoral candidates and tourism professionals looking for a fresh approach to understanding their customers. However, the cognitive paradigm and its applications discussed are equally applicable to other applied fields that utilise psychological theories and concepts.

For tourism researchers and doctoral candidates, the volume aims to contribute to a paradigm shift in the way the field is studied and approached. The authors consider that tourism will benefit by the application of new ideas

Cognitive Psychology and Tourism
Tourism Social Science Series, Volume 27, 3–5

Published under exclusive licence by Emerald Publishing Limited
ISSN: 1571-5043/doi:10.1108/S1571-504320240000027001

found in the mainstream psychology discipline. The importation of these new ideas will enable a better understanding of mental processes, but requires adoption of a new paradigm, termed here the 'Cognitive Wave'. Tourism professionals who apply theory will benefit by identifying how they apply concepts such as attention, emotion, sensations and memory in their work, and critically understand how to measure them. All such professionals do use and measure these concepts, often without being aware of the assumptions they make.

The tourism academy may wish to avoid the coming 'cognitive wave', but failing to innovate and anticipate the future leads to irrelevance. It is the responsibility of an academic to embrace new theories and methods, and at the same time discard ideas that have been disproven. Past giants, on whose shoulders present researchers stand, were insightful and intelligent. They examined those aspects of phenomena accessible to them in their time. However, while it is important to respect them for their ideas, it is vital to evaluate their work against more recent empirical findings in foundation disciplines. Doing otherwise risks myopia and a disconnect between academic tourism research and that found useful in practice.

A word of caution for the reader. The theories and methods used in cognitive psychology have their own associated debates and traditions. Adoption without understanding this background is fraught with problems. Further, the authors acknowledge that they are not classically trained experts in cognitive psychology, despite their decade of work in the area. Thus, they are open to criticism of the detail of their knowledge and in the light of recent theoretical advances and welcome constructive debate in this regard. However, there seemed to be no alternative to bridging the gap between current tourism theory and that of cognitive sciences. Criticism is expected, but wholesale and dogmatic rejection of any possible useful applications or insight is not.

Finally, this volume points to an emergent 'invisible college' of tourism researchers who are adopting the cognitive wave in their work. This volume is designed to stimulate further development of this college and lead to the development of organizational infrastructure, such as regular workshops, conferences and perhaps a journal.

The chapters of this book are arranged into five sections. After this introduction, Chapter 1 introduces the cognitive wave paradigm adopted in this book, discusses the arguments for and against use of a cognitive paradigm and translating concepts into it. Chapter 2 introduces the mental processes central to cognition. Chapters 4 and 5 provide examples of disambiguation, translating tourism concepts and theories into those of cognitive psychology.

In section two, each of the mental processes central to cognition are discussed in more detail. These are sensations and perception (Chapter 6), consciousness (Chapter 7), emotions (Chapter 8), motivation and goals (Chapter 9), novelty (Chapter 10), memory (Chapter 11), memory schema (Chapter 12), prospection and imagery processing (Chapter 13) and coping (Chapter 14). Each chapter discusses the cognitive psychology literature of a process as well as relevant tourism research.

Section three provide applications of these processes to topics of interest to tourism researchers involved in the study of experiences. Chapter 15 discusses experience design and Chapter 16 looks at the co-creation of experiences. The next three chapters examine topics related to experience design, firstly the emotion of awe (Chapter 17), mindfulness, or the experience of the present moment (Chapter 18) and storytelling (Chapter 19).

Section four discusses three methodological techniques used in cognitive psychology, eye-tracking (Chapter 20), electroencephalography (Chapter 21) and skin conductance (Chapter 22). The final section provides a cautionary chapter on the use of the cognitive wave paradigm (Chapter 23), and a final conclusions chapter (Chapter 24).

In resume, this volume aims to advance tourism research and that of cognate fields through the adoption of a new paradigm, with embedded theories, concepts and methods. Each of the chapters highlight relevant existing research and opportunities for further developments. Examples of the application of theory and methods to topics of interest in the tourism, hospitality, events, leisure and services fields are provided. There are many opportunities for developing these and other topics further and in developing the work of the invisible college which underpins the silent shift to the cognitive wave further.

Chapter 2

COGNITIVE SCIENCE AND TOURISM

Noel Scott
Edith Cowan University, Australia

Brent Moyle
Griffith University, Australia

Ana Cláudia Campos
University of Algarve, Portugal

Liubov Skavronskaya
Griffith College, Australia

Biqiang Liu
Griffith University, Australia

Abstract: This chapter introduces the cognitive paradigm. The concept of translation and associated problems is discussed. Several arguments for the adoption of a cognitive-based social sciences are provided. A brief history of the term cognition is provided and the origin of theory in tourism discussed. Justifications for the use of cognitive psychology in the field of tourism are provided. These include the (mostly) hedonic and hence emotional nature of tourism experiences, the novelty of travel outside one's usual environment, the importance of imagining, day-dreams, emotions and desire for travel decisions and the importance of

Cognitive Psychology and Tourism
Tourism Social Science Series, Volume 27, 7–20

Published under exclusive licence by Emerald Publishing Limited
ISSN: 1571-5043/doi:10.1108/S1571-504320240000027002

autobiographical memories because of travel. **Keywords**: Cognition; behaviour; cognitive science; disambiguation; emotions; translation

INTRODUCTION

There is a 'cognitive wave' approaching academic tourism, business and social sciences research (Scott, 2020). This is a wave of new knowledge about how the human brain functions and how its processes take sensations, create knowledge and initiate behaviour. This wave of knowledge originated in cognitive science and neuroscience laboratories in the 1960s and onwards and has begun to break upon the shores of other fields of study. This book provides an introduction to this cognitive literature and its application to tourism.

The need for study of tourism and cognition would on the face of it appear useful. Tourism research is a multidisciplinary endeavour (Tribe, 1997), and many social science concepts have a relationship to cognition. There has been an upsurge in interest in tourism experiences, creative and memorable experiences, emotion, embodiment, transformation, sensations and experience design. Tourism scholars have traditionally been interested in fields such as decision-making, destination image, resident attitudes, cultural differences, interpretation, liminality, savouring, risk, authenticity and so on. To the extent that these phenomena are related to brain functioning, then cognitive sciences would be useful in advancing knowledge. However, there is a 'toothbrush problem', other peoples' theories are treated like toothbrushes – no self-respecting person wants to use anyone else's. It is amusing, but it also points to a conflict that one may be nurturing within the profession to the detriment of science (Mischel, 2008).

Despite this apparent usefulness, there are numerous problems and barriers to the application of cognitive science findings and theories in the study of tourism. The first is the translation problem (Francken & Slors, 2018), which examines the difficulties of relating everyday 'common sense' concepts like memory into more fine-grained concepts such as semantic, episodic, autobiographical, long-term and working memory (Circle 1 in Figure 2.1). There are subsequent problems in identifying appropriate procedures and operationalising these in experiments (Circles 2 and 3). Further, many 'common sense' concepts have multiple meanings in the social sciences derived from different streams of research (Circle 4). For example, many concepts commonly used in tourism are derived from behavioural psychology and operationalised in terms of behaviour (Skavronskaya et al., 2017). This book will address these translation problems in Chapter 3 on disambiguation, and by discussing in each chapter on prominent brain processes, their mechanisms and areas of application to existing fields of study.

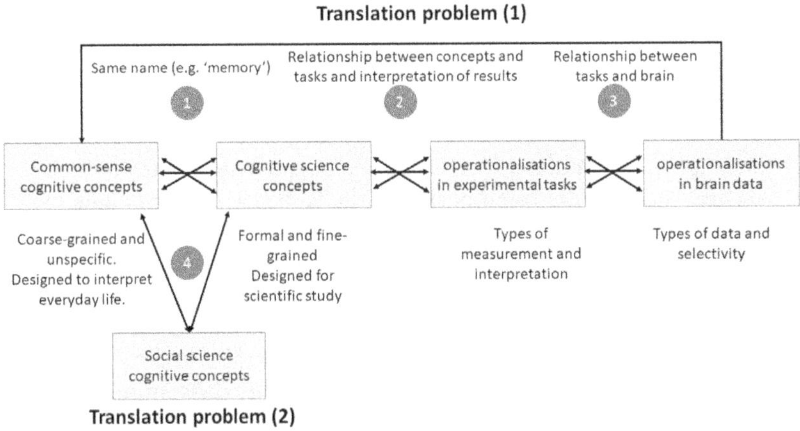

Figure 2.1. The Translation Problem.

A second barrier is due to the applied nature of most tourism research. In many, if not most, tourism journal articles published, the purpose of such papers is to understand real-world problems in a manner that can 'make sense' to a non-scientific public audience. Many academic researchers therefore base their data collection on common sense or social science cognitive concepts without the consideration of 'new ideas' from cognitive science. There is some use of cognitive psychological measurement techniques, such as eye tracking and electroencephalography (Li et al., 2022). Use of such methods may be expensive and time consuming and beyond the reach of many tourism researchers. An alternative is to conceptualise studies and adopt nomological nets that are based on a more detailed study of relevant mechanisms. For example, the study of emotion could use concepts from cognitive appraisal theory (Ma et al., 2013). There is related issue of research priorities. In the climate-challenged world, it might be reasonable to expect that practical solutions do not require new techniques and research should be focused on real-world problems. This can be a result of an anti-science philosophy (it will lead to a big-brother state), or the immediacy of focus on important problems. This book will allay the first prejudice and demonstrate how cognitive science can help to address the psychology of climate change.

A third criticism of use of cognitive science concepts is their 'potential diminution of personhood, agency and personal and political attention conferred by neuroscience–society relations' (Pykett, 2018, p. 163). This concern is due to translational issues as well as neuroscience hyping in the everyday press (Ariely & Berns, 2010). This book will try to avoid hype by discussing applications of 'mainstream' concepts and theories. Several

Table 2.1. Constraints Argument for a Cognitive Social Science.

(1) Since all social processes involve cognitive aspects, social scientists must make assumptions about human cognition in their research practices.
(2) Social scientists' assumptions about the cognitive processes of their research subjects are often based on the subjects' own accounts of these processes and/or the ideas and concepts of 'folk psychology' that people use in their everyday life.
(3) Cognitive scientific studies have convincingly demonstrated that the cognitive processes are not transparent and that one's own understanding of these processes, including social scientists' and their research subjects 'folk psychological theories', is limited and sometimes misleading.
(4) Conclusion: Social scientists' assumptions about cognitive processes of their research subjects should be constrained by the results of cognitive sciences.

philosophical arguments have been used to support development of a cognitive social science such as the constraints ('common sense' psychology constrains understanding Table 2.1; the 'explanatory grounding' argument – social processes are based on mental processes – Table 2.2, Kaidesoja et al., 2019, p. 489).

Certainly, the concepts, theories and methods of cognitive science are being applied in many fields such as education (Thomas et al., 2019) and transformational learning (Taylor, 2001), anthropology – rituals (Hobson et al., 2018); geography (Pykett, 2018); persuasion (Cacioppo et al., 2018);

Table 2.2. Explanatory Grounding Argument for a Cognitive Social Science.

(1) Most social scientists do not currently make use of the knowledge produced in the cognitive sciences when they explain social phenomena.
(2) Cognitive processes are the ontological basis of social processes.
(3) Explanations in the cognitive sciences are deeper than explanations in the social sciences because they bottom out in cognitive processes.
(4) If social scientists ground their explanations in the cognitive sciences, their explanations for social phenomena would become deeper than they are at present.
(5) Conclusion: The social sciences should be grounded in the cognitive sciences.

aesthetic experiences (Kirsch et al., 2016; Pearce et al., 2016); decision-making (Sebastian, 2014); neurourbanism, place and wayfinding (Mondschein & Moga, 2019); theatre (Kemp & McConachie, 2018) and risk (Loewenstein et al., 2001).

For some this wave will be problematic as it challenges the worldview and paradigms common in many fields of research. It disproves the assumptions of conscious rationality that support several research studies and replaces it with the unconscious rationality of emotion. It clarifies and disambiguates many concepts that are related to cognitive processes but have been defined without any reference to them. Ultimately, the reason that it will be adopted is that this knowledge can extend the knowledge of how the human world works.

What Does Cognition Mean?

The term 'cognitive wave' refers to knowledge about cognition; however, the term cognition, like many used in academic discourse, has multiple meanings and associated applications. The concept of cognition is applied in a variety of fields, but in this book, the authors apply definitions from cognitive science (mainly cognitive psychology and neuroscience), the field that studies cognition. 'The word "cognitive" is reported to derive from the Latin word *cognare*, which means "to know." Hence, cognitive psychology is the study of .... knowing or thought'. Cognitive science theories generally assume that 'stimulus encoding operations lead to (and interact with) more elaborate processes, which eventually lead to some response execution' (Balota, 2000, p. 13; Chaney, 2013). The major processes studied include sensation, perception, attention, emotion, memory, consciousness and learning (Braisby & Gellatly, 2012). Each of these concepts is discussed in tourism, business and the social science, but often not based on the experimentally verified findings and theories from cognitive science.

Assumptions of Cognitive Science

Firstly, a key assumption of cognitive science is that the meaning of an object is determined by the thought processes. These thought processes are influenced by the past experiences encoded as memory, emotions, attention, current goals and other internal mental processing characteristics. This is the basis of individuality. The implication of this assumption is that one cannot simply assume that every person has the same response to a theme park, vacation and the like.

On the other hand, the thought processes may seem to be influenced by external factors for several reasons. For example, it may be that many people share common encoded memories (culture). These encoded cultural memories are common across a group *but are a property of an individual*. Alternatively, it may be that people have dissimilar cognition but similar behaviour because there are only so many ways that a person can respond (outcomes limited – only so many activities are available). Finally, it may be that the group is self-selected. Self-selection bias is important in voluntary activities such as leisure tourism.

Assumption 2: Interacting Mental Control Processes

Cognitive models generally are based on the interplay of a reflective system, which is based on conscious deliberation, controlled but effortful, and a quicker automatic system that responds in line with habits and emotions (Kahneman, 2011). Such systems provide a neurological basis for cognitive psychological models such as cognitive appraisal theory (Frijda, 1987; Lazarus, 1984; Roseman, 1984; Scherer, 1988).

Assumptions of Cognitive Science 3 Limitations

The brain capacities and processes place limitations on the ability to understand the world in which one lives. Each of the sense organs has a restricted ability, eyes cannot see in the infrared, hearing has restrictions. Despite this, the amount of sensory stimuli available means that there is a need to limit the mental processes to the most salient at any one point in time. The most salient sensory stimuli may be identified by the use of experiments using psychophysiological measurements techniques such as eye tracking (Li et al., 2020).

Further, the amount of information one can consciously process in working memory is limited and so many 'decisions' are made unconsciously by 'scripts' based on past successful outcomes. This may appear to be a return to Freudian unconscious decision-making, and while sharing some of the basic premises of this early cognitive psychology, it is based on experimentally verified measurement rather than introspectionism.

Additionally, emotions serve as an important information source for decisions and behaviour and thus cognition is not separate from emotion. If one understands from cognitive appraisal theory that emotions are constructed based on a person's goals, then it may better be understood that anger is an indication of a goal incongruent situation. In many situations, a person may not be aware of the reason for their anger. Recognition of that a situation involves an important goal can allow the use of relevant coping strategies (Jiang et al., 2020).

While, for some social scientists, the assumption that the human body has limitations may be well accepted. However, many recent tourism studies continue to ignore this assumption. Arguably, the concept of 'the gaze' is based on attention, sensation, perception and memory processes, but the cognitive aspects of 'gazing' are ignored (Urry & Larsen, 2011). Further, when tourism scholars do acknowledge such limitations, such as in using attention recreation theory, they may attribute the environment with causal powers to redress these limitations rather than the brain's interpretation of these effects (Lehto, 2013). An understanding of how and why the mechanisms behind attention recreation would explain how to improve 'attention restoration'.

Folk Psychology

One of the important implications of the study of cognitive science is that many of common beliefs (assumptions) about how one thinks and makes decisions are incorrect. Thinking about the future (prospection) or past (retrospection) uses the same mental processes as thinking about the present. Memories are mentally recreated when a past event is recalled and not a verbatim encoded record like a picture of what happened. Therefore, it is possible to have false memories. Another implication is that it challenges the enlightenment view that humans are rational and logical. There is a belief that rational thought has led to advances in technology and science, medicine and mathematics. But being emotional does not mean being irrational. Indeed, emotions are unconscious but rational and are the basis for human survival by encouraging (pleasure) or discouraging (fear) action. Many decisions are made based on emotional reactions and subsequent feelings but are rationalised as logical after the fact. Therefore, knowledge of cognitive science is essential to understand people and their actions.

Scope of Cognitive Science

There are some important things that the knowledge of cognitive processes cannot explain except in a very general way and the most important of this is the meaning of an external object. For example, the taste senses enable people to determine how 'hot' a curry is but not if one likes hot curry, that is, if eating a hot curry means enjoying it. The meaning of a curry is determined by the past experiences of curry and if they are pleasurable or not. Similarly, attention is directed to salient objects in the world.

Some indicators of salience are hard wired – colour, sounds and movement attract attention to an object (called bottom-up attention). Other salient indicators are determined by the goals at that time. Therefore, cognitive science can explain that attention is captured by salient objects but may not explain why its salient – its meaning. On the other hand, if one observes that a person's attention is captured by an object (using eye-tracking techniques, for example), then its salient is known. Similarly, emotions are determined by appraisals of a situation based on current goals. Cognitive processes do not operate at the level of meaning, but meaning is inferred from a person's actions by knowledge of cognitive science. A person's goals may not be known, but if strong emotions are observed, then one may infer that a strong goal is involved. Thus, the activity has goal-directed meaning for a person.

Not in Scope and Why Tourism

This book examines the concepts of cognitive science such as sensation, perception, attention, emotion, mental processing of information and memory. It includes only a brief discussion of the biochemistry of the brain (chemical monoamine neurotransmitters) and some aspects of neuronal circuitry by which thoughts are processed. This book does not include the discussion of artificial intelligence or computer science studies although these are important related topics.

There are several reasons why cognitive science is useful for the study of tourism. They are, one, tourism is not a technological or academic activity, but an everyday activity that does not have to be logical or involve sophisticated learnt knowledge (although it can). It is *associated with* relaxation and not doing these types of activities; two, it takes place outside the usual environment which is important as it implies novelty and surprise which are *associated with* emotional arousal as well as attention; three, tourism is mostly an activity that is voluntary and *associated with* important personal goals. Again, this implies that emotions will be evoked, and they will tend to be positive or pleasurable; four, experiences *associated with* stronger emotions are more likely to be recalled more vividly. This is one reason why tourism is found to lead to memorable experiences; five, stronger memories *associated with* important goals and are pleasurable when recalled are recalled more frequently and likely to be interesting to others; six, attention is orientated to salient cues in the surroundings (signs, pictures) which help one to understand how to act in that situation. A lack of salient clues means one may 'gaze' without directed attention.

WHERE TOURISM PSYCHOLOGY CONCEPTS COME FROM

If the ideas of cognitive science are so useful, why are not they used more in tourism research? There are several reasons beginning with cognitive science being a rapidly evolving area of research that has benefitted from technological developments which allow new types of data to be collected (such as physiological responses). This means that cognitive science has significantly advanced much after business and tourism became fields of study. The concepts and ideas used in tourism were (in the early days) adopted firstly in geography, economics and sociology (Butler, 2015) and then transferred to tourism research in a form of Chinese whispers (Skavronskaya et al., 2017). They then were 'locked in' to basic textbooks and seen as accepted wisdom and the basis of the new field. This may explain why all consumer researchers study consumer behaviour rather than consumer cognition.

Therefore, the psychological paradigm used in tourism and in business tend to be based on those in psychology at the time they were established as disciplines. The history of psychology in the last century can be described in simple terms as a succession of introspectionist, behaviourist then cognitive paradigms. Introspectionism was a type of cognitivism in that mental processes directed behaviour. However, the nature of the mental processes was determined by introspectively thinking about the cause of behaviour and was ultimately rejected as unscientific. This led to a rejection of the ability to understand the mind and a focus on observable behaviour. This paradigm involves study of the correlative relationships between stimuli and behaviour only. In the 1960s, cognitive science began using new scientific experimental techniques to understand the mental processes occurring between stimuli and behaviour.

However, and crucially, early geographers and business researchers adopted the 'best available' psychological theories of the time. For example, when Boulding (1956) wrote his book on 'image', the concept is arguably derived from earlier psychological ideas extant in the 1940s. Subsequently, the concept of image was transferred to tourism (Hunt, 1975) where it remains in common use today. A behaviourist approach correlates *tourist-scape* stimuli with outcomes such as emotional reactions or measures of potential future behaviour (intention to revisit). Such research focuses on correlations between antecedent stimuli and consequences of hedonic consumption, rather than subjective mental processes that link them (Alba & Williams, 2013, p. 5). A cognitive approach would instead discuss the appraisal process which explains why a particular emotion is elicited (Ma et al., 2013). Table 2.3 summarises some key differences in the behaviourist and cognitive paradigms.

Table 2.3. Behaviourist and Cognitive Paradigms of Mental Processes.

	Behaviourist	Cognitive
Brain is a	Black box	Information processor
Theory	Correlative	Predictive
Measurement	Subjective – that is, self-report	Objective – that is, skin conductance

The Paradigm Shift in Recent Tourism Research

A shift in tourism research paradigm from behaviourism to cognitivism with the emotions-central perspective is largely determined by changes of the approach to the decision-making process in marketing and consumer research. Starting from the early 1980s, marketing and consumer research theorists began to question the rational nature of the decision-making process, and suggested a more complex perspective, where both cognitive and emotional processes were included (Le et al., 2020; McCabe et al., 2016; Scott et al., 2009). A cognitive approach instead emphasises tourist's goals, prior experiences, attention, emotions, etc., and other mental factors that mediate between perception of an external stimuli and behaviour. Therefore, adopting a cognitive approach can improve on correlations found between stimuli and behavioural responses by considering an individual's mental characteristics. In turn, cognitive appraisal theory of emotions claims that emotions are elicited by subjective evaluation of the experience (Scherer et al., 2001).

Examples of the Development of the Cognitive Paradigm in Experiences

The essence of tourism is creating and delivering exciting and memorable travel and visitation experiences (Andersson, 2007; Hudson & Brent Ritchie, 2009; Scott et al., 2009; Uriely, 2005). Increasingly, destinations and businesses such as attraction operators and service providers are seeking to improve their experiences through better design and management (Ooi, 2005). By enhancing the tourist experience, providers are better able to please their target markets, increase loyalty intentions and improve word of mouth recommendations (Carbone, 1998). The popularity of this approach is shown in several recent examples. The aviation company, KLM, gave personally relevant gifts to its customers who were waiting for a flight connection in order to improve their transit experiences (KLM Royal Dutch Airlines, 2010). This trial was aimed at creating positive emotions such as surprise and happiness. In another example, a rural tourism association in Italy created a series of stories for tourists including one about the town during World War I, and another recreating village life during Mediaeval times. In addition, the Canadian Tourism Commission developed a toolkit to help tourism businesses deliver compelling experiences. These examples reflect the growing importance of the experience economy (Pine & Gilmore, 1999) and of understanding and designing tourist experiences.

Study of Experiences

Academics have used several disciplinary perspectives to understand the nature and characteristics of memorable experiences. Researchers from sociology (Cohen, 1979) and anthropology (Abrahams, 1986) regard a tourist experience as a phenomenon different from the routine experiences of daily life. Marketing and management researchers explore the components of memorable experiences (Schmitt, 1999a), the needs they satisfy (Kim et al., 2012), stages in their formation (Aho, 2001) and how they influence future tourism intentions (Ryan, 2000). A recent review has indicated that the conceptualisation of, and theorising about experiences, methodological development, and exploration of the design and delivery of tourism experiences require attention (Brent Ritchie et al., 2011).

Difficulties in conceptualising tourism experiences are in part due to the subjectiveness of the response to them. Experiences arise out of a person's diverse interpretations of an external stimulus based on their personal, social and cultural background (Ooi, 2005). Indeed, the essence of tourism experience is suggested as 'fundamentally subjective; . . .[and is] shaped by three things – what occurred, the meaning that the service provider applies to what occurred, and the interpretation the consumer gives to what occurred, both during and after the experience' (Brent Ritchie et al., 2011, p. 433). Thus, a better understanding of the antecedents of a memorable experience may help managers to determine the effectiveness of specific stimuli, which, in turn, could enable providers to design stories and meanings and achieve experiential outcomes. This thesis therefore examines how a tourist interprets their experiences and one facet of an experiential outcome: emotion is elicited and differed as a function of their experiences.

Assumption in Social Sciences

A second reason why the behaviourist paradigm remains dominant is more subtle. Tourism is a multidisciplinary field where researchers apply theories and concepts from other disciplines without necessarily having a strong knowledge of it. Many theories and ideas are available for importation. However, the concepts and ideas adopted in tourism do not displace past ideas. Instead, they accumulate, forming a complexity that future scholars must seek to reconcile despite their diverse paradigmatic backgrounds. In part, this is because tourism is seen as a field of social 'science' whereby each idea, concept or theory cannot be disproven, and indeed the principle of disproof may not apply. This complexity helps to explain the plethora of literature reviews published and the despair of doctoral students as they plough through the past literature. In tourism, these two trends indicate a sclerotic layering of ideas, a type of academic atherosclerosis whereby the interdisciplinarity of tourism research has become a weakness. This book seeks to address this problem by providing a basis for comparison and synthesis of prior theories as will be discussed in Chapter 3 on disambiguation. One example of this theory importation and elaboration process is found in the use of a model developed by Mehrabian and Russell (1974), the stimulus-organism-response Model. Mehrabian and Russell developed this model positing that certain environmental stimuli (such as heat or sound) influence an individual's emotional state, which in turn affects approach or avoidance responses (Mehrabian & Russell, 1974). This will be discussed in Chapter 4 and see Donovan and Rossiter (1982) for a discussion of this model.

Consciousness, Rationality and Emotion

The concept of consciousness is complex (Block, 1995), and the term is used in ways which conflate reflexive self-awareness or being aware of one's own mental activity (meaning 1), being aware of bodily sensations (meaning 2) and utilising 'higher' mental functions such as reasoning leading to 'rational' goal-seeking actions (meaning 3).

On any of these criteria, however, most mental activity is unconscious. Action may be undertake based on feelings, intuition or habit (Duhigg, 2013) and not be reflexively self-aware of these thought processes. However, differences between them can be measured experimentally. Indeed, 'feelings-as-information theory' (Avnet et al., 2012; Schwarz, 2011) considers those human sensations termed 'feelings' as the result of mental activity of which one is not conscious. People are also mostly unaware of normal bodily sensations, although they may become aware of those sensations called 'feelings' and abnormal sensations registered as pain (Damasio, 1995). Thus, consumer behaviour models that rely on conscious evaluation of attributes as the basis of decisionmaking are ignoring the mental antecedents of resultant behaviour. There is general agreement that decision-making is a function of brain's mental activity and that one is reflexively aware of only a proportion of brain activity.

Many decision-making models also implicitly assume that such unconscious mental activity is not rational. The concept of rationality is also complex, has a long history and is associated with logic, reasoning rather than intuition, and an absence of emotion. Herbert Simon instead defined rationality in a social science sense as actions or thoughts that:

> …contribute to certain goals, where these goals may be the pleasure or satisfaction of an individual or the guarantee of food or shelter for the members of a society. […] When awareness and intention are present, the function is usually called manifest, otherwise it is a latent function. (Simon, 1978, p. 3)

Using this definition, 'rational' goal-directed decision-making models can be characterised as exhibiting manifest rationality. However, the definition also attributes latent rationality to actions undertaken to achieve a goal even if a consumer is not consciously aware of that goal at that time. This is important since a large body of research has demonstrated unconscious top-down goal-directed actions in studies of attention (Campos et al., 2016; Pieters & Wedel, 2007). Further, as will be described below, cognitive appraisal theory indicates that unconscious goals in part determine the valence and strength of emotion elicited in response to a perceived stimulus. On this basis, emotional decision-making models are rational although the consumer may not be aware of their goal.

TOURISM AS AN APPLICATION AREA

This chapter so far has highlighted that cognitive science is an important area of knowledge of potential use in the study of tourism, albeit the researcher needs to understand its scope and assumptions. This section will briefly discuss some applications where cognitive science has begun to be used.

First, tourism outbound from a country begins usually as a group activity and then progresses to more individual travel. From a psychological perspective, this means that the individual tourist's goals and prior experience become determinants of decisions on where to travel. Cultural differences therefore need to be recognised for a better understanding of how people make decisions and evaluate their experiences of travel and tourism.

Tourism is a phenomenon that is implicitly psychological in nature, and indeed holidays are often used in psychological research as stimuli to investigate hedonic imagery and mental processes (Kane et al., 2012). Thinking about holiday-taking may stimulate fantasy (Rubenstein, 1980) and emotions that play an important role in shaping vacation and destination choices (Decrop & Snelders, 2004). Tourism experience involves fun, amusement, fantasy and sensory stimulation (Babin et al., 1994; Holbrook, 2006) and increasingly 'creativity', as a new experiential offering appealing to tourists' creative skills.

Positive and pleasurable emotions and feelings are important components of tourism experiences (Tung & Brent Ritchie, 2011). Tourism produces hedonic and memorable consumption experiences and promotes and fosters hedonic behaviour (Williams, 2006) that are essentially pleasurable (Floyd, 1997; Gnoth, 1997; Goossens, 2000) and elicit positive emotions (Gretzel et al., 2006; Li et al., 2014). Interestingly, however, whether a holiday experience is considered eudemonic, hedonic or utilitarian is dependent on the goals of the tourism which is implicit in predictions from the cognitive appraisal theory of emotions (Alba & Williams, 2013; Batra & Ahtola, 1991; Pham, 1998). Tourism experiences are multisensory and include tastes, sounds, scents, tactile impressions and visual images (Agapito et al., 2013; Hirschman & Holbrook, 1982, p. 92). Individuals respond to multisensory perceptions of external stimuli by generating multisensory mental images of their experiences (Hirschman & Holbrook, 1982).

Imagining, daydreams, emotions and desires play an important role in hedonic consumption (Goossens, 2000; Hirschman & Holbrook, 1982) such as when visiting a historic battlefield (Chronis, 2012), casino (Wong & Wu, 2013) or theme parks (Bigne et al., 2005). Holidays are memorable and emotions influence memorability (Larsen & Jenssen, 2004; Wirtz et al., 2003). Successful adventure businesses such as rock-climbing gyms, theme parks, museums, hot air balloon rides experience must be personally relevant, novel, surprising and provide learning and engagement (Poulsson & Kale, 2004).

Tourism produces unusual or distinctive events, incorporated in autobiographical memory, often linked to emotion, and such events are therefore more likely to be remembered (Kim et al., 2010). Thus, many concepts studied in tourism are application areas for cognitive science – memory, attention (Campos et al., 2020), beauty (Scott et al., 2020), coping (Jiang et al., 2020), novelty, surprise and unexpectedness (Skavronskaya, Moyle, et al., 2020; Skavronskaya, Scott, et al., 2020).

To conclude, this chapter introduced the cognitive paradigm and emphasised its relevance in the field of tourism. The chapter discussed the translation problem and provided arguments for the adoption of a cognitive sciences approach to better understand the emotional nature of tourism experiences, the importance of imagining and daydreaming in travel decisions and the significance of autobiographical memories for tourism. The chapter highlighted that the cognitive paradigm offered a valuable perspective for comprehending the complexities of the tourism field. The following chapter delved deeper into 'the cognitive wave' to further illustrate the usefulness of the cognitive paradigm in studying concepts such as consciousness, attention, sensation, perception, emotion and memory in the context of tourism.

Chapter 3

THE COGNITIVE WAVE
Major Concepts

Noel Scott
Edith Cowan University, Australia

Brent Moyle
Griffith University, Australia

Ana Cláudia Campos
University of Algarve, Portugal

Liubov Skavronskaya
Griffith College, Australia

Biqiang Liu
Griffith University, Australia

Abstract: This chapter discusses the main psychological paradigms used in the past 100 years, psychodynamism, behaviourism and cognitivism based on an information processing paradigm, and later cognitivism based on complex interactive mental processes. It briefly introduces the main concepts of later cognitive psychology: consciousness, sensation, perception, attention, emotion and memory. Each of these concepts will be discussed in detail in later chapters along with their application to tourism. One basic assumption of cognitive psychology is that the brain emerged through evolution and has survival value. However, this means that the brain is not a unified

Cognitive Psychology and Tourism
Tourism Social Science Series, Volume 27, 21–30

Published under exclusive licence by Emerald Publishing Limited
ISSN: 1571-5043/doi:10.1108/S1571-504320240000027003

designed organ but has layers of development, one building on the others. **Keywords**: Cognitive psychology; consciousness; attention; sensation; perception; emotion; memory

INTRODUCTION

A number of psychological paradigms have developed in the twentieth century to examine subjective experience such as psychodynamism, behaviourism and cognitivism based on an information processing paradigm, and later cognitivism based on complex interactive mental processes (LeDoux et al., 2020). While such labels provide a simplistic view of these developments (Costall, 2006), it is useful to discern a number of 'mainstream' paradigms, so their particular assumptions, focal problems, concepts and theories can be contrasted. For example, a shift from the behaviourism to cognitivism paradigm led to a revival in research on mental states and concepts such as emotion (Barrett et al., 2007). More recently, cognitive psychologists and neuroscientists have used advanced data collection techniques (electroencephalography skin conductance) and computer technology to identify mental events through neuroimaging (Miller, 2003). In this book, the focus is on the assumptions, focal problems, concepts and theories of cognitive psychology.

Origin of Tourism Psychological Theories

As is illustrated in Chapter 4 on stimulus-organism-response theory, tourism is an applied field of study. A range of disciplinary assumptions, theories and concepts are used to examine its phenomena (Jennings, 2001). As an applied field, the theory used in tourism is transferred from other disciplines. As a result of the different disciplines used to study tourism phenomena, a variety of theoretical and methodological paradigms are applied. Further, each discipline applied to the tourism field has a variety of sub-disciplinary approaches, each with its own assumptions, theories and concepts in turn. For example, insights from evolutionary psychology have been used to the study of tourist behaviour in contexts such as risk taking and extremeness aversion (Crouch, 2013; Kim et al., 2021; Kim & Seo, 2018; Kock et al., 2018).

These various psychology paradigms have been adopted in applied fields of study such as management, marketing, leisure, as well as tourism (Scott et al., 2017). However, researchers in these fields may use terms such as consumer behaviour, attitudes, motivation, emotion and cognition without referring to the original discipline's paradigmatic assumptions. In tourism, most psychological research adopts a behavioural paradigm, but will refer to motivation without considering the origins of this concept in drive theory and the related assumption of homoeostasis. A behaviourist approach in tourism research correlates external stimuli found in a servicescape (Bitner, 1992) with outcomes such as emotional reactions, judgements, including perceived value, or measures of future behaviour (intention to revisit), and do not take into account subjective mental processes.

For example, an assumption of cognitive psychology is that the brain emerged through evolution. It therefore has adaptive functions that enable humans to succeed as reproducing organisms. Systems for perceiving, memory and cognition have evolved in a manner that allows adaptation to the environment. The evolutionary nature of brain means that there is no grand design, and the brain may display its evolution through a series of systems that 'work' and which have built one on another in complex ways. Thus, some mental biases may be conserved for their contribution to survival in the past (Cohen & Bernard, 2013). Several tourism researchers have examined the usefulness of evolutionary theory for tourism (Crouch, 2013; Kim & Seo, 2018; Kock et al., 2018).

One of the basic assumptions of cognitive psychology is that the mental processes determine thinking, feeling and behaviour. These mental processes include perception, attention, emotion, mental processing (consciousness, problem-solving, creativity, learning, language, decision-making and reasoning) and memory (Larsen, 2007). Thus, it is fundamentally different to the view that behaviour is determined by external stimuli only. Cognitive psychology is one part of a broader scientific enterprise. Cognitive science is an integration of cognitive psychology, biology, anthropology, computer science, linguistics and philosophy (Hunt, 1989). It represents an interdisciplinary effort to address the same basic issues that confront cognitive psychology.

In an applied field, the lack of discussion of definitions and assumptions can cause considerable confusion among authors and may be a main reason for many scholarly disputes. In addition, researchers may use naive folk psychology knowledge to examine tourism phenomena without defining these terms (Sibicky et al., 2021; Simons & Chabris, 2011). This is the reason why this book places considerable emphasis on disambiguation (Chapter 5).

A large number of existing tourism psychology studies adopt a behavioural approach, which focuses on correlations between an emotion and other variable to measure future behavioural intentions. For example, the behavioural approach to tourism experiences attributes a causal power over emotions to external environment, rather than to a subjective mediating process of cognitive appraisal, and therefore, largely ignores emotions elicitation processes. This leads to a lack of explanation on *how* a particular emotion is elicited and *why* different emotions may emerge from the same tourism experience.

KEY CONCEPTS

The key concepts of psychology are introduced here and discussed in the following chapter.

Consciousness

The concept of consciousness is fundamental to what it means to be human. It is related to the mind and body problem and the extent to which they are separate or the same thing. The mind concerns mental processes, thought and consciousness. The body concerns physical aspects of the brain-neurons and how the brain is structured. Many theories have been put forward to explain the relationship between the mind (defined as the conscious thinking 'you' which experiences your thoughts) and the brain (part of your body).

Behavioural psychologists avoid the mind and body problem by focusing on 'observable actions', namely stimulus and response. They believe that thought processes cannot be studied scientifically and objectively and should therefore be ignored. Radical behaviourists believe that the mind does not even exist. Monism is the belief that ultimately the mind and the brain are the same thing.

Reductionism versus dualism is a debate as to whether mental phenomena be reduced to physical. Reductionists consider that there is no evidence that the mind exists independent of the brain and massive evidence that mind is what brain does (Tryon, 2014). This book assumes a reductionist position and that neural structures and processes are the basis for consciousness (Tryon, 2014).

In the 1970s, cognitive psychologists used a computer analogy (information processing model) arguing that the brain can be compared to computer hardware that is 'wired' or connected to the human body. There are many criticisms of the mind as a computer. The information processing model emphasised encoding, storage and retrieval operations and took it for granted that all operations require a common data format, parallelling the operational requirements of a computer. Their emphasis on amodal knowledge representation has been disproved. But now it is known that, for example, visual imagery process is modal (Krishna & Schwarz, 2014). There is no physical respect in which the brain operates as does a computer.

There are a number of models of consciousness including the global workspace model (Baars, 2005). This model provides a cognitive architecture that identifies consciousness with the global availability of information. According to the global workspace model, information becomes conscious when it is simultaneously made available to a wide range of localised (and individually sub-personal) processes – jointly comprising a 'global workspace'. Other models include the predictive brain model which considers that people create a mental model of what they expect the external world to be and then update it as sensations are received (Nave et al., 2020).

At a neuronal level, conscious awareness involves a distributed, highly integrated, dynamic system of populations of cortical and subcortical neuronal physical, electrical and chemical structures and organs throughout the brain. Nonconscious neuronal activities process and merge signals from sensors into an integrated mental representation of an event or object. Chemical neurotransmitters play an essential role communicating among cells across the brain participating in these processes and they are key to the flexibility of connectivity and overlapping activity (Statler, 2015).

Executive function refers to a suite of higher-level cognitive skills and capacities that generally promote successful human functioning. Attention, task switching, working memory and inhibitory control are usually described as executive functions, though there is debate about the precise definition of the term (Banich, 2009). Executive function involves some degree of updating information, shifting focus between targets or mental sets, and inhibiting irrelevant or distracting information (Miyake et al., 2000). Executive function has three characteristic features: it is effortful, operates consciously and engaged in service of novel goals as opposed to rote or overlearned ones (Miyake & Friedman, 2012). The executive function abilities evolved to help one deal with novel challenges (Berkman, 2018).

One important distinction is between unconscious and conscious cognition. Dual process theory indicates that not all cognitive processing is accompanied by consciousness in this sense of informational access. Unconscious processes are fast, automatic, intuitive and unreflective. These occur without the ability to report on them. The cognitive processes that give rise to conscious awareness are, by contrast, slow, effortful and deliberate. However, the unconscious processes posited in contemporary cognitive theories are not identical to the concept of the Freudian unconscious described in psychodynamic theory (Freud, 1961).

Evans (2003, 2008) refers to these as System 1 and System 2, and suggests that these systems differ in terms of evolution, structure and function. When one makes a 'gut' decision, it might be a decision that is based on System 1. When it is a more conscious, deliberative decision, it might be thought to involve System 2 processes more heavily.

Consciousness is a core concept in cognitive psychology. In this book, consciousness is considered a function of neural circuits although these neurons are not located only in the brain as is discussed in embodied cognition. Much of the behavior is due to subconscious mental processes.

Sensation

Mental activity is grounded in sensory experience. At present, there is a consensus that one cannot understand human cognition without taking into account that humans interact with the world through their senses and do their thinking within a body; but there is much less consensus about what this implies and how it is best conceptualised (Krishna & Schwarz, 2014). Embodied perception refers to a conglomerate of theoretical claims postulating that the body, its movements and the interaction with the environment fundamentally shape people's perception of the world. All these claims have in common that they criticise 'disembodied' notions of perception that are characteristic of the information-processing approach.

Perception

Sensations (the activation of sensory organs: eyes, ears, nose, skin and taste receptors) act as the initiator of the individual's perception of the surrounding world, a process through which sensory inputs are selected, organised and interpreted, resulting in a 'conscious sensory experience' (Goldstein, 2010, p. 8). Conceptually driven processes operate from the top down – from long-term memory to sensory memory – to identify the stimulus. Data-driven processes operate from the bottom up (from sensory memory to long-term memory) to achieve the same goal.

Attention

Attention is focused mental engagement involved in processing external sensations or internal memories and thoughts (Davenport & Beck, 2001) and which influences perception, memory and learning (Dayan et al., 2000). The quantity of sensory information the brain receives exceeds its processing ability and attentional processes direct attention those aspects of the environment which are perceived as particularly salient. Attention is related to goal achievement, memory processes and behaviour. In general, attentional focus is affected by top-down goals and bottom-up perception, where top-down goals are related to endogenous mental activities, and bottom-up perception, respectively, refers to exogenous stimuli from the external environment (Dijksterhuis & Aarts, 2010). Goals 'guide behaviour through attention' to particular parts of the flow of information (Dijksterhuis & Aarts, 2010, p. 467), allowing mental goals translated into overt behaviour (Monsell & Driver, 2000) and alerting person to goal-relevant information (Dijksterhuis & Aarts, 2010).

Attention may also be directed internally as is found in the concepts of engagement (Brodie et al., 2011), immersion (Ellis et al., 2018) and narrative transportation (van Laer et al., 2014). Higgins considers: 'Engagement is a state of being involved, occupied, fully absorbed, or engrossed in something—sustained attention' (2009, p. 102). Similarly, a high level of immersion is characterised by [in part] 'high focus of attention on a limited stimulus field' (Ellis et al., 2018). Narrative transportation has been related to immersion (van Laer et al., 2014) and conceptualised as co-activation of attention, imagery processing and emotions (Appel & Richter, 2010).

Emotion

The study of emotions is difficult, due to the complex nature of this phenomenon (Harré & Parrott, 1996). First, there is need to differentiate the terms of *affect, emotions* and *mood,* as they are often inconsistently used in the literature (Bagozzi et al., 1999). *Affect* is commonly viewed as an umbrella term for specific mental processes, such as emotions, moods and attitudes (Bagozzi et al., 1999), while *mood* state is allocated from any particular object or event and refers to an affective state with a low level of intensity (Huang et al., 2012). Emotions may result in intense feelings and be associated with a specific factor, such as an event or person.

Emotion Appraisal

Emotions are mental events elicited by an appraisal of relevant stimuli and create measurable bodily changes such as sweating, raised heartbeat and so on. Over the past 30 years, the cognitive appraisal theory describing the formation of emotions has become accepted by most psychological and cognitive science researchers. Appraisal theorists maintain that the critical determinant of any positive emotion is the resultant evaluation and interpretation that arise after comparing an actual state with a desired state (Lazarus, 1991a, 1991b, 1991c). People appraise or evaluate characteristics of events in terms of respective goals or motives using structured appraisal dimensions (Scherer et al., 2001). Appraisal dimensions are inherent aspects of emotions themselves and appraisal theories specify appraisal dimensions that distinguish discrete emotions and illustrate their variance by dimensions and corresponding appraisal patterns (Scherer, 1997a, 1997b).

The study of emotions has a long history (Gendron & Barrett, 2009) with numerous streams of research. Over the past 30 years, cognitive appraisal theory describing the formation of emotions has become accepted by most psychological and cognitive science researchers (Ellsworth & Scherer, 2003; Frijda, 1986; Lazarus, 1991a, 1991b, 1991c; Roseman, 2001; Watson & Spence, 2007). Appraisal theorists consider that emotions are subconscious mental events elicited by an appraisal of relevant stimuli on a limited number of dimensions. Desmet and Hekkert write about emotional experiences that:

> arise from encounters with products that are appraised as having beneficial or harmful consequences for the individual's concerns, that is, his or her major goals, motives, well-being, or other sensitivities. (2007, p. 61)

Similarly, Volo writes that

> knowledge of the emotional tourists' space is essential in designing services which enable providers to relate with tourists in personal and memorable ways creating moments of engagements [that]....amplify tourists' emotions. Indeed, it is argued that congruity with tourists' goals realization can elicit positive emotions while incongruity can generate negative emotions. (2017, p. 38)

Coping

Coping is defined as cognitive or behavioural effort to manage a situation that exceeds the resources of the individuals. One is able to reduce his/her emotional distress and induce more positive emotional states via coping. Negative emotions arising in service delays or failures can lead to either problem-focused or emotion-focused coping behaviours. Problem-focused coping behaviours refer to people's activities/actions to alter or manage the cause of a negative experience, such as switching to another airline or lodging a complaint with the airline company or seeking compensation like a refund or free accommodation. Emotion-focused coping behaviours refer to people's activities/actions to regulate and control their own emotions, such as disengaging themselves from the situation by using a laptop or mobile phone while waiting or seeking emotional comfort by complaining to a friend.

Emotions and Feelings

Often the terms 'feeling' and 'emotion' are used interchangeably but are distinguished here. Feelings are the result of an 'emotional body state' being signalled back to the brain, not only to the limbic system but also to the somatosensory system (Damasio, 1995). An emotion of happiness may result in a smile and thereafter the state of the muscles around the mouth, the effect of neurotransmitters released and so on are subsequently reported back to the brain reinforcing a feeling of happiness. This is why bodily feedback or unconscious priming, such as asking a person to smile, can successfully induce emotion (Briñol & Petty, 2003). More generally, emotions and feelings have a recursive relation, each making the other more likely (Han et al., 2007). This leads to the idea that feelings provide information to the brain as is discussed in feelings as information theory. Emotions are motivational and informational, primarily by virtue of their experiential or feeling component. There is a limited discussion of feelings in the tourism literature (Buda et al., 2014), with most studies focuses instead on perception of external physical stimuli (touch, taste, smell and vision) rather than perceptions of conscious or unconscious emotional signals (Agapito et al., 2014).

Memory

Memory is a mental process by which information is encoded, stored and retrieved (Atkinson & Shiffrin, 1968), although memory may also refer to the information itself, the memory that one recalls and remembers (Winstanley, 2006). Cognitive psychology recognises two major types of memory having different capability and capacity. Short-term memory (working memory) can store information for a certain period of time but with limited capacity, while long-term memory has a large capacity and contains memories of experiences and information that have accumulated over a lifetime. Two types of long-term memory are distinguished: semantic and episodic memory (Pearce & Packer, 2013). Semantic memory is memory of general knowledge about the world (Braun et al., 2006), while episodic memory refers to specific memories such as personal experience (Pearce & Packer, 2013) and events (Braun et al., 2006). Autobiographical memory is a subset of episodic memory and has defined as recollected events that belong to the person's past (Pearce & Packer, 2013; Rubin, 2005; Tung & Ritchie, 2011). The spreading activation model describes how an entire network of ideas (schema) is searched to retrieve specific information from our memory.

Memory is not a perfect system and can be a subject to false information (Loftus, 2005; Roediger & McDermott, 2000). People may be unable to recall a prior event, may remember events differently from the way it happened, or remember an event that never happened (Braun et al., 2006; Braun et al., 2002; Roediger & McDermott, 2000, p. 149). Memories are malleable and can be distorted by outside sources, such as advertising or stories or due to later experiences. Attention given to new stimuli determines the amount information that becomes encoded for storage (Fivush, 2011).

CONCLUSION

Cognitive psychology seeks to unveil the mental processes occurring between stimuli and behaviour, thus assuming that mental processes include perception, attention, emotion, mental processing (consciousness, problem solving, creativity, learning, language, decision-making and reasoning) and memory (Larsen, 2007). Elucidating the key concepts and principles in cognitive psychology is important for tourism studies applying them and discussing key areas for future research (Scott, 2020; Skavronskaya et al., 2017).

This chapter has touched upon several related concepts: consciousness, sensation, perception, attention, emotion and memory. It aims to introduce the 'Cognitive Wave' (Scott, 2020) – discussing concepts and theories from cognitive psychology and to demonstrate its usefulness for understanding mental process underlying behaviours human exhibit, especially for transfer of psychological paradigms to tourism research.

Chapter 4

UPDATING TOURISM THEORY
The S-O-R Model

Noel Scott
Edith Cowan University, Australia

Brent Moyle
Griffith University, Australia

Ana Cláudia Campos
University of Algarve, Portugal

Liubov Skavronskaya
Griffith College, Australia

Biqiang Liu
Griffith University, Australia

Abstract: This chapter presents a case for updating the use and development of psychological theory in tourism through an examination of the psychological stimulus-organism-response (S-O-R) model. The historical development and the criticism of the model has been addressed. A revised version of the model is provided based on importing new theoretical knowledge from the original discipline. This chapter argues that with an explicit focus on tourism psychology, there is some evidence to suggest that tourism theory may benefit from adopting theory from the parental discipline rather than inventing its own. Future research can address other examples of imported theories and to further examine mechanisms for this

Cognitive Psychology and Tourism
Tourism Social Science Series, Volume 27, 31–43

Published under exclusive licence by Emerald Publishing Limited
ISSN: 1571-5043/doi:10.1108/S1571-504320240000027004

knowledge transfer and update from original disciplines relevant to the study of tourism. **Keywords**: Theory; stimulus-organism-response model; cognitive appraisal theory; emotions; goal; arousal

INTRODUCTION

The first two chapters of this book discussed numerous problems and barriers associated with developing tourism theory, including the translation problem (Francken & Slors, 2018) (see Chapter 2), as well as the need for paradigm shifts and other issues associated with the ongoing development of tourism theory. Theory is critical to the representation of the social world, as it provides the grounds used to understand, explain or predict phenomena (Gerring, 1999; Seelos, 2010). Tourism and cognate phenomena, such as events and hospitality, are considered fields of research and derive theory from many disciplines (Jafari & Ritchie, 1981; Tribe & Liburd, 2016). Tribe proposes that tourism is a field of study 'embracing and adapting theories from its contributory disciplines and neighboring fields' (2010, p. 13).

A discipline is characterised by a well-defined object of study and a collection of related phenomena, concepts used to refer to them, theories to describe their interrelations and specific methods designed to study and test theories. Over time, foundation disciplines, such as economics, geography, anthropology, sociology and psychology, undergo their own paradigm shifts and develop new theories and concepts. Theory development in these disciplinary areas may occur inductively from the analysis of empirical data using, for example, grounded theory (Strauss & Corbin, 1997), or deductively where research is applied to test theory (Bricker et al., 2015). Theory evolution and obsolescence are a recognisable part of established disciplines. Over time new theories and concepts are developed and these then can be applied to study tourism phenomena. This continual development of theory in foundation disciplines, such as psychology, implies that it is necessary to periodically check back on the status of theories applied in the parent (sub) discipline which informs tourism research and scholarship.

Psychology is a discipline that is applied in tourism research to understand tourist cognition and behaviour (Koseoglu et al., 2019; Pearce & Packer, 2013), using concepts such as motivation (Weiler et al., 2012) and emotion (Ma et al., 2013). Psychology is not a unitary area but has a number of sub-disciplinary fields that have been used to study tourism (Crouch et al., 2004) including behavioural psychology (Weiler et al., 2018), social psychology (Tang, 2014), environmental psychology (Fridgen, 1984), positive psychology (Filep & Laing, 2019) and cognitive psychology (Glenberg et al., 2013; Skavronskaya et al., 2017). Each of these sub-disciplinary fields has an active research programme with new concepts and theories successively supplanting others and becoming mainstream.

Each of these subdisciplines of psychology has its own focal problems, assumptions, units of analysis and associated methodology (Skavronskaya et al., 2017). In tourism, most existing studies examine consumer behaviour and are based on theories and concepts from the environmental, behavioural or socio-psychological sub-disciplines. Each of these sub-disciplines is dynamic and indeed 'mainstream' psychology has developed over the past century from introspectionism, through behaviourism to a cognitive paradigm (Scott, 2020). This chapter presents an argument for updating the use and development of psychological theory in tourism through an examination of a particular example. It focuses on the psychological stimulus-organism-response (S-O-R) model and contrasts it with mainstream theoretical advancements from cognitive psychology. The chapter examines the historical development of S-O-R theory and explores the efficacy of this model in tourism. It provides insights on how to advance existing tourism theory through engagement with advances in parent subdisciplines.

THE STIMULUS-ORGANISM-RESPONSE MODEL

The S-O-R model was developed by Mehrabian and Russell (1974) as part of a comprehensive environmental psychology framework to examine how a person responds to their environment. Environmental psychology is described as an emerging interdisciplinary field of environment and behaviour, or 'human-environment relations' (Stokols, 1978). Environmental psychologists assumed that health and happiness derive from the harmonious and supportive nature of such relationships (Cicerali et al., 2017). Mehrabian and Russell (1974) developed the S-O-R model positing that certain environmental stimuli (such as heat or sound) influence an individual's emotional state, which in turn affects approach or avoidance responses.

The S-O-R theory has incrementally changed over the past 50 years to adapt it to the consumer and retail context and subsequently tourism. Later authors such as Baker (1986) and Bitner (1992) extended the scope of this model to include a variety of concepts: social factors (people in the store, customers and employees), design factors (visual cues such as layout, colour, clutter and cleanliness) and ambient factors (non-visual cues such as smells, sounds and lighting effects). In the process, several of its original theoretical assumptions have been revised. Crucially, these adaptations have been based on theoretical developments in the original discipline of psychology (not the subfield of environmental psychology).

Environmental psychology was developing at the same time as the academic study of tourism and there appears to be an overlap between tourism and environmental psychology in concepts such as destination image (Fridgen, 1984), place identity and place attachment (Suntikul & Jachna, 2016). Since its inception, the S-O-R model and its underlying concepts and assumptions have been applied in several studies of restaurant dining, events, theme parks and destination marketing (Bakker et al., 2014; Donovan & Rossiter, 1982; Jacoby, 2002; Kucukergin et al., 2020). As previously mentioned, the S-O-R model was originally developed to account for the effect of external factors, such as colour, temperature and sound on humans. Thus, its original focus was to examine the effect of external stimuli on perceptual processes that usually do not involve conscious mental awareness.

One important assumption of the S-O-R model is that, unlike previous behavioural models, environmental stimuli were specifically considered to affect a person mentally, and evoke individuals' emotional states, which then in turn determine behavioural responses. The emotional impact of natural and built environments was conceptualised by Mehrabian and Russell (1974) in terms of three basic mental dimensions: pleasure, arousal and dominance. Therefore, the S-O-R model provided a new perspective on emotion as a key motivator for consumption, clearly a departure from the traditional view that regarded consumers as rational and cognitive decisionmakers and neglected emotions in consumer behaviour (Hirschman & Holbrook, 1982).

The Model in Consumer Behaviour Research

Several authors adopted the S-O-R model to various contexts. The first was Belk, a consumer behaviour researcher who was struggling with how 'consumer behaviour depends upon the situation' (Belk, 1975, p. 427). At that time, the model of the consumer was that behaviour was determined by innate characteristics (personality, general skills and intellect), but it was found that these factors did not predict purchase outcomes well. Belk found that behaviour was dependent on the 'situation', noting that 'there is ample evidence and speculation that consumer behaviour is a function of the interaction between the individual and the situation' (1975, p. 427).

Belk (1974) conducted research on buyer behaviour and how a purchase (say of a brand of cracker biscuit) was dependent on a situation. Belk found that the variance in the brand of biscuit cracker purchased could be reduced if the 'situation' were defined. As a result, his studies specified the situation or 'consumption setting' using statements such as 'you want to buy some things for a picnic you are planning with friends'. Belk attempted to make the case that a particular situation (having a picnic) provides a reason for why a particular type of product is chosen. This is true to the extent that every respondent reacts to a situation in the same way. Belk claimed that a 'picnic with friends' is a situational factor. However, if attributing a person with an internal mental function, this 'situation' may be interpreted as a person's goal in a usage case. For example, a person's particular goal in this consumption situation (to impress a girlfriend versus to celebrate with the footy team) would also affect the brand purchased.

Belk (1975) however appeared to assume that stimuli (a servicescape or even a destination) could be objectively evaluated like colour or temperature. Subsequently, this assumption has been 'forgotten' so that now the objective effects of a stimulus on internal states are taken for granted. Therefore, the literature using S-O-R and indeed most tourism research on emotion has built this assumption into it. This approach may be useful for simple purchase and 'usage situations' but ultimately relies on assumptions about the homogeneity of a consumer's goals.

Donovan and Rossiter (1982) adopted the S-O-R model to examine the atmosphere of a retail setting. The advantage of the model was that previously store atmosphere and other physical in-store variables such as aisle width, brightness and crowding were seen as a component of store image. These authors considered physical variables as antecedents of store atmosphere. Donovan and Rossiter also considered that 'store atmosphere effects are basically emotional states' (1982, p. 35). While Donovan and Rossiter (1982) maintained that the environment (as store environment) affected behaviour, they did update S-O-R theory to remove the dominance dimension in measurement of emotion based on the work of Russell and Pratt (1980), who found that the two dimensions, pleasure and arousal (pleasant-unpleasant, arousing-sleepy), were adequate to represent people's emotions.

Further Development of the Model

Bitner (1992) was next to use the S-O-R model. The study adopted the assumptions of previous authors that particular physical settings are associated with recurring behaviour patterns and thus their social behaviours can be predicted. Here the action of a person is associated with (although not necessarily determined by) the situation. Similarly, Bitner adapted the environmental assessment dimensions of novelty (mystery) and complexity, writing that 'preference for or liking of a particular environment can be predicted by three environmental dimensions: complexity, mystery, and coherence' (1992, p. 63). While perhaps applicable in the particular area of study, these are dimensions that are not necessarily universally applicable to study consumer situations. However, Bitner was able to recognise that the consumer's goal was important, writing 'Each individual comes to a particular service organization with a goal or purpose that may be aided or hindered by the setting' (1992, p. 61).

Critically, the servicecape model developed by Bitner (1992) included moderators of the relationship between stimuli and organism. These include personality as previously noted in the S-O-R model, as well as the person's 'plan or purpose for being in the environment' as discussed by Donovan and Rossiter (1982). Bitner goes further and notes that 'what the individual notices and remembers about the environment, as well as how he or she feels about it, is influenced by the purpose for being there' (1992, p. 65). Bitner also identifies expectations: 'What an individual expects to find in an environment also affects how the individual responds to the place' (1992, p. 65). In effect, the study identified expectations and goals as moderators of emotional responses. Later writers adopted this model although also extended the type of elements in a servicescape (Rosenbaum et al., 2011).

Consistent with Bitner's extension of the S-O-R model, later studies examined the relationships among external environmental stimuli, perceptions and emotions, assuming that cognitive perception of the external environment (perceived servicescape/atmospherics and perceived service quality) elicits customer emotions (Tuerlan et al., 2021; Wirtz & Bateson, 1999). However, how perceptions elicit emotions is not discussed or subsequent cognitive processes articulated (Massara et al., 2010).

The Model and Research on Emotions

The Bitner (1992) model is applied in several tourism studies without modification. Jang and Namkung (2009) applied the model to examine how the physical servicescape of theme restaurants affects customers' emotional states. This study reiterates the S-O-R assumption that 'emotions are caused by customers' exposure to specific stimuli' (2009, p. 146). These results indicate that emotions mediate between perceived quality and consumer behavioural intentions. Other studies also assume a causal link between stimuli and emotions (Chen et al., 2022; Flavián et al., 2021; Hwang & Lee, 2018; Su, Cheng, et al., 2020; Su, Hsu, et al., 2020; Su & Hsu, 2013; Tsaur & Lo, 2020; Tubillejas-Andrés et al., 2020).

Indeed, a recent literature review of the literature of emotion in tourism found that 42 studies among the selected 178 hospitality and tourism management articles attempted to establish an external stimulus–emotion relationship (Tuerlan et al., 2021). These studies regard stimuli, in line with S-O-R, as direct antecedents of emotion without considering how individuals' emotions are evoked through (cognitive) mental processes. Essentially, these studies assume that stimuli, such as a building or a certain environmental attribute, is emotion-arousing and that a stimulus automatically elicits emotions of the same type (Tuerlan et al., 2021).

Later papers developed the S-O-R model in unexpected directions, modifying the existing parameters to suit the research context or the type of activity engaged in, but still assuming that the environment will determine the emotional arousal experienced as per the standard model. Su et al.'s (2020) study extends the model on the basis that emotions experienced and activity type will both directly influence the storytelling intentions of tourists. Thus, the study considers that an adventure activity situation is related to a story telling activity, noting 'a tourist engages in a particular type of activity, eliciting their affective experience (arousal) and ultimately a response (storytelling)' (Su et al., 2020, p. 3). The results demonstrate a correlation between type of activity (relaxing or challenging), emotional arousal and storytelling intentions and indicate that arousal fully mediates between tourism activity type and storytelling intentions.

These results extend S-O-R theory, whereby the stimulus now appears causally antecedent to storytelling behaviour. Later in the paper the authors note that emotional arousal only increases the likelihood of storytelling. This statement appears to transform the theory into a probabilistic model, although these implications are not discussed. The application of the model to a destination further complicates the picture (Kucukergin et al., 2020). A destination provides a number of usage situations and does not allow identification of a specific goal for visiting. A set of destination attributes are provided for a person to choose as their reason for visiting, but there is no assessment of whether these attributes have any relationship to a person's goal. Thus, a respondent will answer a question as to whether the mountainous scenery in a destination is beautiful and whether or not it is important for a particular person. While use of the S-O-R model in tourism research may have been useful, there is a need to consider recent developments in psychology to determine how it could be improved or superseded.

Criticism and Developments of the Original Model

Since its development, the S-O-R model has been subject to at least four major criticisms in the original discipline of psychology. The first is its assumption of a passive reaction to stimuli. While the model was an advance of stimulus-response models where people are machines which react automatically to stimuli (Lazarus, 1998, p. xvii), the nature of the 'organismic' reactions unclear. The S-O-R suggests that when a person is exposed to external stimuli, 'inner organism changes' precede behavioural responses. However, the model only discusses personality as linking stimuli and emotional response.

The second is that the model assumes stimulus situations have an objective meaning: the model assumes that a stimulus has only one interpretation by the organism, but stimuli ambiguity is inevitable. The interpretation of a situation is found to depend on various factors, including the stimulus context, some form of social priming, the level of arousal, experience, motivation and thinking (the search for a proper interpretation). Third, the model is linear and implies a unidirectional environment–behaviour relation. Fourth, the model assumes that emotions are caused by stimuli novelty and complexity (as will be discussed below, this is found to be incorrect). Two people experiencing the same stimuli may express vastly different emotions.

There is a need for an understanding of the cognitive or mental processes which underpin the elicitation of emotion. Here, a cognitive appraisal model of emotions can be used to replace this assumption.

Since the development of the S-O-R model, much progress has been made on the process of elicitation of emotions in cognate subdisciplines of psychology. Indeed, cognitive psychology has developed an impressive body of research on emotion based around cognitive appraisal theory (Scherer & Moors, 2019; Watson & Spence, 2007). There is general agreement in this literature that goal congruity determines emotional valence (pleasantness-unpleasantness) and while novelty, goal importance and other dimensions determine physiological responses such as degree of arousal (Chapter 8 on emotion and Chapter 10 on novelty). Certainly, in situations involving non-habitual responses, an appraisal process mediates between stimulus and emotional arousal. The level of novelty of an experience for a person undertaking an activity, such as mountaineering willingly (therefore goal congruent), affects the degree of arousal. A person who has climbed a mountain many times will not be aroused as an inexperienced climber.

Traditional S-O-R vs. Updated Stimulus

This section critically analyses each component in the original S-O-R model prior to introducing a revised version of it based on the theoretical advancements from the original discipline. Its first component presented below is a stimulus. According to the S-O-R model, the environment can be described as consisting of stimuli, such as physical objects. These objects may be characterised as having an informational 'load' that is mentally assessed using two parameters, their novelty and complexity (Mehrabian & Russell, 1974). The mental mechanism by which these characteristics are assessed is not identified in the model. The organism reacts to these stimuli with emotions that can be differentiated on the dimensions of pleasure, arousal and dominance; keeping in mind that later dominance was dropped (Russell & Pratt, 1980). The emotional response is moderated by a person's personality, which is a fixed characteristic that moderates emotional responses to particular stimuli. Finally, following the exposure to stimuli and the development of consumer emotion, responsive behaviour emerges as either approach or avoidance (Figure 4.1).

While much of the tourism literature adapts classical S-O-R theory or in some cases adapts it based on idiosyncratic assumptions, other scholars in consumer research applied more recent concepts from cognitive psychology. Jacoby (2002) provides some theory to support the S-O relationship by accepting cognitive psychology findings that subconscious processes, such as implicit learning and learning without awareness, may be involved. This was based on habits in determining automaticity of daily life (Bargh, 2002).

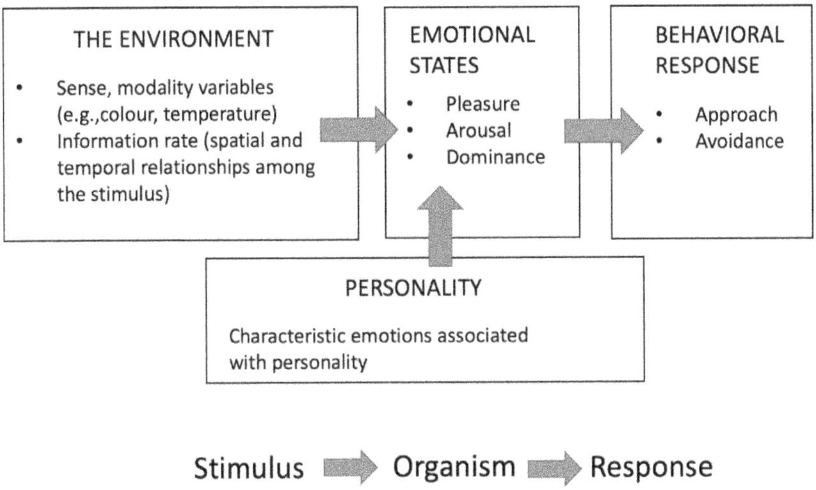

Figure 4.1. Original S-O-R Model of Approach and Avoidance.

Jacoby (2002) argued that rather than mediated by conscious thought and reflection, most psychological phenomena (including how one perceives and interprets the outside world) are unconscious. Concomitantly, subconscious processes determine the emotions resulting from stimuli. Bigne et al. (2020) applied this theoretical development to overcome the limitations of the traditional S-O-R model and accommodate automatic processing. Thus, for Bigne et al. the organism component includes prior experiences, knowledge, beliefs, attitudes, predispositions, intentions, values, cognitive networks, schema, scripts, motives, the individual's personality and feelings. This would explain why the in-store aroma of coffee, a conditioned response, leads to a positive emotion (Walsh et al., 2011).

Improvements Using Cognitive Psychology Theory

Another major issue of the S-O-R model is providing an explanatory mechanism that relates perception of a stimulus environment and the elicitation of an emotion. While Bigne et al. (2020) have recognised that the appraisal of a situation may be subconscious and lead to the elicitation of an emotion, there is no psychological mechanism specified. Without such a mechanism, the results of S-O-R studies are based on correlation without explanation. In many cases, authors make the claim of causality based on theory, something that it cannot conceptually support. Other scholars have made a general claim that the relationship between stimulation and behaviour is mediated by psychological structures and processes (Loureiro et al., 2021) but not specified a mechanism.

In the wider psychology literature, it is noted that an individual takes action to regulate emotions for their well-being, with positive emotions motivating the continued pursuit of successful goals (Fredrickson, 2001, 2004). Cognitive appraisal theory has been developed to determine how the appraisal of an environment leads to emotion (Watson & Spence, 2007) (Chapter 8 on emotions). The theory was initially proposed by psychologists to understand subjective causes of the elicitation of a specific emotion for the sake of predicting and adjusting behaviours, especially negative ones (Lazarus, 2001; Roseman & Smith, 2001; Smith & Ellsworth, 1985). In its simplest form, the appraisal theory of emotion is 'the claim that emotions are elicited by evaluations (appraisals) of events and situations' (Roseman & Smith, 2001, p. 3).

Appraisal is a 'cognitive process, the way [an] individual defines and evaluates relationships with the environment' (Lazarus, 1991a, p. 3). The quality and intensity of an emotional response depends on the appraised relationship between the individual and the environment. Appraisals are 'the results of the information-processing tasks that indicate the implications of the situation for the interests and goals of the individual and therefore determine the form that emotional reaction takes in a given situation' (Johnson & Stewart, 2005, p. 5). Thus, appraisals start the emotion process, initiating the physiological, expressive, behavioural and other changes that comprise the resultant emotional state (Lazarus, 1991b; Roseman, 1984; Scherer et al., 2001). An individual's appraisal of a situation depends on conditions both internal (personality, beliefs, goals) and external, such as product performance, responses to others (Lazarus, 1991a). In all, an appraisal process implies a cognitive mechanism – a subjective construal of personally relevant information that generates the variability in emotional reactions.

People generate various patterns or profiles on appraisal dimensions, and different emotions may be elicited by the same situation. The multi-stage appraising process is illustrated in three steps: the antecedents of the appraisal process: people's knowledge about the situation and perception of the personal relevance information that influenced the knowledge; the process of appraising personally relevant information within the organism of appraisal dimensions; and the consequences of the appraisals and emotions. These consequences include the elicited discrete emotion, people's emotional coping and action coping adaptation to the situation (Lazarus, 1993). See Chapter 8 on emotions and Chapter 10 on novelty for more information on appraisal dimensions and the mechanism of cognitive appraisal.

The Stimulus-Cognitive Appraisal Response Model

This section restructures the S-O-R model to introduce the assumption that stimuli are mentally evaluated before a response is evoked. This is a simple model as it avoids processes such as reappraisal and coping, feelings as information and so on. Also, the model assumes that emotions provide the foundations or motivation for behaviour. The revised model is specified in Figure 4.2.

The stimulus-cognitive appraisal response model indicates that a stimulus that does not attract attention is not responded to. This is different to the S-O-R model where attentional processes are not discussed. Next the model explicitly includes subconscious mental appraisal (cognitive appraisal) which are based on a small number of dimensions, goals congruence and relevance, novelty, certainty, etc. This approach allows a researcher to predict the emotional response based on an appraisal of how important a goal is to a person and if the situation is goal congruent or not. The revised model is termed the stimulus-cognitive appraisal response model. This model is presented as an alternative to the S-O-R model through the inclusion of cognitive appraisal process into the model. It should not be contentious, as mounting literature has already validated the usefulness of cognitive appraisal theory, discussed above (Hosany, 2012; Hosany & Prayag, 2013; Hosany et al., 2015). The model is simply an application of mainstream cognitive thinking.

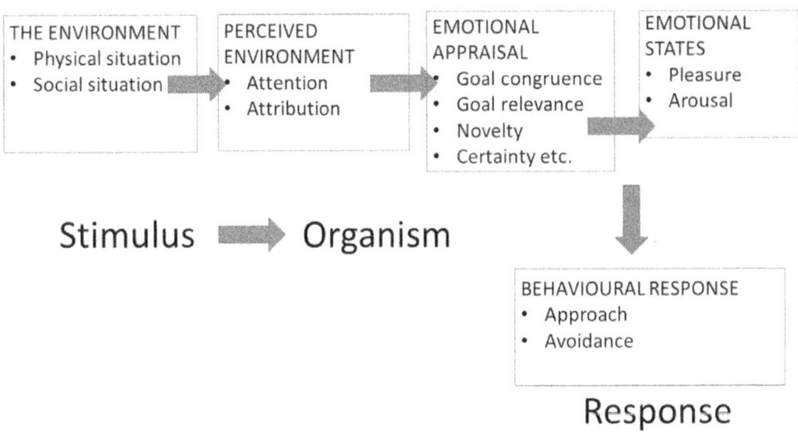

Figure 4.2. Stimulus-Cognitive Appraisal Response Model.

CONCLUSION

This chapter examined how the S-O-R model evolved and was adapted and used in various fields, including tourism research. The development of the S-O-R model over a 50-year period has been based on importation of concepts and theories from psychology. This is evident in the use of implicit or subconscious processes by Jacoby (2002) and subsequently by Bigne et al. (2020). The ensuing discussion highlights that modification of the S-O-R theory in an ad hoc (post-disciplinary) manner without supporting theory, as in Su et al. (2020), appears to introduce inconsistencies. Cognitive appraisal theory was then introduced from the original discipline provides a way forward for S-O-R theory. Cognitive appraisal theory has already been shown to relate a situation and emotions elicited through appraisal on a number of dimensions (Hosany, 2012; Hosany et al., 2020; Ma et al., 2013).

This chapter advocates, with an explicit focus on tourism psychology, that tourism theory may benefit, as recommended by Larsen (2007), from adopting theory from the original discipline rather than inventing their own. This in turn suggests that at least in the study of tourists and their cognition and behaviour, researchers may benefit from adopting more recent psychological theory. This is consistent with the call by Pearce and Packer to 'discard some past approaches, focus more on contemporary studies and anticipate links to the new technologies and futures of psychology study' (2013, p. 402).

A more general point for future discussion is why it has taken so long to adopt theories and ideas from cognitive psychology into mainstream tourism thinking. The basic concepts of goal-directed behaviour were discussed by Holbrook and Hirschman (1982) when they emphasised the importance of task and involvement and intervening cognitive processes in emotional situations. Interestingly, a study of the distinct knowledge domains of tourism research indicates that consumer behaviour is declining in importance 'arguably, due to no groundbreaking additions to its theoretical pillars and no observable change in its ontological or epistemological stands' (Koseoglu et al., 2019, p. 10). One reason suggested for the lack of use of more recent psychological theory is its lack of accessibility. A solution to this problem is noted by Pearce and Packer who write:

> ...increasingly, major figures in the psychology field write more popular and accessible, but still research-derived, versions of their work in key monographs. The approach of monitoring the accessible work of well-established psychology scholars backed by detailed contemporary journal review articles offers stimulating educational pathways for many tourism students and scholars. (2013, p. 4)

Further research is needed to examine efficient mechanisms for this knowledge transfer to occur. This chapter has addressed only one example of imported theory; other studies would contribute to further corroborate this chapter's claims. This study has focused on psychological theory which is in itself a series of sub disciplines. It is important to 'match' the concepts used to study a tourism phenomenon of interest to an appropriate theory in a particular subfield. This requires a detailed knowledge of the current developments within these subfields which creates a problem for those in an applied field. For example, the study of emotion is a core area of cognitive psychology. Alternatively, personality is a focus of personality psychology. Cognitive psychology does not address the concept of an attitude except through schemata.

In a similar manner, this study has examined only one aspect of psychological theory. Further research to review recent theoretical developments in other original disciplines relevant to the study of tourism (social sciences) may be useful in applying useful knowledge from the core disciplines to tourism. Such integration may help to provide insightful knowledge to tourism stakeholders. Overall, this chapter presents a foundation for discussion of tourism theory development and also the nature of tourism as a field of study. Further, the approach to development of tourism theory based on disciplinary knowledge rather than radical non-disciplinarity is illustrated by the adoption of cognitive appraisal theory to address failures of the S-O-R model. The chapter supports the conceptualisation of tourism knowledge as the result of disciplinary knowledge applied to this field (Tribe & Liburd, 2016).

Chapter 5

CONCEPTUAL DISAMBIGUATION

Liubov Skavronskaya
Griffith College, Australia

Noel Scott
Edith Cowan University, Australia

Abstract: There is need for definitional clarity when using psychological concepts in tourism research. The same term used in 'folk psychology', or behavioural or cognitive psychology can have quite different definitions and associated paradigmatic assumptions. This chapter argues that an understanding of mainstream cognitive psychology is needed when using concepts that are within its explanatory scope. Emotion is one such concept that has been extensively studied in cognitive psychology. The chapter discusses emotion and related terms such as feelings, novelty, unexpectedness and surprise to provide conceptual clarity. This will benefit researchers studying psychology of tourism emotions, tourism behaviour and decision-making. **Keywords**: Conceptual clarification; folk psychology; feelings as information; emotions; novelty; facial feedback hypothesis

INTRODUCTION

Disambiguation is the removal of uncertainty in the meaning of a term: distinguishing between similar things, meanings, names, etc., to make the meaning or interpretation more clear or certain. In an academic sense, it is a process of checking to see that a concept is used in a manner consistent with its definition. This of course raises the question of which definition to compare a particular term with. One problem found in the tourism literature is that researchers from different disciplinary perspectives use the same concept but mean different things.

Cognitive Psychology and Tourism
Tourism Social Science Series, Volume 27, 45–55

Published under exclusive licence by Emerald Publishing Limited
ISSN: 1571-5043/doi:10.1108/S1571-504320240000027005

Disambiguation is a common problem in research. Here it is formally considered as determining if two terms define the same concept within the same theoretical structures and with the same paradigmatic assumptions. Disambiguation is a process of comparison and analysis of concepts with similar names to determine their conceptual similarity. There are at least two important advantages of using the same definitions of theoretical concepts – first, theories, researchers and practitioners can agree on their meaning, and second, theoretical abstractness enhances the development of generalisations about the phenomenon to which the same concepts apply.

How a tourism researcher deals with the identification of a phenomenon as an example of a concept is based on parameters such as the research paradigm they ascribe to (Deshpande, 1983), and the theory and conceptual definitions and operationalisations used. No paradigm, theory or operationalisation has de facto privilege in every situation or in studying a particular type of phenomenon. However, it is generally considered good academic practice to specify the parameters a researcher uses when writing an academic journal article. This can allow the reader to understand and better critically assess the article.

In some research communities, the parameters and intension of a concept may be generally assumed to be known and accepted (Tribe, 2010). However, if a new theory or concept, embedded as it is in its paradigmatic, theoretical and methodological assumptions and definitions is adopted into the field of tourism, it is important that the intension and parameters of that concept are discussed and understood by the researchers using it. This mutual comprehensibility must underlie interdisciplinary research. Failure to do this can lead to ambiguity and confusion in interpretation of the findings of the research.

As noted before, there is no absolute privileged research paradigm or theory. This may be thought to imply that no final truth can be identified, and no best theory adopted. While many paradigms accept that absolute truth cannot be attained, the realist paradigm (Platenkamp & Botterill, 2013) assumes that progress can be made in the theoretical understanding of the world and, therefore, some theories (based on current evidence) may be considered as providing a 'better' explanation and hence should be used instead of 'lesser' theories. This approach does not mean that this better theory will always prove to be the most useful. Instead, it seeks to use the best possible theory where there is a consensus in the original discipline that it is the most explanatory one (here called mainstream theory). For example, this volume argues that cognitive psychology provides several mainstream theories that can be used to study phenomena such as emotion.

One exception is in research studies that are interested (only) in the commonly used *meanings* of an emotion-related term used by everyday people (folk definition) in specific cultural situations. In such situations, researchers studying a phenomenon may adopt common everyday definitions of terms, by assuming that a particular intentional definition such as those used in cognitive psychology studies and the everyday usage of the term are the same. Again, this is not to say that the academic definition is privileged, just that confusion and ambiguity should be avoided by using the same word to refer to different phenomenon. However, caution is recommended in the adoption of everyday definitions of terms such as consciousness, attention, emotion, memory, etc., as there are many examples where everyday understanding of these phenomena can be seen to be incorrect. For example, Simons and Chabris (1999) have demonstrated that a person looking at a video may not be aware of a gorilla walking through a group of people. This experiment demonstrates that one may not see what is in front of you and that folk understandings of how the perception and attention works are incorrect. Similarly, folk theories about how the memory works have been disproven such that in the legal system, there has been a recognition that the ordinary understanding of everyday mental processes does not necessarily match how the brain is shown to operates, including false memories (Roediger & McDermott, 2000). This chapter provides examples of disambiguation of terms commonly used in tourism, such as emotion, feelings, novelty, surprise, unexpectedness, motivation, among others.

USAGE OF THE TERM 'EMOTION'

Academic tourism research on emotion typically uses definitions and assumed conceptual mechanisms based either on folk-psychology and everyday beliefs or using dated theoretical frameworks. For example, many studies assume that certain types of events (a visit to a theme park) elicit certain emotions (excitement) automatically (Tuerlan et al., 2021). The implicit assumption is that the person has experienced excitement because it was caused by the theme park, thus it was an inevitable result of a visit. This is in line with common usage. For example, sadness is defined as 'the feeling of being unhappy, especially because something bad has happened' (Macmillan Education, n.d.). This definition of 'unhappiness' may be interpreted as 'something external' that has the property of 'bad' causes one to 'feel unhappy'. This is incorrect as people may interpret a 'bad event' in many ways and further, a 'bad event' in the past may be seen as amusing in retrospect. An alternative interpretation is that 'something external' has occurred that one appraises as 'bad' (non-goal congruent) and as a result sadness is felt. The dictionary definition does not indicate any such appraisal has occurred.

The cognitive appraisal theory of emotion is consistent with the second interpretation. It considers that the emotion of sadness is an outcome of the appraisal of a situation: for instance, during a recollection of an event during a holiday, sadness may be elicited if a tourist appraises this past event as desirable but no longer available (goal incongruency) (Roseman & Smith, 2001). However, if a desirable future event is no longer available due to the actions of an external agent (a trip has been accidently cancelled by a tour agent or an airline cancelled a flight), then anger may be elicited. Typically, a layperson is unaware of the mental processes that lead to elicitation of emotions and therefore may simply associate their emotion with the event, hence attributing causality to it (Sibicky et al., 2021; Simons & Chabris, 2011).

An example of using a layperson meaning of an emotion related term in research is found in a study by Choi and Choi (2019, p. 4). The study uses a definition of 'fun' based on a four-dimensional scale (social vigour, psychological zest, emotional spark and flow) developed by Tasci and Ko (2016) who in turn identified these dimensions based on dictionary definitions of fun, the literature on fun and voluntary descriptions of fun provided by online respondents. This leads to inclusion of items such as 'Makes me feel social', and 'Involves active people' in this scale. These items associate different contexts where people often experience fun with the experience of fun itself. In contrast, cognitive appraisal theory considers fun as a positive emotional state elicited by goal congruence. Thus, the use of a layperson definition of emotion in research may lead to potentially erroneous conclusions and confound the antecedent (eliciting condition) with the consequent (emotional response patterning, including subjective feeling) (Roseman & Smith, 2001).

Disambiguation of Emotion-Related Concepts

Emotions have been recognised as core building blocks of tourism experiences (Bastiaansen et al., 2019) and as a central to many topics that tourism scholars study. In studying emotions, the concepts of emotion, feeling, affect and mood are often used interchangeably, with different definitions based on the paradigm and theoretical position taken by the researcher. In the business and marketing literature as well as in cognitive psychology, affect is an umbrella term for specific mental processes, such as emotions, moods and attitudes (Bagozzi et al., 1999).

As discussed in Chapter 8, the concept of emotion has a distinct and identifiable meaning in cognitive psychology (Ma et al., 2013; Skavronskaya et al., 2020). Here emotions are mental events elicited by an appraisal of relevant stimuli on goals and other dimensions and create measurable bodily changes such as sweating, raised heartbeat and so on (Li et al., 2015). In turn, mood is a state that unrelated to any object or event and refers to a longer-term affective state with a low level of intensity (Huang et al., 2012). Compared to emotion, mood is considered longer lasting (up to days) but lower in arousal intensity

(Bagozzi et al., 1999). Mood is different to emotion, as mood does not have an object of reference or relate to any particular event or situation (Frijda, 1993).

The emotion elicitation process leads to the secretion of neurotransmitters, such as serotonin and other biochemicals, that in turn elicit measurable bodily changes governed by the autonomic nervous system, such as heartbeat, blood pressure and increased sweating (Moyle et al., 2019; Statler, 2015). These 'emotional body states' are then registered as sensations and signalled back to the brain as perceptions that are called 'feelings' (Damasio, 1995). This explanation is called 'feelings as information theory' (Figure 5.1).

Disambiguation of Feelings

The term feeling is used to describe diverse phenomena such thoughts as 'I feel like I am being misunderstood'; emotion, 'I feel happy'; or a wish, 'I feel like taking a holiday' (Damasio, 2004). 'Feelings as information theory' indicates that feelings provide information to the brain and explains how bodily feedback or unconscious priming operates. For example, the facial feedback hypothesis notes that asking a person to smile can induce an emotion of happiness (Briñol & Petty, 2003; Coles et al., 2022). The impact of feelings increases with their perceived relevance and applicability and decreases with the consideration of alternative inputs (Liu et al., 2020). More generally, emotions and feelings have a recursive relation, each making the other more likely (Han et al., 2007).

The elicitation of an emotion occurs subconsciously (Roseman & Smith, 2001) and one only becomes conscious of it through bodily sensations. These

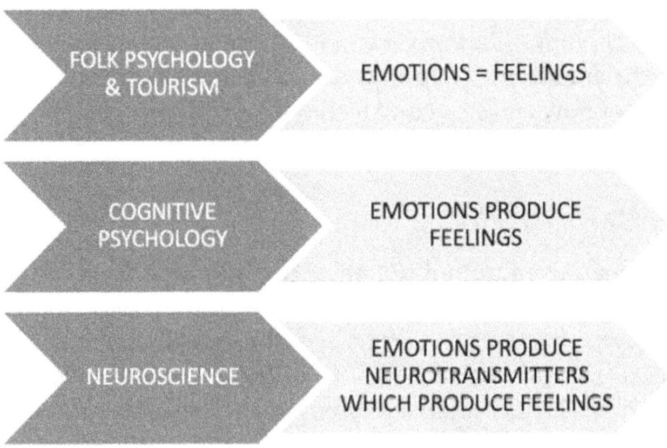

Figure 5.1. Disambiguation of Emotions and Feelings.

feelings or sensations (goosebumps, stomach churning worry) are then consciously perceived and provide information that can play a critical role in decision-making (Schwarz & Clore, 1996). When people make decisions, they may employ these feelings as a source of information to form judgements by asking themselves 'how do I feel about this?' and may adopt the shortcut heuristic of 'what feels good must be good' (Schwarz, 2011). Pleasant feelings are viewed as evidence of liking, satisfaction and happiness, and unpleasant feelings are viewed as evidence of disliking, dissatisfaction and unhappiness (Schwarz, 2011). In feelings as information theory, feedback from the feelings sensed during an experience helps a person determine whether to continue or not (Schwarz & Clore, 1996).

Feelings are also evoked during prospection or retrospection. For example, recalled memories of a place are appraised and elicit emotions, which then create feelings, using the same process as used to appraise the current situation. The same process is also used to appraise mental imagery about an imagined future. Thus, when people think about visiting a future destination as goal congruent, they develop positive anticipatory feelings towards it (Liu et al., 2020). Thus, feelings are different from emotions.

There is a limited discussion on causal relationship between emotions and feelings, especially from cognitive appraisal perspective in the recent tourism literature. For instance, Godovykh and Tasci (2021) discuss conceptual and methodological aspects of emotions, moods, feelings, and affect in tourism, and emphasise fundamental differences between these concepts, including their causes, duration, intensity and outcomes from a psychology and neuroscience perspective. Tuerlan et al. (2021) systematically reviewed the research on emotions in tourism and hospitality literature and identified inconsistency in the way authors discuss the emotion elicitation process by drawing insights specifically from the appraisal theories of emotions in cognitive psychology. This review addresses common misunderstandings concerning the emotion elicitation process which in turn leads to conceptual inconsistency of how emotions and feelings are defined or applied in tourism.

Disambiguation of Novelty

Tourist trips feature more novelty than daily by definition as they involve travel outside the usual environment. In tourism studies, the concepts of novelty, unexpectedness and surprise are often used interchangeably (Skavronskaya et al., 2020). Novelty has been identified as partially mediating positive emotions (Mitas & Bastiaansen, 2018). Several studies have identified that novelty is critical for the memorability of tourism experiences (Blomstervik & Olsen, 2022; Chandralal & Valenzuela, 2013; Kim, 2014); however, they do not provide an explanation of the mechanism of how novelty and memorability are related. This chapter disambiguates each of these three concepts and evaluates them as part of a process of cognitive appraisal with memorability of the experiences as one of the outcomes. All three concepts have neutral valence and are related to the level of arousal (Lazarus 1991a; Roseman & Smith, 2001). The concepts can be distinguished as follows.

Novelty

Novelty has many definitions. From cognitive psychology perspective, when the brain detects novel stimuli in the environment, attention is directed towards it (Bradley, 2009). Detection of novelty has high survival value and is integral to critical mental cognitive processes such as attention, emotion and memory (Lisman & Grace, 2005). The process of novelty detection in the environment allows one to identify important new information. The attention is directed towards unexpected novel stimuli, and this occurs subconsciously. Novelty is not a property of an object but is determined by experience. Detecting novelty requires comparing real-time perceptions with expectations based on contents of memory, to determine if the stimulus has been previously experienced (Tulving & Kroll, 1995). Novel information is more likely to be retained and encoded in memory (Mather, 2013).

Novelty's connection with emotions can be explained through the process of cognitive appraisal. As shown in Figure 5.2, an appraisal of a stimulus as novel leads to arousal and hence more intense emotions and feelings (Barto et al., 2013). Novelty, when coincident with positive goal congruence, may elicit delight and positive surprise (Ma et al., 2013), pleasure and enjoyment (Mitas & Bastiaansen, 2018). Novelty is also an antecedent of negative emotions. When associated with goal incongruence, novelty creates strong negative emotions, such as a feeling of anger (Bradley, 2009). Therefore, it is independent from emotional valence (Förster et al., 2010).

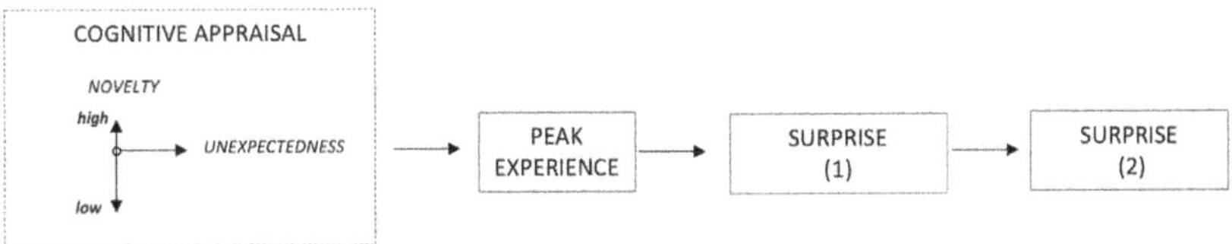

Figure 5.2. Cognitive Appraisal of Novelty, Unexpectedness and Surprise.

Surprise

Terminological ambiguity concerning the use of surprise in tourism arises from differences in the accepted definitions of surprise in early cognitive psychology literature and that current in the discipline. Early theorists in cognitive psychology defined it as an emotion (Meyer et al., 1997; Plutchik, 2001), while current researchers consider surprise as an unconscious cognitive state that affects behaviour through the direction of attention (Reichardt et al., 2020). Here, surprise is a result of a discrepancy between an incoming perception and previous expectations derived from the knowledge recorded and encoded in memory (Lorini & Castelfranchi, 2007). Therefore, surprise is a result of cognitive appraisal of novelty. Interpreted further through feelings as information theory, surprise is also a feeling which is the result of the appraisal process communicated to the brain as conscious feelings through sensations (Reisenzein, 2000; Schwarz & Clore, 1996). Thus, one becomes aware of surprise and recognises it by feeling surprised.

Surprise in the stated senses is neutral in valence (Foster & Keane, 2015; Noordewier & Breugelmans, 2013). When surprise is felt, the conscious brain recognises the unexpectedness of the experience (Meyer et al., 1997); therefore, unexpectedness accompanies surprise.

Unexpectedness

Unexpectedness and surprise are closely related in folk psychology; however, they are distinct concepts. Early cognitive psychology studies define unexpectedness as antecedent to the emotion of surprise; when a stimulus is unexpected, it will produce surprise (Roseman, 1996); however, recent studies state that unexpectedness is a more complicated phenomenon and is associated with different stages of neurocognitive processes, not only surprise (Foster & Keane, 2015).

According to cognitive appraisal theory, surprise and novelty initiate a series of mental processes that commence with the experience appraisal as exceeding a perceived threshold value of unexpectedness (Reichardt et al., 2020). Consistent with a cognitive appraisal of novelty, once this threshold is exceeded, the unexpectedness signal continues to interrupt ongoing information processing and is thought to reallocate 'processing resources' to the investigation of the unexpected event (Foster & Keane, 2015).

The reallocation of resources through the cognitive process culminates in the evaluation of this event with the immediate reaction and/or the updating or revision of the disconfirmed expectations or knowledge (Meyer et al., 1997). Thus, unexpectedness can be considered as a *threshold degree* of novelty which triggers an appraisal process and leads to surprise (Figure 5.2).

Unexpectedness can also be thought of as a degree of novelty which, when exceeded, leads to surprise. When a stimulus is appraised as having high novelty and is consciously recognised as the feeling of surprise. Similar to surprise and novelty, unexpectedness is neutral and can have both positive and negative valence. All three concepts are involved in the process of cognitive appraisal and have a neutral valence and are therefore useful to understand positive and negative experiences in tourism.

Figure 5.2 demonstrates how novelty, unexpectedness and surprise are involved in the process of cognitive appraisal. In cognitive psychology, these concepts are mental phenomena involved in cognitive appraisal theory emotion elicitation (Roseman & Smith, 2001).

Other Examples of Disambiguation

There are a number of other examples on how concepts not directly related to cognitive processes can be 'translated' and interpreted though cognitive psychology terms already discussed. For instance, several terms related to motivation, such as the concept of intention, can be interpreted as an outcome of an appraisal process in a form of prospective readiness to take a goal-oriented action. This is consistent with emotion as being motivating. Thus, motivation can be interpreted as a persistent goal-oriented state of readiness resulting from an appraisal process (Lang & Bradley, 2013; Lazarus, 1991). Lazarus (1991) emphasised that understanding motivational states is critical to differentiate joy from relief and sadness from distress based on understanding whether an event is related to reward-maximising versus punishment-minimising motives. At the same time, through a cognitive appraisal framework, risk can be explained as results of an appraisal of prospective events in terms of possibility of lack of coping potential and its consequences (Lazarus, 1974, 1985). It may manifest itself first as a feeling and then be identified as 'risk'.

The definition of the concepts such as authenticity (Newman, 2019) and value (Scherer, 1999) also may be considered through the lens of cognitive appraisal process. Thus, authenticity can be interpreted as an indicator of a goal-congruent experience. An authentic object is important and therefore seeing it may relate to achieving a goal. Value in the appraisal process context can be conceptualised as an outcome of appraisal process in a form of summative judgement based on the emotions experienced, when a stimulus is appraised positively on one or more appraisal dimensions (goal relevance, goal congruence, novelty). Table 5.1 summarise definitions discussed in this chapter.

Table 5.1. Cognitive Appraisal Theory – Conceptual Disambiguation.

Concept	Disambiguation
Emotion	Subconscious mental events elicited by an appraisal of relevant stimuli that create measurable bodily changes such as sweating or raised heartbeat and so on
Feeling	Conscious result of an 'emotional body state' being signalled back to the brain and use as a source of information
Fun	A positive valence response to a goal congruent situation with the degree of arousal unspecified
Novelty	An appraisal dimension influencing the intensity of emotion through comparing real-time perception with prior experiences
Unexpectedness	Threshold degree of novelty which triggers the appraisal process and leads to surprise; a conscious recognition that novelty or a feeling of surprise has been experienced
Surprise	Result of appraisal of novelty; unconscious (emotional) state; conscious knowledge of experienced emotion (feeling)
Intention	Outcome of appraisal process in a form of prospective readiness to take a goal-oriented action or a behaviour preparation for establishing, changing or abandoning relationship with the environment
Risk	Appraisal of prospective events in terms of possibility of lack of coping potential and its consequences
Authenticity	Appraisal of goal-congruent experience associated with pleasure with a capacity to affect people's judgements and behaviour, and significant for person's well-being
Value	An outcome of appraisal process in a form of summative judgement, when a stimulus produces values for one or more appraisal dimensions (e.g. goal relevance, goal congruence, coping potential, expectancy)
Motivation	Goal-committed action-oriented state driven by emotion

CONCLUSION

This chapter argues that maturity of the cognitive paradigm and the progress of emotions research in tourism allows utilisation of specific theories to study important nuances of emotions utilisation and differentiation. The coverage demonstrated the usefulness of the cognitive appraisal theory of emotions and feelings as information theory not only to differentiate closely related concepts, such as emotions and feelings, or surprise and novelty, but also to clarify the meaning of the concepts such as motivation, authenticity, risk and intentions, based on the understanding of the cognitive appraisal mechanism.

Disambiguation based on cognitive psychology theories emphasises the importance of individual interpretation of the experiences and significance of understanding momentary and long-term personal goals, values and prior knowledge for developing a holistic view on emotions elicitation and consequent behaviour and intentions in tourism context. Overall, cognitive psychology-consistent disambiguation can contribute to tourism research in two ways: to distinguish between closely related concepts and to clarify the meaning of the concept in a theoretically based and well-defined manner. Overall, this discussion demonstrates that cognitive appraisal theory providing a powerful framework to better understand emotions in tourism and can provide much needed conceptual clarity on mental processes which underpin behaviour and experience in tourism context. During the appraisal process, unexpectedness may be considered as high novelty and derived by comparing current perceptions with expectations, memories and previous knowledge (Lazarus, 1991b; Skavronskaya et al., 2020). Thus, unexpectedness can be explained as conscious recognition that novelty or a feeling of surprise has been experienced. Once unexpectedness elicited by novelty reaches a certain threshold, the unconscious state of surprise is evoked (Reichardt et al., 2020).

SECTION 2
PROCESSES

Chapter 6

SENSATION AND PERCEPTION

Noel Scott
Edith Cowan University, Australia

Biqiang Liu
Griffith University, Australia

Abstract: Sensory experiences and subjective perception serve as unique sources of tacit information that subconsciously influence people's cognition, emotions and behaviour. This chapter begins with an overview of key themes of sensation and perception in cognitive psychology, including tracing the history of perceptual processes, introducing different types of senses and perceptual processes and cognitivism-informed theories in this area. Following this, the discussion focuses on the application of sensation and perception within tourism field, discussing implications for research on sensescapes. This chapter concludes with an agenda for future research which delves into the role of multi-sensory and interaction of senses in touristscapes. **Keywords**: Sensation; perception; review; sensory experience; future research; embodiment

INTRODUCTION

The environment is multi-sensory in nature. Though visual sensations dominate, the accompanying sounds, smells, tastes and touch are all processed to create perceptions of the environment. The important role of the senses in shaping core human functioning is evident from ongoing discussions since the early days of philosophy (Aristotle, 2001; Plato, 2003). Recently, paradigm shifts in the scientific foundations of psychological research have prompted empirical investigations into the physiological underpinnings of

Cognitive Psychology and Tourism
Tourism Social Science Series, Volume 27, 59–72

Published under exclusive licence by Emerald Publishing Limited
ISSN: 1571-5043/doi:10.1108/S1571-504320240000027006

human perception, highlighting the significance of the senses in shaping our understanding of the world (Krishna, 2012).

Sensation and perception are two separate processes that are very closely related. Sensation is input about the physical world obtained by the sensory receptors – it is biochemical and neurological in nature. Perception is the process by which the brain selects, organises and interprets these sensations (Krishna, 2012).

SENSATION AND PERCEPTION

The process by which the sensory organs – those involved with hearing and taste, for example – respond to external stimuli is called sensation. Sensation is deemed the foundation of how individuals interact with their surroundings, and thus a starting point for the process of deriving meaning from these interactions (Agapito, 2020). There are five types of sensation channels: vision, hearing, touch, olfaction and taste (Figure 6.1). Each sense organ requires a minimal amount of stimulation to detect a stimulus. This is called the absolute threshold.

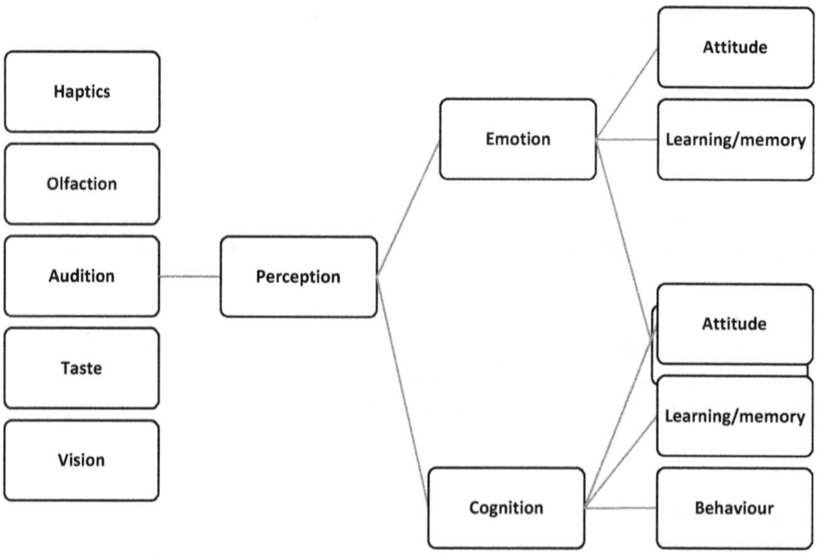

Figure 6.1. A Conceptual Framework of Sensory Processing.
Source: Adapted from Krishna (2012).

Sensations

Vision: Light enters the eye through the pupil, a tiny opening behind the cornea. The pupil regulates the amount of light entering the eye by contracting (getting smaller) in bright light and dilating (getting larger) in dimmer light. In the pupil, light passes through the lens, which focuses an image on a thin layer of cells in the back of the eye, called the retina. It is in the retina that light is transduced, or converted into electrical signals, by specialised cells called photoreceptors. Next, this electrical signal travels down the optic nerve.

After passing through the thalamus at the base of the brain, this signal makes its way to the primary visual cortex, where information about light orientation and movement begin to come together (Hubel & Wiesel, 1962). Information is then sent to a variety of different areas of the cortex for more complex processing. Some of these cortical regions are specialised for processing faces (fusiform face area) and body parts (extrastriate body area).

Hearing: People can get a large amount of information from the basic qualities of sound waves. The amplitude (or intensity) of a sound wave codes for the loudness of a stimulus; higher amplitude sound waves result in louder sounds. The pitch of a stimulus is coded in the frequency of a sound wave; higher frequency sounds are higher pitched.

Touch: Tactile stimuli – those that are associated with texture – are transduced by special receptors in the skin called mechanoreceptors. Just like photoreceptors in the eye and auditory hair cells in the ear, these allow for the conversion of sensations a form the brain can understand.

Olfaction (Smell): Both olfaction and gustation (taste) require the transduction of chemical stimuli into electrical signals. Unlike any of the other senses discussed so far, the receptors involved in the perception of both smell and taste bind directly with the stimuli they transduce. Odorants in the environment, very often mixtures of them, bind with olfactory receptors found in the olfactory epithelium. The binding of odorants to receptors is thought to be like how a lock and key operates, with different odorants binding to different specialised receptors based on their shape.

Taste: Taste works in a similar fashion to smell, only with receptors found in the taste buds of the tongue, called taste receptor cells. The binding of these chemicals with taste receptor cells results in the perception of the five basic tastes: sweet, sour, bitter, salty and umami (savory).

Some Senses Communicate Intrinsic Value. Sensations such as touch, smell and taste have an intrinsic valence, positive or negative, while others such as sight and hearing do not. The sensations that evolved later (for example, vision and hearing) do not contain intrinsic valence, although they are commonly labelled with valences generated from body states (Damasio &

Carvalho, 2013). Congruity theory highlights that individuals are more likely to develop positive attitudes towards products when the elements composing the consumption environment are perceived as consistent. Specifically, sensory studies show that when the arousal qualities of at least two sensory modalities are congruent, approach behaviours are improved (Agapito, 2020; Krishna, 2012).

Perception

The psychological process of making sense of the stimuli is called perception. It involves a series of steps involving what to perceive (what to pay attention to) and the interpretation of sensations. Both orientation (attention) to sensations and interpretation can be affected by goals. In addition, perceptions can also be affected by one's beliefs, values, prejudices, expectations and life experiences.

Amodal Information Processing. In the 1970s, the development of general models of information processing provided an integrative framework for the conceptualisation of human memory, judgement and decision-making (for authoritative contemporary treatments, see Lachman et al., 1979; Wyer, 1974). In a nutshell, the approach assumed that people acquire information through their senses and that this information is translated into a common code that is independent of the modality in which the information was initially acquired (Krishna & Schwarz, 2014). These amodal representations were assumed to be the building blocks of knowledge (semantic memory), on which the mind operates with general procedures that are also independent of the initial modality. Once stored, knowledge representations were assumed to be context independent, although context could influence how easily a stored representation could be accessed (Glenberg, 2010).

Multimodal perception: It is now known that experience is multimodel (vom Lehn, 2006). Real-world experiences typically involve multimodal sensations, wherein various senses combine to create a unified perceptual experience. Interestingly, one responds more strongly to multimodal stimuli compared to the sum of each single modality together, called the super additive effect of multisensory integration. The multimodal sensory stimuli are processed separately, and the results of those processes are qualitatively different from those of unimodal stimuli.

Top-down and Bottom-up Processing. While sensory receptors are constantly collecting information from the environment, it is ultimately how they interpret information that affects how one interacts with the world. Perception refers to the way sensory information is organised, interpreted and consciously experienced. Perception involves both bottom-up and top-down processing. Stimuli have been defined as what is 'out there' in the environment. In other words, the external cues that stimulate the sensory receptors in turn are responsible for transmitting sensory information to the brain via electric signals. Top-down processing plays a role in perception via previous knowledge, memory and anticipation, which individuals bring to the situation (Goldstein, 2010). For example, current goals can direct attention to salient objects.

Mere Exposure Effect. When novel stimuli are repeated, people tend to rate the repeated stimuli as more likeable than nonrepeated stimuli (Glenberg, 2010). One well-supported explanation of the mere exposure effect is that repetition leads to more fluent processing, and that fluency leads to increased liking. In contrast, Lee (2001) provided an alternative discourse – uncertainty reduction explanation. Lee (2001) posits that individuals misattribute perceptual fluency to liking. The uncertainty reduction explanation posits that individuals prefer stimuli that are familiar and is consistent with findings that stimuli judged old were preferred to those judged new.

The Stroop Effect. The 'Stroop effect' is one of the best-known phenomena in cognitive science and indeed in psychology more broadly. It is also one of the longest standings, having been reported by Stroop (1935) in the published version of his dissertation. In its basic form, the task is to name the colour in which a word is printed, ignoring the word itself. When it is a colour word printed in a mismatched ink colour, this is very difficult to do and results in slow, error-prone responding (MacLeod, 1991).

Implicit Attitude and Implicit Association Test. Explicit attitudes are deliberate, intentional and available to conscious awareness. Implicit attitudes were defined by psychologists Greenwald and Banaji (1995, p. 8) as 'introspectively unidentified (or inaccurately identified) traces of past experience that mediate favorable or unfavorable feeling, thought, or action toward a social object [concept]'. Such attitudes can influence how humans see, think and react to the world, even without the individual having awareness of their existence, activity or influence (Nosek & Banaji, 2009).

Both the Stroop and implicit association tests are interference tasks. It was proposed by Greenwald et al. (1998) to measure individual differences in implicit social cognition. The test is a simultaneous classification task where participants have to classify pairs of stimuli into mutually exclusive categories. Participants' pre-existing associations between stimuli and categories can interfere with the classification task. As a result, responses are faster when the pre-existing associations match the pairing of categories on a response key. The test scores reflect the interference effect from pre-existing associations on the classification task (Schimmack, 2021). As a result, the implicit association test and Stroop tests can indicate the existence of introspectively unidentified attitudes and hence aspects of schema.

Mind Body Connections

The human mind is dependent upon the body in its cognitive processing. Indeed, parts of the world different from those of the biological bodies can sometimes be processed as if they are part of them, as 'embodied' (Iriki et al., 1996). According to a more thorough formulation: given an external object X other than the biological body, 'X is embodied if some properties of X are processed in the same way as the properties of one's body' (De Vignemont, 2011, p. 82). Considering this definition, the 'embodiment' phenomenon would correspond to a specific modality of presentation of the properties of an object and would derive from the same way as the properties of the object are processed (Iani, 2021).

Embodiment discusses the profound connections between body and mind. Cultural geography and sensuous geography recognise the need to understand this mind and body interaction. From the perspective of cognitive science, it is known that bodily states may affect cognition, and that cognition may affect bodily states. Bodies provide sophisticated information-bearing channels, beyond vision or audition. A well-adapted mind uses any available, reliable signals. Similarly, mechanisms have evolved wherein mental states (such as fear) can affect physiological functions.

In embodiment cognition theory, sensory experiences and bodily feelings serve as unique sources of tacit information that subconsciously influence people's cognition, emotions and behaviour. This means sense-specific systems are involved in the cognitive processing which gives rise to 'embodied' mental representations. For instance, observing a given object (a bottle) will immediately activate the congruent motor information required to interact with it, such as a grasping movement and the appropriate hand shape (Goldinger et al., 2016).

The related term, 'situated cognition' indicates that cognitive processes change (either qualitatively or quantitatively) based on a person's goals and the immediate context (Goldinger et al., 2016). There are a number of approaches to embodiment and the theory has been used in a number of fields such as 'the feminist gendered body' (Alessandroni, 2017). In feminist literature, grounded cognition might mean that women think differently due to differences in male and female bodies. Embodiment has been related to aesthetic evaluation of stimuli such as dance (Kirsch et al., 2016; Papies et al., 2017).

Affordances: Affordance studies are often used in embodied theories. The concept of affordances originated in ecology, and can be seen as a way to combine the mind-body and environment (Buckley & Akhoundogli, 2019). The notion of affordance was first introduced by psychologist James Gibson (2015) to describe the potentialities for action afforded to individuals by the features of the environment. Affordances can be summarised as action possibilities offered from the environment to the animals that can seize these offering. A step affords stepability to an adult, not to a toddler (Pucillo & Cascini, 2014). This concept has been used to study service design, not from a psychological approach but more as a metaphor (Tomej & Xiang, 2020).

Grounded Cognition. The central theme of the grounded approach is that cognition emerges from interactions between classic cognitive processes: body and physical environment, and the social environment. Grounded cognition rejects traditional views that cognition is computation on amodal symbols in a modular system, independent of the brain's modal systems for perception, action and introspection. Instead, grounded cognition proposes that modal simulations, bodily states and situated action underlie cognition (Barsalou, 2008). Cognition is not a module in the brain that can be studied effectively in isolation but must be studied in the context of these other domains. Another central theme is that cognition supports situated action (Clark, 1998; Glenberg, 1997). Cognition is not an end but typically guides effective action in the world. Building on Barsalou's (2008) work, Krishna (2012, pp. 344–345) attempted to develop precise definitions of grounded cognition in terms of how it interacts with bodily states, situated actions and mental simulations, respectively.

Metaphors. For many consumer researchers, embodied cognition is probably synonymous with one particular stream of research, namely work that illuminates how sensory experience in one domain can influence cognition, emotion and behaviour in a different domain in ways that follow familiar metaphors (Krishna & Schwarz, 2014). Depending on theoretical perspective, metaphors are either seen as the unique driver of the observed effects (Lakoff & Johnson, 2008; Lakoff et al., 1999) or as expressions of the multimodal nature of mental representations (Anderson, 2010; Barsalou, 2008; Williams et al., 2009). The current interest in metaphors breaks with a long tradition in philosophy and linguistics that treated metaphors as rhetorical devices best reserved for poets (for a review, see Ortony, 1993). For example, the abstract concept of importance is closely associated with bodily sensations of weight (Schneider et al., 2011). In line with an embodied simulation account but contrary to a metaphor-enriched perspective, Schneider et al.'s (2011) study suggested that the connection between importance and weight goes beyond metaphorical mappings (see also Schema Chapter 12).

Presence. The construction of body awareness appears to be mediated by specific brain mechanisms involving a region of the cortex known as the temporoparietal junction. Damage to this brain region can generate distorted body awareness, such as feelings of a substantially elongated torso. Altered neural activity in this region through artificial stimulation can also produce an out-of-body experience, in which one feels like the body is in another location and one sees it and the world as if from the ceiling of the room. Remarkably, comparable brain mechanisms may also generate the normal awareness of the sense of self and the sensation of being inside a body. In the context of virtual reality, this sensation is known as presence (the compelling experience of actually being there). The normal localisation of the self may be equally artificial, in that it is not a given aspect of life but is constructed through a special brain mechanism.

Sensation and Perception in Tourism

Many of the tourism studies of senses accept that tourist experiences are intimately involved with their sensory perceptions of the destination. One experiences the surrounding landscape through the stimulation of senses (Rahman et al., 2017). Senses are deemed the foundation of how individuals interact with their surroundings, mediating the process of deriving meaning from these interactions (Agapito, 2020). A more complex idea of the tourism experience suggests that all five senses or a multisensory approach should be considered in, for example, gastronomic experiences and destination products (Pedro et al., 2020). All tourism destinations have unique and distinct sensory characteristics, which play an important role in creating memorable tourism experiences and tourist loyalty (Agapito et al., 2013; Pedro et al., 2020).

In the tourism context, Agapito et al. (2014) suggest that the sensory dimension of the tourist experience appears to be strongly linked to positive, unique and memorable experiences. They have therefore analysed the sensory dimension of the tourist experiences. In a rural context, they found that most references senses were to sight (26%), hearing (23%), taste (19%), smell (17%) and, finally, touch (15%). Each sense is, however, stimulated by different and specific environmental settings and provides different sensory information (Agapito et al., 2017). The entire sensory component (the multisensory stimulus) allows tourists to achieve an experience more truly meaningful and personalised (Kim & Fesenmaier, 2017a) and therefore more memorable. These personalised experiences suggest specific stimuli to provide memorable moments. Tourism experiences that include a multisensory offer may therefore contribute to the reinforcement of positive emotions, satisfaction and the creation of long-term memories and destination loyalty (Agapito et al., 2013; Diţoiu & Căruntu, 2014; Pawaskar & Goel, 2014). The multisensory dimension of the tourist experiences has been termed a 'sensescape'. Tourist experiences are not merely visual, but multisensory and tourism should involve various sensescapes, such as soundscapes, smellscapes, tastescapes and the geography of touch in addition to landscapes (Liu et al., 2018).

Embodiment: In embodiment cognition theory, sensory experiences, and bodily feelings serve as unique sources of tacit information that subconsciously influence people's cognition, emotions and behaviour (Barsalou, 2008; Parrinello, 2012). One embodies other people's actions. When humans observe someone else executing an action, their brain covertly simulates the performance of the action being observed. In other words, the brain creates an internal representation of the observed motor programs that, in the absence of overt movements, is usually called 'action simulation'. Several studies have suggested that such a simulative mechanism is rapid and even obligatory (Iani, 2021). Embodiment may be seen as acting within the framework of mental simulations (Papies et al., 2017). These simulations draw upon multiple systems activated by a cue (Barsalou, 2020).

In the tourism field, Small et al. (2012) focuses on embodied experiences of disabled people. The paper concludes that the tourism industry and community must understand the multi-sensory nature of the tourist experience if quality accessible experiences are to be available for tourists with vision impairment. In addition, provision of multi-sensory experiences also enhances the experiences of sighted tourists. Sun and Lv (2021) discuss how a dark experience affects people. This is based on the idea that evaluative processing of general affective metaphors has been shown to bias perceptual judgements of stimuli. The mind-to-body pathway involves the dark experience exerting a measurable effect on participants' sensory expression in photographs and sketches. They then discover the body-to-mind pathway in which the expressions of darkness can be decoded by others. Xie and Sun's (2018) study expands the understanding of embodiment theory in dark tourism with implications for product design and marketing.

Embodied cognition effects on tourist behaviour (Kock & Ringberg, 2019). This notion is anchored in (bodily) feelings-as-information theory according to which people consult their affective states subconsciously to make judgements about objects (situations, people, the self or the environment) unrelated to the bodily senses (He et al., 2022). Traditional sensory marketing uses the smell of citrus fruit to signal freshness and cleanliness, whereas in embodied cognition, the sensory experience of the citrus smell is incidentally and metaphorically projected upon others as being morally purer. That is, incidental embodied experiences differ from tourism research which only examines the direct sensory dimensions of tourist experiences, such as sensory destination aspects. Kock and Ringberg (2019) provide numerous examples. Carrying weight makes hills seem steeper and distances seem greater (Körner et al., 2015), highlighting the body's role in information processing.

Technological Embodiment. Although the integration between technology and a tourist's body is one of main features of technological embodiment (Tussyadiah et al., 2018), no previous studies have empirically tested this technological feature of virtual reality, nor analysed the mechanisms through which technological embodiment affects the customer experience. In the hotel industry, Flavián et al. (2021) study analyses the affective process by which technologies, with different levels of embodiment (virtual reality headsets, smartphones and desktop computers), can be effective in the hotel pre-experience stage. Tussyadiah et al.'s (2018) study provides a theoretical reflection on the phenomenon of embodiment relation in technological mediation and then assesses the embodiment of wearable augmented reality technology in a tourism attraction. The findings suggest that technological embodiment is a multidimensional construct consisting of ownership, location and agency. Flavián et al. (2019a) examined the effects of technological embodiment on customer engagement and behavioural intentions towards the destination through sensory simulation.

Landscapes: This dimension of sensescapes describes the visual channel and has attracted more scholarly attention than the other four human senses. Tourism studies have been systematically centred on the visual component of the tourist experience (Pan & Ryan, 2009), mainly focusing on the western-based sociological concept of tourist gaze developed by Urry (1990). The tourist gaze remains a key concept in tourism research, and its value choice is reflected in the destination choice and is reflected in the tourists' perceptions and behaviours, which form the destination's image (Samarathunga & Cheng, 2020). However, many studies using the gaze concept appear to assume that meaning lies within the object gazed upon, which is inconsistent with the cognitive approach adopted here (Zhang et al., 2022).

Smellscapes: An over-emphasis on the visual disregards that the unique character of a place can be imparted by its aromas (Dann & Jacobsen, 2003). Although traditional conceptions of the senses have often dismissed smell as vulgar, sub-human or lacking any aesthetic importance, some of the earliest anthropological studies recognised the importance of smell in understanding culture and social change (Śliwa & Riach, 2012). Olfactory sensations are extremely difficult to describe and categorise. Interviewees in Everett's food study particularly commented on the importance of smell in their appreciation of place, 'sometimes you can walk into a shop and walk past fruit, and you won't smell fruit, it's like, next! I think everything here is local produce, you can smell everything, you know' (2008, p. 344), thus supporting Dann and Jacobsen (2003) who insist that the senses must be considered in the appreciation of the landscape, and that 'smellscapes' is a way of experiencing the city and countryside.

Soundscapes: Soundscapes are an important component of parks and outdoor recreation (He et al., 2018). The importance of sound to tourists lies in at least two aspects: one, sound can be the core element of tourist attractions (music) and it is an intangible resource that tourists consume when travelling; and, two, sound helps tourists to establish the sense of a place and enhances their perception and understanding of a natural destination. The unique soundscape of destination is not only a key part of its attraction but also an element that can enrich the tourism experience (Kankhuni & Ngwira, 2022). Jensen (2016) noted that sounds in tourism should be revalued and researched by utilising the localised, sensuous and embodied act of listening.

Tastescapes: Much research on smell and memory has been done within the basic sciences, psychology and recently in marketing (Krishna, 2012). Tastescapes have been associated with the field of food tourism (Everett, 2019), but little attention paid to this issue from cognitive psychology. Everett (2019) deploys the term tastescape as an entry point to delineate the evolution of food tourism research. Inspired by the idea of a therapeutic landscape, Su and Zhang (2022) propose the notion of tastescapes to highlight that well-being is a multisensory condition, and tea tasting can be conducive to well-being and health.

Touchscapes: Personality analogies often dwell on haptic characteristics (Krishna, 2012). In studies of altruistic behaviour, Peck and Wiggins (2006) gave people brochures for an arboretum that either had a 'touch' element associated with it (a tree bark or feather in the front page of the brochure) or did not. They found that people's willingness to donate to the arboretum increased with a touch element in the brochure for people who had a high autotelic need for touch. People who have a high autotelic need for touch engage in touch because it is fun, interesting or enjoyable, an experience that is more hedonic than instrumental. There was no difference in donation willingness with and without the touch element for those who had a low autotelic need for touch.

Perception of Time. The sense of time has no direct sensory system that specifically supports one's appreciation of the duration of events and activities (Pearce, 2020). Nevertheless, it is widely recognised that people do have skills in assessing how much time has passed, even if one does not fully understand how the relevant information is processed. In a tourism setting, how tourists perceive, use and respond to time in their holiday experiences provide opportunities for recasting the design of many experiences (Pearce, 2020). Li et al.'s (2022) recent study postulates time perception is an inevitable factor influencing tourist behavioural decisions during the course of travel; however, their study is based on social psychology. Instead, drawing on work in cognitive psychology and neuroscience, Pearce (2020) considers ideas about the perception of time and temporal features of tourists' social episodes and articulates six principles for furthering the consideration of time in the better design of contemporary tourism experiences. These are specifying duration, managing, and filling downtime, preparing for emergencies, tailoring tourist sequences, encouraging engagement and implicating time in memorability.

Sensory Marketing and Design

Krishna defines 'sensory marketing' as 'marketing that engages the consumers' senses and affects their perception, judgment, and behavior' (2012, p. 332). From a practical standpoint, sensory marketing may trigger subconscious cues which tourist perceptions of abstract notions of the product (its sophistication or quality) are sensitive to. Despite the widespread popularity of explicit marketing appeals to consumers, implicit subconscious cues have untapped potential in influencing consumers attitude and behaviour. These sensory triggers may result in a consumer's recognition of (desirable) brand attributes, different from those provided in advertising. An understanding of sensation and perception as it applies to consumer behaviour may help to understand these sensory triggers (Krishna, 2012; Krishna & Schwarz, 2014).

Digital sensory marketing has emerged as information and communication technologies have evolved (Petit et al., 2019). Integrating new technologies into multisensory online experience, other interfaces involving sensory triggers (including touch screens, together with a range of virtual, and augmented solutions) are increasingly salient and available to people when interacting online (Petit et al., 2019). With the advancement of the field of human–computer interaction, online environments are likely to engage more of the senses and be more connected with offline environments in the near future. This expansion will likely coincide with an increasing engagement with the consumer's more emotional senses, namely touch/haptics, and possibly even olfaction (Agapito et al., 2013). Forward-thinking marketers and researchers are thus expected to appropriate the emerging techniques to deliver richer online experiences for tomorrow's consumers (Petit et al., 2019).

Sensory Design. The importance of addressing multisensory aspects in the design of tourism experiences and places has become apparent in a recent strand of research (Agapito, 2020). Recent comprehensive action-oriented frameworks in the book of *Design Science in Tourism* (Fesenmaier & Xiang, 2017) involve a sustainable approach to tourism experiences and place design, embraces the senses as one of its foundational aspects. They conclude experience as a response to the environment is generally considered a systematic process where sensation is the first order of experience. How people 'interpret' the stimuli and 'make' meaning from them and generally is framed within the mechanisms of perception. Thus, it is argued, travelers' perceived senses are not direct records of the world around them. Rather, they are 'constructed' internally along with constraints imposed by the construction of the nervous system and its functional abilities (Fesenmaier & Xiang, 2017).

Affordances in tourism: There is limited research discussing the application of affordances within the area of tourism design (Tomej & Xiang, 2020). Recently, Fesenmaier and Xiang (2017) listed affordance as one of the six components of the 'experience production system'. However, it appears that the concept has not yet received a careful, systematic examination in terms of how it is connected to various aspects of design in tourism. Tomej and Xiang (2020) present a case study wherein they used the concept of affordance to analyse the service experiences in an organised holiday for a group of tourists with and without visual impairment.

CONCLUSION

Literature reviews are available on specific senses, on using senses for marketing purposes, or on topics such as wine (Agapito, 2020; Brochado et al., 2021). This chapter provides a systematic overview of research incorporating the sensory component of tourism from a cognitive psychology perspective. Agapito's (2020) review finds that regardless of the variety of research methods used and multidisciplinary approaches in the research addressing the sensory component of the tourism experience, it can be observed that the articles in the dataset are predominantly developed from the perspective of tourists. Moreover, in general, only one stage of the tourism experience (pre-visit, onsite or post-visit) is addressed. A few exceptions (Agapito et al., 2017; Ballantyne et al., 2011) perform follow-up research. Notwithstanding, it is acknowledged that future research seeking to inform more effective tourism experience design around the senses should analyse how sensory-informed experiences evolve across different stages of the tourism experience and involve different stakeholders.

As in other chapters, the study of the senses has been undertaken from a variety of disciplinary perspectives, the gaze concept from sociology. These studies appear to assume that meaning lies within the object gazed upon, which is inconsistent with the cognitive approach adopted here. Despite this, there may be some benefit to reexamine the concept of the gaze from a cognitive psychology perspective. This may help, for example, to understand the consequences of 'gazing' such as an inability to engage with the viewed surroundings. Viewed from a cognitive perspective, 'gazing' involves difficulty in interpreting unfamiliar stimuli due to a lack of cues and past experience. This would appear to be a useful focus of further enquiry.

By addressing contemporary technological developments, neurophysiological instruments, multisensory components can be cannibalised by technology to enhance tourism experiences in an inclusive manner (Flavián et al., 2019b). However, the application in tourism remains limited. There is a surge of interest among a body of researchers who focus on the effectiveness of virtual reality on advancing understanding of sensory components. Alongside the advancement of immersive technologies, an analysis of the research methods employed in the articles indicates that sensory data collection has primarily been conducted through questionnaires and interviews. These methods could be complemented by using physiological measurement devices and mobile devices to better understand how senses interact with each other, and how different profiles of individuals react to different stimuli in diverse tourism contexts (Kim & Fesenmaier, 2017b). Technology can also aid the design of experimental research and longitudinal studies to examine the evolution of sensory-informed experiences across different stages of the tourism experience.

While a lot of work which predominantly focuses on single channel of five senses of humans has been done on sensory marketing in the last two decades, there is still a great need for additional research on the interaction of multi-senses: sensory congruence, sensory conflict and synaesthesia. In a retail setting, the congruence between the arousing qualities of scent and music improves approach behaviours (Mattila & Wirtz, 2001). Krishna et al. (2010) focus on the effect of the congruence between semantic associations of smell and touch. In contrast, many other studies have also examined vision–touch discrepancy and found strong or complete dominance of vision over touch. Little tourism work has examined sensory conflict even though many sensory conflicts exist within the consumption domain, which is an area for future research (Krishna, 2012).

Finally, synaesthesia, a new theoretical framework for understanding intersensory development, has been used to describe that sensory stimuli induce extra cross-modal or intramodal percepts (Spector & Maurer, 2009). Recent findings from cognitive psychology, functional brain imaging and electrophysiology have shed considerable light on the nature of synaesthesia and its neurocognitive underpinnings, which warrants scholarly attention from tourism (Grossenbacher & Lovelace, 2001).

Chapter 7

CONSCIOUSNESS

Noel Scott
Edith Cowan University, Australia

Liubov Skavronskaya
Griffith College, Australia

Abstract: This chapter clarifies the definition and reviews recent theories of consciousness, such as the dual systems framework, global workplace theory, and higher-order theory. Aspects of conscious experiences are useful in studying tourism decisionmaking, memory, emotions, tourism behaviour, and other topics related to tourism psychology. For instance, studies on conscious experiences of decisionmaking are useful to better understand destination choice and tourists' emotions elicitation, unconscious biases help to understand tourism attitude formation, while explicit and implicit cognition research provides insights to destination image attributes. Future studies may consider using objective and subjective methods of studying consciousness in the tourism context. **Keywords**: Consciousness; global workspace theory; priming; rosy view; dual-process theory; memorable tourism experiences (MTEs)

INTRODUCTION

'Consumers are not disembodied minds floating through space, but are instead biological machines whose behaviors are intimately tied to physical states'. (Williams & Poehlman, 2017, p. 17)

Consciousness from an everyday phenomenal perspective is the subjective experience of the present moment (Mannell & Iso-Ahola, 1987), or what it is like to perceive, feel, or think from a first-person perspective (Chalmers, 1998). It is the awareness of some of the content of mental processes and the

Cognitive Psychology and Tourism
Tourism Social Science Series, Volume 27, 73–83

Published under exclusive licence by Emerald Publishing Limited
ISSN: 1571-5043/doi:10.1108/S1571-504320240000027007

ability to describe them to others. Consciousness was a focal topic of study for psychodynamic psychologists, while later behaviourists considered consciousness beyond the scope of their research. Cognitive psychologists revived the study of consciousness but began by applying an information processing model. This early cognitive information processing model (Massaro & Cowan, 1993) used a computer analogy of data input, processing, and output. This model has difficulty with the complex the brain's interactive and recursive feedback processes. It also incorrectly assumes that the mental structures the brain uses have some overall unifying architecture and common data format rather than are the outcome of evolutionary processes. The information processing analogy is useful to examine how some higher-level processes are a function of lower-level neural structures. For example, 'Spatial attention is comprised of neural mechanisms that boost sensory processing at a behaviourally relevant location while filtering out competing information' (Fiebelkorn & Kastner, 2020, p. 221).

Today, consciousness is a core concept in cognitive psychology which views it as a function of neural circuits. Damasio (1998) divides the phenomenon into two categories: core and extended consciousness. The former is when an organism is wakeful and has alert awareness of the present. The latter is a sense of self, one's history and thoughts. Thus, consciousness may be thought of as a series of increasingly complex layers of processes based on neuronal circuits in different regions of the brain.

Anatomically, consciousness is dependent on three main brain structures: the brainstem, thalamus, and cerebral cortex. The brainstem is the oldest (in evolutionary terms) part of the brain and is responsible for wakefulness, homoeostatic regulation, and mediation of information from the senses. The thalamus receives sensory data and directs it to other parts of the brain for further processing and interpretation. The cerebral cortex works to maintain wakefulness, selective attention, and a sense of self built from memories and expected future experiences. There are disagreements as to whether neuronal processes necessary for core consciousness are located only in the brain or are more distributed throughout the body (embodied cognition).

Definitional Ambiguity

Applied in everyday life or in the non-cognitive psychology literature, the term 'consciousness' is used with a variety of meanings. For many authors, saying that consumers are conscious of a process denotes that they can verbalise that process and mentally represent it in a series of coherent thoughts (consciousness = awareness of internal processes) (Williams & Poehlman, 2017). However, for others, saying that consumers are conscious of a process suggests that they can manipulate that process via volitional, top-down, executive processes (conscious = control). For some, labelling a process as conscious indicates that it involves cognitive (as opposed to affective) processes (conscious = cognitive). This approach incorrectly seeks to distinguish emotional processes as non-rational and separate from 'normal' mental processes which are logical and rational. Still other authors define conscious processing in terms of effort or resource intensive mental activity (conscious = effortful or intentional). For example, intentionality (goal driven actions) has sometimes been considered one of the hallmarks of conscious thought although it is now known that individuals are able to elicit and pursue goals unconsciously. Many authors equate conscious thought with conscious decisionmaking (conscious = deliberate decision); however, individuals have been shown to make decisions unconsciously. Finally, researchers traditionally have viewed flexibility and adaptive integration of information as defining features of consciousness (conscious = flexible), but emerging evidence points to the existence of an adaptive, flexible unconscious.

Approaches to the definition of consciousness may be summarised under the three groups with different meanings. Consciousness is used in ways which conflate being aware of bodily sensations (meaning 1), reflexive self-awareness or being aware of one's own mental activity (meaning 2) and utilising 'higher' mental functions such as reasoning flexibly and deliberately leading to 'rational' goal seeking actions (meaning 3). On any of these criteria, however, most mental activity is not conscious. Instead, most actions are based on mental processes such as emotions, intuition, or habits of which one is mostly not reflexively self-aware. For example, 'feelings-as-information theory' (Avnet et al., 2012) considers those human sensations termed 'feelings' as the result of unconscious emotional processes. Individuals are also mostly unaware of normal bodily sensations although they may become aware of abnormal sensations registered as pain (Damasio, 1995). Furthermore, a mental goal can affect decisionmaking processes and behaviour without an individual being self-aware of the goal (Dijksterhuis & Aarts, 2010). Therefore, to better understand what consciousness is, one also needs to understand the unconscious processing. The next section discusses other important characteristics of unconsciousness processing from cognitive science literature.

COGNITIVE PSYCHOLOGY OF CONSCIOUSNESS

There are a number of models of consciousness.

Models of Consciousness

The dual systems framework is a useful model to understand conscious and unconscious states. The dual processing model highlights two mental 'information' processing systems, system 1 and system 2 (Kahneman, 2011). System 1 is an intuitive, rapid, automatic, and effortless process, while system 2 is a rational, slower, deliberate, and effortful process. The model derives from cognitive psychology and neuroscience and used in studies of attitude change and persuasion (Jun & Vogt, 2013; Shu & Scott, 2014) as well as decisionmaking (Walls et al., 2011). The dual processing model relates decisionmaking to cognitive psychology and neuroscience and provides a neural basis of it (Mograbi, 2011).

Recent cognitive and neuroscience literature associates system 1 with non-conscious processing and system 2 with conscious processing (Smith, & DeCoster, 2000). However, there are problems associated with mapping the notion of intuitive processing onto the distinction between conscious and non-conscious processes (Epstein, 1994; Kahneman, 2002). Kahneman (2002) argues that two system thinking categorises mental phenomena through multidimensional space with three principal axes: the distinction between automatic and controlled processing; the distinction between non-conscious and conscious processing, and relative degree of emotional content. The third axes links consciousness with emotions and feelings, and considers emotions are an integral part of most decisionmaking processes, rather than external influences acting on rational processes (De Sousa, 2013; Pfister & Böhm, 2008).

Global Workplace Model: Another useful model to understand consciousness is the global workspace model (Baars, 2005). The global workspace model provides a cognitive architecture that identifies consciousness with the global availability of information. According to the model, information becomes conscious when it is simultaneously made available to a wide range of processes – jointly comprising a 'global workspace' (Shanahan, 2005). This theory holds that when one is conscious of (essentially consciously paying attention to) integrated perceptual stimuli represented in working memory, other brain areas responsible for monitoring goals or emotional appraisal, access and process this information in a substantially modular fashion. A theatre stage is used as a metaphor to describe the workings of this model. The stage represents a dynamic mental picture of the self and surroundings held in the working memory and where the part of the stage lit by spotlights represents the focus of attention on certain stimuli. The brain represents the external world on the stage as a play by combining external perceptual stimuli and internal memories, scripts, and so on in a cohesive 'show'. The audience can interactively appraise the stage show and provide comments on it, for example, by producing neurotransmitters that stimulate bodily feelings which in turn become new input stimuli. Such reflexive processes are motivating and may influence the direction of the play. Overall, the global workspace provides a central information exchange, receiving its input from processes that currently command attention, and broadcasting widely across the brain. The global workspace model provides a theoretical basis for the advances in neuroscience with the development of other theories of consciousness, including higher-order theories of consciousness.

Higher-Order Theories of Consciousness: Higher-order theories of consciousness account for the distinction between unconscious and conscious brain processing and explain the difference in terms of a relation obtaining between the conscious state in question and a higher-order representation of a perception of that state, or a higher-order thought about it (Rosenthal, 2004). According to higher-order theory, both types of mental operations involve first-order manipulations, and what makes cognition conscious is a higher-order observation of the first-order processing (Lau & Rosenthal, 2011). Higher-order theories can be considered as an intermediate position between the GWM and early sensory views (Brown et al., 2019). The former posits that specific high-level processes are required, however, do not necessarily lead to global broadcasting and advantages in performance in simple tasks. Furthermore, most proponents of higher-order theories do not stipulate consciousness as equivalent to metacognition. Overall, the dual processing model deals with issues of consciousness and rationality but does not specifically address elicitation of emotion. It is being argued in the literature, that compared with other existing theories, higher-order theories more readily account for everyday experiences, such as emotions and episodic memories (Brown et al., 2019).

Predictive Brain: Evidence increasingly suggests that simulations implement patterns of prior experience to construct one's current experience, whether that experience is oriented in the past, in the here and now, or in the future. Simulation is the mechanism by which the brain capitalises on prior learning to efficiently navigate the situation at hand (Clark, 2013; Fernandez Velasco & Loev, 2020; Hoemann et al., 2017; Oettingen & Reininger, 2016).

Mental Biases and Misattributions: Unlike conscious thought, unconscious mental processes operate faster and effortlessly. One impact of faster processing and effortlessness is that associative processes are beyond deliberate control or monitoring. The parsimonious use of cognitive capacity does not allow for evaluating associations as true or false (Gawronski & Bodenhausen, 2006) and makes misattribution common (Wörfel, 2021). A second effect is caused by an unconscious striving for congruence. To operate efficiently, the unconscious mind relies, if possible, on already existing memory traces. This renders it more likely that individuals interpret novel information in a manner that supports existing beliefs (Baars, 1993). Implicit stereotypical beliefs toward social groups, for example, alter how individuals interpret ambiguous behaviour (Levy, 2014).

Unconscious Mental Activity and Decisionmaking

A third effect is that unconscious decisionmaking is more effective. The points illustrated above may create the impression that the unconscious hinders effective decisionmaking, because it confounds perceptual input and is biased in favour of existing beliefs (Wörfel, 2021). As mentioned previously, several important mental activities happen unconsciously. It is a common folk belief that conscious thought leads to behaviour (Williams & Poehlman, 2017, p. 231). However, one may undertake action based on feelings, intuition or habit and hence not involve reflexive self-awareness. There is also a body of literature demonstrating that a goal can affect decisionmaking processes and behaviour without the consumer being self-aware of the goal (Dijksterhuis & Aarts, 2010). Different ways of presenting a choice alternative can affect choice (Krishna & Schwarz, 2014). Incidental exposure to sensory experiences in one domain can affect choice in another, substantively unrelated domain (Krishna & Schwarz, 2014). Thus, consumer behaviour models that rely on conscious evaluation of attributes as the basis of decisionmaking ignore the mental antecedents of much behaviour. There is general agreement that decisionmaking is a function of the brain's mental activity and that one is reflexively aware of only a proportion of brain activity. The processes that enable much of the thinking are unconscious and unavailable for verbal report. Just as individuals do not have conscious access to the workings of the cardiovascular system or the autonomic nervous system, they also do not have access to the processes that lead to a decision. One has awareness only of the product – a mental representation of the world and its evaluation.

There is growing evidence that much of social judgement and behaviour occur without conscious awareness or intent, but in an implicit or unconscious fashion. The identifying feature of implicit cognition is that experience influences judgement in a fashion not introspectively known by the actor (Greenwald & Banaji, 1995). One of the reasons for this focus on implicit processes is that often theoretical models based on laboratory findings are found not to apply in the field. One of the main problems of these models as they were studied under more naturalistic conditions was the role played by deliberate conscious choice processes (Bargh, 2002). Today a number of studies examine implicit memory, implicit goal pursuit, implicit consumption, implicit theories, unconscious decisionmaking and make use of the concept of implicit attitude (Wörfel, 2021). There are also some examples from tourism, for instance Lee and Kim (2017) conduct single-target implicit association test to evaluate individuals' explicit and implicit cognitions of image attributes in three tourism destination countries to better predict behavioural intentions in tourism. Next section focusses on other examples where studies on consciousness might be useful for advancing tourism research.

Consciousness and Tourism Research

There have been several studies that have sought to apply research on consciousness in consumer behaviour that can be useful for tourism. For instance, Williams and Poehlman (2017) discuss the strong emphasis attributed to control of behaviour by access conscious, or that individuals willful and aware of the reasons for their actions. Duerden et al. consider conscious experiences and reactions are the primary determinants of behaviour and argues that experiences involving intrinsically motivated behaviour commonly found in recreation and leisure, are conscious 'as they involve instances where objective environmental stimuli attract and hold an individual's attention long enough to produce subjective reactions' (2018, p. 200). The concept of consciousness is a developing topic in tourism research.

Conscious Experiences and Decisionmaking: In consumer research, discussion of decisionmaking usually emphasises either rationality and objectivity or the emotional, sensual, and mystical aspects of human nature (Hirschman, 1985). Rational decisionmaking refers to mental activity that is effortful, goal directed, and that a person is aware of undertaking, while emotional and intuitive decisionmaking is considered non-rational. These attributions about whether human nature is 'objective and rational or sensual and emotional....ultimately address the nature of consumers' consciousness' (Hirschman, 1985, p. 223).

Such models also implicitly assume that unconscious mental activity is not rational. The concept of rationality is also complex, has a long history, and is associated with logic and an absence of emotion. Herbert Simon defined rationality in a social science sense as actions or thoughts that:

> contribute to certain goals, where these goals may be the pleasure or satisfaction of an individual or the guarantee of food or shelter for the members of a society". [..] When awareness and intention are present, the function is usually called manifest, otherwise it is a latent function. (Simon, 1978, p. 3)

Using this definition, 'rational' goal directed decisionmaking models can be characterised as exhibiting manifest rationality. However, the definition also attributes latent rationality to actions undertaken to achieve a goal even if a consumer is not consciously aware of that goal at that time. This is important since a large body of research has demonstrated unconscious top-down goal directed actions in studies of attention (Campos et al., 2016; Pieters & Wedel, 2007). Further, cognitive appraisal theory indicates that unconscious goals in part determine the valence and strength of emotion elicited in response to a perceived stimulus. On this basis, emotional decisionmaking models are rational although the consumer may not be aware of their goal. Thus, studies on conscious experiences of decisionmaking can provide insights to tourism decisionmaking and destination choice, as well tourists emotions elicitation process.

Unconscious Biases: Unconscious processing allows one to integrate an immense amount of information, to weigh vast numbers of attributes, and arrives at superior decisions for complex matters (Dijksterhuis, 2004). Conscious and unconscious arousal have a significant and divergent effect on attitude, which might provide useful insights for tourism attitude studies. Several studies have examined the effect of the subconscious on decisions. Boshoff (2017) examined subconscious gender effects on service failure and recovery, while Bettiga et al. (2017) distinguished between conscious arousal and unconscious arousal (Sharma et al., 2020).

Rosy View and Halo Effect: Studies understanding how tourists evaluate events and experiences have explored cognitive biases such as the rosy view which mitigates negative emotions in people's and magnifies positive experiences. These are specific examples of cognitive biases as discussed in Kahneman and Tversky's (2013) prospect theory. For example, Lee and Oh (2017) use the rosy view to justify that sharing experiential purchases leads to more positive behavioural outcomes than sharing material purchases. The rosy view is used to study food experiences (Sthapit et al., 2017), sports events (Schnitzer & Barth, 2019), and social media (Lee & Oh, 2017). It has also been applied to prospection (Sie et al., 2021). One plausible explanation for the positive bias is that although negative experiences reduce the enjoyment of the moment, these disappointments are fleeting (Mitchell et al., 1997) and people reinterpret their memories in ways consistent with their original expectations (Klaaren et al., 1994).

Priming: Priming is one of the most well-established phenomena in cognitive and neuroscience literature believed to occur unconsciously (Lohse & Overgaard, 2019), and considered to be one of the strongest pieces of evidence for unconscious mental processes and unconscious perception (Dolan, 2002). In general, priming refers to a preceding stimulus that affects a subsequent performance (Roediger, 1990). The individual is not aware of the effect and, sometimes, even the preceding stimulus remains below the threshold of conscious awareness (Strahan et al., 2002). Prime exposure activates knowledge structures that alter perception, motivation, evaluation, and behaviour (Bargh, 2006; Wörfel, 2021). For example, store atmospherics can be the result of unconscious priming.

Conscious Episodes and Tourism Experiences

Memorable tourism experiences (MTEs) have become an important topic for tourism research. As previously mentioned, cognitive psychology distinguishes consciousness as the process of moment-by-moment experience and memory as the outcome of an experience (Mannell & Iso-Ahola, 1987). The tourism literature examines the concept of experience, but it is still poorly defined and is used to refer to both the experience process and the psychological outcomes (Gao et al., 2012). In cognitive psychology, memorable experiences are defined as a remembered segment of consciousness (Baars &

Franklin, 2003). Dreams can be broadly defined as mental experiences about which memories can be described after waking (Pagel et al., 2001). There is agreement in the psychological literature that the stream of consciousness is divided into episodes (Bastiaansen et al., 2019).

Thus, research on conscious episodes and the stream of consciousness can further clarify the definition of MTEs. For instance, Bastiaansen et al. (2019) explaining the mechanism of stream consciousness by breaking down experiential episodes into their constituent elements, including sensory, cognitive, affective, behavioural, and interpersonal elements. Scholars argues that mental models stored in episodic memory serve to segment the stream of consciousness into experiential episodes (Figure 7.1, adapted from Bastiaansen et al., 2019).

These episodes can occur on different time scales, but usually are only seconds long. Episodes occurring at longer time scales (so-called extended experiences) are not related to a single event or activity; rather, they together form the narrative of an episodic memory trace, with specific emotions linked to it (Bastiaansen et al., 2019). This distinction raises the question of precisely when and how an experiential episode translates into a memorable experience. The management implications of the bridge between experiencing and memorable experiences give this question urgency and where consciousness can add value. Thus, by reviewing the studies on consciousness from cognitive science literature and applying them to tourism context, one knows that external stimuli and internal mental states and processes make up the continuous stream of consciousness. This stream is being partitioned into episodes through the activation of mental models, or schemata in episodic memory. If an experiential episode elicits sufficient (strong) emotions, it is stored in episodic memory and as such it transforms to MTEs over time.

There are other examples of how the studies on consciousness are useful for tourism. For instance, Hosany et al. (2020) developed an emotion apps model that views consciousness/unconsciousness as a continuum and

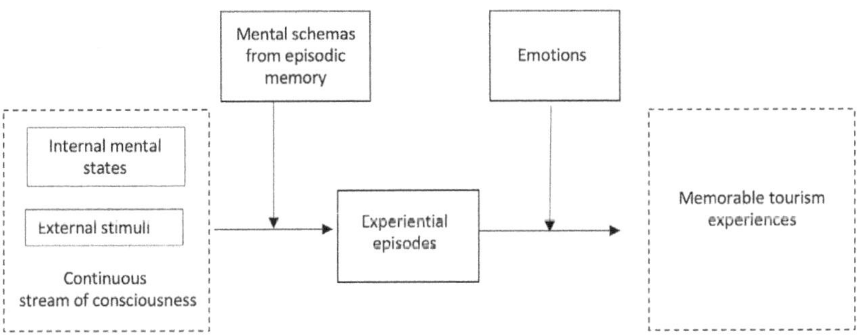

Figure 7.1. Continuous Stream of Consciousness and MTEs.

consider that the individual often does have some (but incomplete) ability to interpret some aspects of her or his unconsciousness. Furthermore, Ellis et al. (2015) proposed a process-based structured tourist experience framework based on fast and slow thinking or dual-process theory (Kanheman, 2002). Structured experience theory describes the flow of tourists' immediate conscious experiences across different tourism encounters such as dining experiences, sporting events, theatrical performances, concerts, and museum visits. This is critical to optimise probability of positive tourist experiences, which directly related to satisfaction, emotion elicitation, and MTEs.

Prospection and Retrospection

The brain uses the same mental processes to think about future events (prospection), current events, or past events (retrospection) (Suddendorf & Corballis, 2007). However, the process of imagining future events draws on memories and therefore may be less detailed or vivid than the current experience. This is why the dream beach holiday destinations are associated with prototypical attributes, such as a beach, blue sky, and sea and palm trees, perhaps with a hammock, sailboat, and seagull (Kane et al., 2012). Similarly, the memories of the future are constructed based memories of past emotions are reconstructed based on subsequent reinterpretations of past events (Levine, 1997; Schacter & Addis, 2007). For example, participants' recollection of their emotional responses to an event can be altered if their current goals used to appraise the event have changed from the time the event had occurred (Levine et al., 2001). Such a reconstructive process provides a mechanism for savouring holidays (Filep et al., 2013).

Implicit Association Test: Psychological research on consciousness can also contribute to tourism research methods. For instance, researchers using the implicit association test identify two types of cognition towards objects of interest: conscious (explicit) and unconscious (implicit) (Greenwald & Banaji, 1995). In tourism, for instance, the implicit association test can improve studies of destination image and attributes (Lee & Kim, 2017) by facilitating exploration of individuals' explicit and implicit cognitive perceptions towards a destination country and to predict tourists' behavioural intentions. Furthermore, Tse and Tung (2020) applied this test to study tourist stereotypes.

CONCLUSION

This chapter reviewed contemporary theories of consciousness with an emphasis on cognitive functions and conscious experience. The discussion contributes to research by clarifying the definition of consciousness from a first-person perspective, and discussing aspects of subjective conscious experience, such as memories, emotions, feelings, decisionmaking, rather than neuroscientific aspects on consciousness. This chapter considers consciousness as a moment of awareness. There are three definitions of consciousness that tourism researchers need to be aware of. The first aspect is reflexive self-awareness, the second meaning of consciousness is being aware of bodily sensations, and a third one is utilising 'higher' mental functions such as reasoning leading to 'rational' goal-seeking actions. Understanding these nuanced aspects of how consciousness is defined can improve tourism research. For instance, if one thinks about consciousness as an experience of feeling and understanding sensory input, of a temporal sequence of autobiographical events or memories, it helps a better understanding of memorable tourism experiences by emphasising the importance of understanding tourist's past experiences to improve experience design and MTEs outcomes.

Research on consciousness is useful in several ways. For instance, tourism concepts that can benefit from utilisation of the theories of consciousness from cognitive and neuroscience include but are not limited to decision-making in tourism, mental time travel, mindfulness, tourists' emotions, and memorable experiences. Modern cognitive science defines consciousness as self-awareness of one's own mental activity and a 'higher' mental function such as reasoning, leading to rational goal-seeking actions (Dijksterhuis & Aarts, 2010). The approach taken here is like that in other applied areas such as consumer behaviour studies (Wörfel, 2021).

Studying different aspects of conscious experiential episodes improves the definition of MTEs by better distinguishing between the experience as a process and the psychological outcomes of such experiential episodes, including memory. It also has management implications by connecting experiencing (present moment) and memorable experiences (outcome). This is important as often memorable experiences predict future revisit or recommendation decisions. Secondly, understanding conscious experiences of embodiment helps to better distinguish emotions from feelings and to better understand the embodiment phenomena, which is critical in studies on immersive virtual tourism experiences. Last, but not least, applying cognitive models of consciousness improves understanding of rational and irrational aspects of the decisionmaking process in tourism.

Research on the mechanism of consciousness explains that behaviour is often the result of unconscious mental processing, not conscious volition as in many situations the brain retrospectively produces a reasonable explanation for behaviour. Thus, the brain may generate a reason for an action unrelated to the actual basis of the decision. Future studies can focus on using methods in experiment design that make a clear contrast between conscious and unconscious processing or between different conscious experiences in tourism context. Objective and subjective tourism data can be combined by correlating neural activity with subjective experience to identify the neural networks that are specifically involved in conscious processing.

Chapter 8

EMOTIONS

Vikki Schaffer
University of the Sunshine Coast, Australia

Lee Kannis-Dymand
University of the Sunshine Coast, Australia

Liubov Skavronskaya
Griffith College, Australia

Noel Scott
Edith Cowan University, Australia

Abstract: Emotion is a key cognitive process that is central to being human. This chapter discusses various psychological approaches to understand emotions and introduces cognitive appraisal theory and its appraisal dimensions. By reviewing recent studies on emotions, this chapter recognises the theory as one of the most promising ones to advance research on emotions. The significance of appraisal theory for tourism is in the ability to explain why the same stimulus can generate different emotions in different people or in the same person at different times. Cognitive appraisal theory is able to assist in the prediction of emotions and subsequent behaviour. This chapter concludes by outlining potential topics where the theory can be particularly useful in tourism. **Keywords**: Emotions; cognitive appraisal theory; agency; novelty; memory; goals

Cognitive Psychology and Tourism
Tourism Social Science Series, Volume 27, 85–96

Published under exclusive licence by Emerald Publishing Limited
ISSN: 1571-5043/doi:10.1108/S1571-504320240000027008

INTRODUCTION

Emotion is a key cognitive process that is central to being human. Tourism services are usually designed to be pleasurable and therefore to elicit positively valanced emotions. However, negatively valanced emotions may be associated with a failure to meet personal goals for a holiday or leisure activity. Understanding how and why emotions are elicited is critical to understanding tourists' decisionmaking, attitude, behaviour, and satisfaction. Emotions are useful in the study of experience as they are seen as central to memorability.

As will be found in many of the concepts discussed in this chapter, there are several different approaches to emotion, some from folk psychology and others from different branches of psychology. This chapter examines emotions through the lens of the cognitive paradigm. Cognitive psychology defines emotions are mental states resulting from appraisals of experiences. Within cognitive psychology, there are several theories of emotional appraisal. Cognitive appraisal theory in particular emphasises that the emotion elicited is based on a limited number of appraisal dimensions (Johnson & Stewart, 2005; Roseman, 2001; Roseman et al., 1990). In other words, emotions are produced by a cognitive evaluation of the eliciting circumstances. The core assumption of all cognitive theories of emotion is that stimuli per se, do not elicit emotions, but stimuli as it is appraised. This helps to explain why the same stimulus can generate different emotions in different people, or in the same person at different times.

Chapter 2 in this volume provided a general outline of cognitive psychology and cognitive appraisal theory was briefly introduced. This chapter discusses the various approaches to emotion commonly used in tourism research; defines cognitive appraisal theory and its appraisal dimensions; and relates the literature of emotion to tourism and how it can be useful. It argues that knowledge of the cognitive emotion elicitation process and appraisal dimensions can improve studies of tourists' decisionmaking, memory formation, and other processes.

PSYCHOLOGICAL THEORIES OF EMOTIONS

There is an extensive history of intellectual discussion of emotion potentially dating back to the Greek philosopher Aristotle. Originally, this was philosophical in nature, concerning the mind and body, cognition, and rationality. It is convenient here to begin the history of scientific studies of emotion with the work of Charles Darwin (1809–1882). Darwin in his book *The Expression of the Emotions in Man and Animals* (1872) proposed eight families of emotion and their functions. Darwin wrote that an emotional state is embedded (or encoded) in behaviour and can serve as a universal display of information to others (Gendron & Barrett, 2009). For example, there are specific configurations of facial muscle movements that correspond to different emotion categories in a one-to-one manner (Ekman et al., 1972; Izard, 1971; Tomkins, 1962), such as smiling. Furthermore, Darwin assumed that 'expressions' are habits or reflexes that occur without the conscious intent of the organism (they are automatic). Following his work, Tomkins (1962) developed a psychobiological basic emotion theory.

Basic emotions theorists (Ekman, 1992; Izard, 2008; Plutchik, 2001; Tomkins, 1962) believe that there are a number of discrete emotions that are largely universal in animals (Volo, 2021). These theorists identify certain biologically privileged emotions that are automatically elicited by objects and events. Thus, emotions of the same kind share some common biological cause and show the same pattern of behaviour, bodily activation, and facial actions (called 'expressions'). This allows virtually all people to easily recognise an expression on another's face. Emotional states involve feelings, action tendencies, appraisal, motor expression, and physiological activity.

Various emotion classification schemes have been developed, with two of note being Mehrabian and Russell (1974) and Plutchik (1980). The three dimensions of pleasure, arousal, and dominance were found to capture more information about the emotional character of consumption experiences than did the eight basic emotional categories recommended by Plutchik (Havlena & Holbrook, 1986). Classification schemes, such as the consumption emotion scale (Richins, 1997), pleasure-arousal-dominance model (1980), and positive affect negative affect scale (Watson & Tellegen, 1985), do not explain how an individual's consumption experience leads to one specific emotional response rather than another (Johnson & Stewart, 2005). These models are not intended to explain the formation process for emotions, but rather a method for measurement.

Constructivist emotion theories such as those of Barrett (2017) and Siegel et al. (2018) postulate that emotion categories are not 'unavoidable consequences of the genes, rather they emerge out, depend on context and are conceptual categories' (Barrett, 2017; Siegel et al., 2018; Volo, 2021, p. 4). Constructivist theories include two main components: a mental representation of bodily changes or sensory body input related to an 'object' and a process of accessing memory and prior knowledge that enables the brain to create meaning (Barrett, 2017; Lindquist et al., 2013). Barrett (2017) focusses on the effect of the social environment (through a person's past experiences and schema) as determining emotions. In this view, emotions are based on a person's memories, thus variation is not uncommon. Social pressure determines emotional expressions in a process like coping. Overall, constructionists emphasise variation in emotional elicitation due to a person's experiences and challenge the basic emotion theory assertion that there is some dedicated circuit for each emotion and instead considers the brain constructs an instance of emotion by categorising the present moment using 'emotion concepts' from one's culture.

Appraisal Theories of Emotion

Appraisal theorists, like constructivists, assume that emotions are elicited by an individuals' assessment of an event or situation (Lazarus, 1991; Roseman, 1991; Scherer, 1984; Scherer et al.,2001). They agree with constructivists that the quality and intensity of an emotional response changes depending upon an appraisal of the relationship between the individual and the situation. However, these theorists differ from constructivists as certain criteria, called appraisal dimensions, are used to assess the situation. Further, that it is the combination of appraisals on these dimensions that determine the emotion elicited. Appraisals are defined as 'the results of the information-processing tasks that indicate the implications of the situation for the interests and goals of the individual and therefore determine the form that emotional reaction takes in a given situation' (Johnson & Stewart, 2005, p. 5). While appraisals may not be the single cause of emotions, it is evident that they are pivotal in 'generating and shaping emotions' (Uusberg et al., 2019, p. 270).

Appraisals start the emotion elicitation process, initiating the physiological, expressive, behavioural, and other responses that comprise the resultant emotional state (Roseman, 2001; Scherer & Moors, 2019). The appraisal process implies a cognitive mechanism, a subjective construal of personally relevant information that generates the variability in emotional reactions (Frijda et al., 1989; Lazarus, 1998; Moors et al., 2013; Ortony & Clore, 2015; Ortony et al., 1988; Roseman et al., 1990; Scherer, 2022).

Cognitive appraisal theory was developed to further explain variations in emotional responses and the mechanism of emotion elicitation associated with stress and coping processes. Lazarus and Folkman (1984) developed a transactional model of stress to explain responses to stressful events based on a two-pronged cognitive appraisal process involving primary and secondary appraisals. Primary appraisal occurs when an individual interprets an event based on motivational relevance and importance to personal goals (Bagozzi et al., 1998; Bagozzi & Utpal, 1999; Perugini & Bagozzi, 2001). The purpose of secondary appraisal, or emotion-focussed coping, is to lessen emotional distress (Folkman et al., 1986). Several coping strategies may be adopted such as avoidance, or wishful thinking. During secondary appraisal, the extent to which the situation either inhibits or facilitates achievement of personal goals is assessed along with an appraisal of the resources or options for coping (see Chapter 14 on coping).

Different dimensions have been used to examine the structure of the appraisal process. While some cognitive appraisal theories propose a categorial and discrete number of appraisals, other appraisal theories have a dimensional approach proposing the potential for multiple appraisals (Moors et al., 2013). For instance, Lazarus (1991) introduced six dimensions: goal-relevance, goal-congruence or incongruence, type of ego-involvement (self or event), blame or credit, coping potential, and future expectancy. Scherer et al. (2001) distinguished among appraisals of relevance, consequences, coping potential, and normative significance. Ellsworth and Scherer (2003) identified seven major appraisal dimensions: novelty, intrinsic pleasantness, certainty or predictability, goal significance, agency, coping potential, and compatibility with social or personal standards. Watson and Spence (2007) developed a core set of appraisal dimensions: goal relevance, goal congruence (also called motive consistency), certainty, coping potential (also called control and/or power), agency (also called cause) (Moors et al., 2013), surprise, uncertainty (in relationship to expected experience) or novelty (Ma et al., 2013; Skavronskaya et al., 2020), and norm/self-compatibility (Scherer, 1984). Cognitive appraisal posits that each emotion has a unique appraisal pattern. For instance, the appraisal structure of delight involves appraising an event outcome as highly relevant to a goal; appraising the event as congruent with the goal and surprised about the outcome of the event.

Five appraisal dimensions are relevant in a consumption context goal congruence, goal importance, agency, certainty, and novelty. The valence of the emotion is determined by goal congruence. An appraisal of goal congruence leads to positive emotions and non-goal congruence leads to negative valanced emotions. Appraisals on the goal importance and novelty dimensions affect the strength of emotional arousal leading to mild positive emotions, including pleasantness or highly aroused emotions such as elation or delight (Table 8.1).

In psychology, individuals are considered to take action to regulate their emotions and maintain their sense of wellbeing. Positive emotions encourage continued pursuit of successful goals (Fredrickson, 2001; Garland et al., 2010), while coping strategies such as shifting goals may be used to moderate negative emotions (Lazarus, 1993) (see chapter 14 on coping).

Often the terms 'feeling' and 'emotion' are used interchangeably but are distinguished here. To clarify the variability, feelings may be distinguished as an intention 'I feel like going' are related to motivation; feeling as a sense 'feels rough' are related to sensation and perception (Chapter 6); or feeling as emotional representation such as to 'feel happy'. The later sense is that used as information theory where feelings provide information to the brain (Avnet et al., 2012; Schwarz, 2011). Other researchers have provided a neurobiological view of feelings (Damasio, 1995; Damasio & Carvalho, 2013). Here, feelings are the result of an appraisal process triggering the production of neurotransmitters leading to an 'emotional body state'.

Emotion processes result in secretion of chemical monoamine neurotransmitters that evoke sensations in our bodies. These sensations in turn provide input that enable the brain to recognise that one has experienced a

Table 8.1. Appraisal Dimensions and Definitions.

Appraisal Dimension	Definition of the Appraisal Task	Effect
Goal congruence	Evaluating the situation in terms of (expected) goal success vs. failure, or whether it is consistent vs. inconsistent with values or ideals	Differentiates between positive and negative emotions (valence)
Agency	Inferring whether there is a person (self or event/circumstance) that is responsible for, or in control of, the situation	Differentiates emotions that focus on the self, another person, or an object, from emotions that do not reference an agent
Certainty	Determining whether the outcome is known or certain	Differentiates between outcome-related emotions and anticipatory emotions
Novelty	Evaluate the unexpectedness or suddenness of a situation or an experience	Differentiates on emotional arousal
Goal relevance	Evaluate the importance and value of the desired state in the current situation	Differentiates emotional intensity

particular emotion, and thus influence, subsequent behaviour (Moyle et al., 2019). This allows emotional states to be signalled back to the limbic and somatosensory systems in the brain. An emotion of happiness may result in a smile and thereafter the muscles around the mouth, the effect of neurotransmitters released, and so on are subsequently reported back to the brain reinforcing a feeling of happiness. This is why bodily feedback or unconscious priming, such as asking a person to smile, can successfully induce emotion (Briñol & Petty, 2003).

More generally, emotions and feelings have a recursive relationship with each affecting the other (Han et al., 2007). Emotions are motivational and informational, primarily by virtue of their experiential or sensational component. The various appraisal dimensions commonly used in cognitive appraisal theory shed more light on the subject.

Goal Congruence. Goal congruence, sometimes referred to as motive consistency (Roseman et al., 1990), outcome desirability (pleased/displeased) (Ortony et al., 1988), intrinsic pleasantness (Scherer, 1993), or pleasantness

(Frijda, 1987), determines the valence of emotional responses and differentiates positive and negative emotions. Goals are 'abstract benefits sought by the consumer that are available through the (abstract or concrete) features of a product class that offer fulfilment of these goals' (Huffman & Houston, 1993, p. 194). For example, a goal sought by a diner in a high quality restaurant could be to have an enjoyable meal as a celebration of family reunions or other special occasions (Bowden & Dagger, 2011). Appraising goal congruence results in an assessment of whether achievement of a personally relevant goal is facilitated or hindered in the current situation (Watson & Spence, 2007). An evaluation response results with the motivation pertaining to a personal goal (Frijda, 1987). If an individual perceives the situation as moving them closer to a desired goal, positive emotions are more likely to be elicited and accompanied by approach tendencies. The goal congruence dimension is the most basic, and accounts for the variabilities among different emotions (Ruth et al., 2002).

The nature of the person's goal is also significant to the emotion elicited based on the situation. If it is a helpful goal, the pursuit of which is good for the person's wellbeing, the goal congruency evaluation is likely to elicit a positive emotion (happiness, joy, or gratitude). However, if it is a goal that the person may seek to avoid, the goal incongruence evaluation is more likely to generate a negative emotion. Therefore, appraisal theories list goal significance as a sub-dimension of goal congruence to refine the variance among emotions of the same valence. This is in line with outcome desirability (appetitive [desirable] or aversive [undesirable]) identified as a dimension by Roseman (1984). People's appetitive goal congruence is like the disconfirmation of their inner-generated motivations and outcomes.

Agency. Appraising agency involves the evaluation of the various entities (people, objects, products, etc.) in a situation. The process of appraising agency involves assessing whether the person or object is perceived as causal or responsible for the outcome in the situation. Agency is believed to be more relevant in situations involving negative emotion than those involving positive emotion (Peeters & Czapinski, 1990). A diner may experience the emotion of anger if a service failure is appraised as the fault of the waiter. A sense of guilt may manifest when individuals attribute a negative outcome, like losing a game, to their skill. However, positive affect, such as gratitude and pride, are empirically differentiated from happiness/satisfaction by different appraisal agencies of other people, self, and circumstance (Soscia, 2007).

An adventurer will probably feel proud if they believe their success is due to their ability, grateful if the achievement is with another's help, or satisfied if they attribute the accomplishment to a tool or device that is of good quality and helpful to the completion of the adventure. The difference in agency determines the target of emotions (Lazarus, 1991; Roseman, 1984). It is an appraisal of responsibility (Manstead et al., 1989), or as the locus of cause in attribution theory that suggests the elicitation of key emotions, such as pride, shame, and anger can be distinguished solely on the basis of internal-external attribution of responsibility (Weiner, 1985).

Certainty. The appraisal of certainty, also known as likelihood or outcome probability, involves the assessment of the extent to which the situation is perceived to lead to a particular outcome. Past events are perceived to be certain since the outcomes of events have happened already, while future events are uncertain. Certainty in a goal outcome can arise from it coming to pass in the current situation, as well as through the knowledge or experience about the goal having been achieved previously in a similar situation. Alternatively, uncertainty may arise from a lack of experience or knowledge about achieving the goal in a given situation. The emotions generated by appraisals of certainty serve a motivational function to convince individuals about the likely success of continuing to pursue the goal or changing to another goal, either because the current goal has been achieved or is unlikely to be achieved (Bagozzi et al., 1998). The certainty dimensions distinguish anticipatory emotions that refer to whether an outcome is known or uncertain, such as hope, anxiety, and fear. With post-consumption emotions, the dimension of certainty predicts emotions of joy or relief based on pre-consumption perceptions of the situation as certain or uncertain.

Novelty. This dimension allows evaluation of the extent to which an experience departs from an individual's expectation. Novelty is interpreted by people as the suddenness or unexpectedness of an experience, and this high degree of unexpectedness is associated with delight (Scherer, 1993). There requires no or low level of unexpectedness for the elicitation of moderate emotions (see chapter 10 on novelty).

Dimensions Affecting Emotional Intensity

Two types of appraisals have been linked to the intensity of emotion: goal relevance, also known as motivational relevance (Frijda, 1987; Lazarus, 1991; Scherer, 1993) or importance (Ellsworth & Smith, 1988a, 1988b); and goal realization, also known as the degree of goal congruency (Johnson & Stewart, 2005), or the extent to which the event is consistent with personal goals (Lazarus & Smith, 1988), or the consistency of the outcome with expectations for the present time (expectation) (Scherer, 1993), or prospect (Ortony et al., 1988).

Goal relevance is the assessment of the significance of the event or situation to the person's goals or needs. The appraisal of a goal's importance affects the intensity of emotion because it is associated with the value or desirability of the state that is sought, or the severity of potential consequences of failure (Perugini & Bagozzi, 2001; Sonnemans & Frijda, 1995). Emotional reactions are more intense because of the expectation of more valuable gains or more painful losses. As Scherer (1997a, p. 141) noted, 'events that give rise to joy are evaluated as highly anticipated'. Goal relevance also impacts on the intensity of enjoyment because it is related to interest in the experience (Ellsworth & Smith, 1988a, 1988b). Interest is cognitive in nature and usually linked with motivation or goals (Krapp, 1999). In addition, interest is a general response to situations perceived as subjectively important and motivates high levels of attention activity, hence increasing chances of effectively coping with the event and eliciting an intensified positive emotion such as enjoyment. Therefore, goal relevance includes two sub-concepts as importance and interest in this thesis. In the case of a goal congruent situation, the goal relevance determines the intensity of the positive emotion and helps specify the discrete positive emotion.

Goal realisation relates to emotional intensity as well, with respect to the degree to which the situation meets expectations or approximates the desired state (Ortony et al., 1988). It is called prospect realisation and defined as, 'the degree to which an anticipated event actually occurs' (Ortony et al., 1988, p. 84). If the situation exceeds expectation, the resulting positive emotion is likely to be more intense than if the situation merely met expectation or matched the desired state.

Emotional intensity within the context of appraisal theory is similar to the concept of arousal used in dimensional theories of emotion (Johnson & Stewart, 2005). However, while dimensional theories do not explain the factors leading to the intensity of emotions, appraisal theories specifically address and predict the degree of emotional intensity that is likely to be experienced. In addition, appraisal of intensity is independent of other appraisals (Scherer & Ceschi, 1997). The appraisal dimensions of goal relevance and realisation can distinguish emotions of the same valence but of different intensity, such as joyous and happy, angry and irritated, hopeful and expectant, anxious and afraid.

Satisfaction

In marketing research, concepts multiply often without concern for their overlap in definition with others in different fields. Satisfaction, a concept commonly used in the marketing and tourism research, has been defined as 'result of the interaction between a tourist's experience at the destination area and the expectations he had about that destination' (Pizam et al., 1978, p. 315). This definition does not explain what satisfaction is but only how it may be formed. Here it can be 'translated' satisfaction into the concepts of cognitive psychology. Duerden et al. (2018) distinguishes an experience (say a hotel dining experience) as a current state of our mental processes (termed erlebnis in German) and the recalled memory of that hotel dining (termed erfahrung in German).

In the theory of expectation congruency, satisfaction is conceptualised as dependent on an appraisal of the extent of expectancy disconfirmation, comparing the expectations and outcomes of consumption experience (Dubé & Menon, 2000). This process essentially corresponds to the appraisal of the direction and degree of goal (Scherer & Ceschi, 1997), the dimensions related to valence and intensity of emotion in appraisal theory. In addition, the ratings of satisfaction are consistently found to be related to the ratings of the degree of positive and negative affect, which are likely to relate to the degree of goal congruence and valance appraisals (Mano & Oliver, 1993). According to Kumar and Oliver, 'disconfirmation may be evocative of various consumption specific appraisals consistent with those in appraisal theories' (1997, p. 17). The specific emotions resulting from these appraisals are shown to describe the affective content of the satisfaction response (Oliver, 1993). In essence this means that satisfaction as an immediate outcome of an experience may be considered a feeling that is followed by a cognitive evaluation (Bagozzi & Utpal, 1999).

There are only a few empirical studies in marketing that have tested appraisal dimensions within consumption experiences, and most of these were conducted in an experimental research situation. Appraisals of goal congruence and agency were adopted to examine the consumption emotions of gratitude, happiness, guilt, anger, pride and sadness, and their impacts on behavioural intentions in experimental scenarios (Soscia, 2007). Soscia (2007) found that gratitude towards others, but not happiness, predicted more repurchase intention and positive word of mouth, while guilt caused by oneself inhibited complaint behaviours and negative comments. Appraisals of goal congruence and controllability/power over the situation were found to be significant in predicting emotions during a service encounter by analysing video vignettes with positive and negative experiences during a restaurant encounter (Dalakas, 2006). This study indicated that the customer might still experience positive emotions when an event is inconsistent with the customer's desire if there is a perceived ability/power to cope with the undesirable event. The appraisals of goal congruence, goal relevance, and coping as potential determinants of consumption emotions, such as anger, sadness, and joy/satisfaction were examined in an experimental study (Nyer, 1997). Nyer (1997) found that goal congruence situations lead to positive emotions, while goal incongruence events lead to negative emotions. Further, goal relevance as a moderator was found to increase the intensity of the emotions experienced. The study argued that complaining behaviour, as consumers' coping mechanism to unfavourable consumption, facilitates the emotion of anger.

In summary, appraisal theories have been used in the marketing field although not in a systematic way. Researchers have suggested integration of more appraisals to further understand the distinct emotions elicited and their effects on satisfaction, as well as behavioural intentions. For example, 'an examination of additional factors that impact emotional responses, such as goal importance and certainty appraisals, might provide increased precision in understanding and predicting reactions to consumption experiences' (Johnson & Stewart, 2005, p. 25).

Tourism Applications

Emotions are an important component of a tourism, contributing to the formation of experiences in parallel with cognitive evaluations and hence satisfaction (Hudson & Brent Ritchie, 2009). Emotions such as satisfaction, as well as delight, hope, pride, and relief, are important outcomes from a purchase in terms of their effects on loyalty intentions, such as positive recommendation and repurchase (Bagozzi & Utpal, 1999). The majority of tourism studies of emotion have examined recalled emotions and using behaviourist approaches (Tuerlan et al., 2021). For example, the willful recall of positive emotions has been termed 'savouring' (Filep et al., 2013).

In tourism, cognitive appraisal theory has been used to study the important role of emotions in determining memorable tourism experiences (Hosany & Gilbert, 2010; Hosany et al., 2020). Hosany (2012) investigated emotions of happiness, positive surprise and love among tourists and their correlations with appraisal dimensions of goal congruence, certainty, agency, novelty, and self-compatibility. Ma et al. (2013) examined the appraisal processes eliciting delight. A first path involves appraisals of appetitive goal congruence and unexpectedness on the dimension of novelty; and a second path to delight is through high degree of goal realisation. A third path that elicits delight is through relevance to the personal wellbeing or special needs under the condition of appetitive goal congruence; and a fourth path is through personal interest and goal congruency. One study has examined the effect of negative emotions from flight delays (Jiang et al., 2020). Another has examined the relationship between emotion and storytelling using cognitive appraisal theory and script theory (Manthiou et al., 2017).

Tourism studies have used cognitive psychology to examine coping (Zheng et al., 2021), complaint behaviour (Loo et al., 2021), and pre-consumption emotions (Le et al., 2020). Ma et al. (2017) were able to distinguish between elicitation of delight or satisfaction based on goal importance. A recent review discusses measurement issues (Hosany et al., 2020). There is a limited discussion of feelings in the tourism literature (Buda et al., 2014; Liu et al., 2020; Shin et al., 2018; Williams, 2006), with most studies focusses instead on perception of external physical stimuli (touch, taste, smell, and vision) rather than perceptions of conscious or unconscious emotional signals (Agapito et al., 2014).

CONCLUSION

This chapter discusses the concept and theories of emotion from cognitive psychology and how it may be applied to tourism. While tourism research emphasises the growing demand for more dynamic and emotional tourist experiences (Kim & Fesenmaier, 2015; Nawijn et al., 2013), the use of recent theory of emotion and related phenomena is only beginning. Cognitive psychology acknowledges the dynamic nature of the appraisal mechanism whereby an appraisal process is followed by re-appraisals, which follow changes in the environment and in internal variables, and incrementally shape emotions over time (Gao & Kerstetter, 2018). It appears that emotions play a critical role in the operations of mental processes. Indeed, some scholars consider emotions as largely determining the contents and focus of consciousness throughout the life span (Izard, 2008). Emotions provide motivation to achieve future goals and provide guidance on current goal-relevant circumstances. Emotions provide subconscious, but rational feedback about goal achievement through the mediation of feelings (Izard, 2008). Dimensions, such as goal valence, goal importance, and novelty are central to appraisal theories and provide a useful first approximation of the type of emotion elicited.

The emotion elicitation process is related to coping and feelings processes. Together these form a cohesive account of how emotions are elicited through an appraisal process, may be revised through coping mechanisms and lead to feelings and hence provide situational feedback. This comprehensive account of the emotional process must be useful to tourism researchers due to their interest in the effect of both positive and negative experiences. This chapter has discussed satisfaction as a result of emotional process. Further research on satisfaction and its relationship to goal strength would be of use. A number of authors have examined risk a 'feeling' (Loewenstein et al., 2001; Luna-Cortés, 2021). This is also an area of research related to entrepreneurship.

Furthermore, cognitive appraisal theory is able to map out the beginning of emotions elicitation process (a subjective reaction to the experience) which other theories of emotions are unable to capture (Scherer et al., 2001), and to contribute to the studies on the long-term consequences of emotional experienced in tourism, such as memorability, satisfaction, or loyalty. Memories and their associated emotions about an event are recreated. Experiences after an event may explain how a recalled event can evoke a different feeling compared to the original (Scherer et al., 2001). The appraisal process is important for tourism not only to understand the emotional experience of the present moment or emotions in recalled experiences but also to predict emotional reaction to future tourism experiences to better understand tourism decisionmaking (Han et al., 2007).

Chapter 9

MOTIVATION AND GOALS

Ana Cláudia Campos
University of Algarve, Portugal

Noel Scott
Edith Cowan University, Australia

Abstract: Goals play a critical role in everyday life and shape life and lifestyles. This chapter reviews and discusses the main approaches to the study of goals, addressing main theoretical endeavours that have emerged from psychological research, and emphasising the contributions from cognitive psychology to the topic. It also discusses literature on goals and motivation as found in tourism studies, highlighting the theoretical underpinnings used in this field. The discussion concludes with the need to update, in tourism studies, knowledge from cognitive psychology and propose how a better understanding of the cognitive foundations of goal-oriented behaviour can contribute to change tourism practice. **Keywords**: Goal-oriented behaviour; motivation; drive; executive function; self-determination theory; theory of planned behaviour

INTRODUCTION

The study of motivation and goals is central to the social sciences and to understanding why people do what they do (Eccles & Wigfield, 2002). Motivation is considered the driving force of all behaviour (Fodness, 1994), providing an explanation to why a person in a situation chooses one course of action rather than another in an energised and repeated way (Gollwitzer & Oettingen, 2012), and goals are what controls and provides the direction of the motivation (Dijksterhuis & Aarts, 2010). Goals guide everyday life and shape the lifestyle (Hommel, 2022). Goals' capacity to produce response is

Cognitive Psychology and Tourism
Tourism Social Science Series, Volume 27, 97–113

Published under exclusive licence by Emerald Publishing Limited
ISSN: 1571-5043/doi:10.1108/S1571-504320240000027009

cognitively important to explain it (Gollwitzer & Oettingen, 2012). It would therefore seem important to understand what goals and motivation are.

As has been found for the concepts examined in other chapters, there is a long history of the study of motivation and goals within various disciplines. For example, sociologists consider goals as end-results towards which an individual or collective action is directed, shaped, and motivated by a particular social arrangement at a particular point in space and time (Robertson, 1981). Sociologists and social psychologists examine social effects on behaviour and consider that goals are generally socially determined. In addition, motivation is not analysed through the lenses of the mind and its cognitive processes dynamics although these constantly change an individual's motivational conditions. 'Social psychological theories have often treated motivation as separate from cognition, and have often approached it in a somewhat static manner' (Kruglanski et al., 2018, p. 207).

The importance of goals in human activity has contributed to the high interest from different psychology sub-fields in their study. These sub-fields include developmental, social, cognitive, and neuro-psychology (Benn et al., 2014). In each of these fields, theories are often based on the dominant psychological paradigm of the time. Theories including drive theory, expectancy value theory, and cognitive model theories are discussed in this chapter (Graham & Weiner, 1996). More recently, the cognitive paradigm suggests that the notion that goals are rewarding to attain indicates that goals control behaviour through cognitive as well as motivational processes (Bargh & Huang, 2009; Custers & Aarts, 2010; Kruglanski et al., 2002). Accordingly, any behaviour or outcome that is represented in terms of a result of more concrete action can potentially operate as a goal if one is motivated or encouraged by the external environment to attain it.

An abundance of recent evidence (reviewed more fully later) demonstrates that goals, as a result of practice and skill automation, can be activated unconsciously by goal-related features of the environment (Aarts & Custers, 2012; Bargh, 2002; Dijksterhuis & Aarts, 2010; Kruglanski et al., 2002) resorting to a complex array of cognitive, affective, and behavioural information (Aarts & Custers, 2012). Priming by the environment can influence a person's goals at a point in time. But this is because goals are mentally represented as desired behaviours or outcomes (Aarts & Custers, 2012; Bargh & Gollwitzer, 1994; Dijksterhuis & Aarts, 2010). These more recent findings suggest that there is a need to examine the use of the concepts and theories of goals and motivation relevant in the tourism context.

THEORIES AND CONCEPTS

Early theorists such as William James (1842–1910) considered behaviour to be driven by instincts, which aid survival. These 'instinctive drives' or inner urges created needs or motives. Instincts are biologically determined, existing in all species, and drive behaviour (James, 1981/1890).

Early Theories

Instincts are inherent motives; they are fixed, rigid, and predictable patterns of behaviour that are not acquired by learning. Hunger is a typical example of a drive, motivating a person to seek food. According to instinct theory, humans primarily react to environmental stimuli, precluding explorative and planned behaviour. This approach does not explain the effect of learning on behaviour.

Another early theory considers that deviations from homoeostasis create physiological needs (Hull, 1943). Here a 'drive' is a state of arousal or tension triggered by a person's physiological or biological needs, which might be food, water, or sex. These needs result in psychological drive states that direct behaviour to meet the need and, ultimately, bring the system back to homoeostasis. This theory is related to the concept of optimum level of arousal, deviations from which lead either to boredom or anxiety (Zuckerman, 1979).

Early Cognitive Theories

Humanistic theories of motivation were based on the idea that people also have strong cognitive reasons to perform various actions. Such a perspective runs counter to behaviours caused by automatic responses governed by innate mechanisms involving drives, needs, and reactions. Cognitive motivation was believed to be affected by the information available and past experience.

Incentive theory suggests that people are motivated to do things because of outside (extrinsic) or inside (intrinsic) rewards. Behavioural learning concepts such as association and reinforcement play an important role in this theory of motivation (Killeen, 1982). According to expectancy value theories (Eccles et al., 1983), motivation is determined by what a person expects to get (expectancy), the likelihood that one will get it and the subjective value of a goal. Thus, cognition is presumed to play a key role in motivated behaviour. An expectancy is the result of experience. Self-efficacy is a concept related to fulfilment of a goal as an individual's belief in her own capability to complete a task may include a previous successful completion of the exact task or a similar task. In this sense, self-efficacy involves expectancy beliefs (Bandura, 1997). Bandura (1994) theorised that an individual's sense of self-efficacy plays a pivotal role in motivating behaviour. Bandura also argues that motivation derives from expectations that people have about the consequences of their behaviours, and ultimately, it is the appreciation of the capacity to engage in each behaviour that will determine what one does and the future goals set.

Self-determination theory refers to people's ability to make choices and manage their own life, leading to feelings of self-control which are considered to impact on psychological health and wellbeing (Deci, 1975; Deci & Ryan, 2000, 2015). Self-determination assumes that there are two types of motivation, autonomous and controlled. In order to understand behaviour it is fundamental to examine which type is at work in a situation. Additionally, self-determination theory hinges on three fundamental psychological needs: autonomy, competence, and connection. Studies based on this theory challenge the assumption that reinforcement of task performance depends on external rewards.

Theory of planned behaviour (Ajzen, 2015; Armitage & Conner, 2001) uses concepts such as subjective norms, along with attitudes and perceived behavioural control in predicting behavioural intention (Ajzen, 2005). The theory assumes that behaviour typically serves an individual's goals. This may work well for some behaviour, such as wearing a seatbelt, where the dominant goal (preventing injury in an accident) is self-evident and there is little variability in goals across individuals. However, for other behaviour, such as exercising or travel, goals can vary greatly from one individual to another and can fluctuate across time and context.

To understand and predict a behaviour's occurrence, its motivational context must be taken into account (Ajzen & Kruglanski, 2019). The behavioural predictions based on the theory of planned behaviour are successful only when the behaviour investigated explicitly or implicitly contain active behaviour-relevant goals that drive those behaviours. The theory assumes that a person's expressed attitudes will predict subsequent behaviour. Attitude represents a summary evaluation of a psychological object captured in such attribute dimensions as good-bad, harmful beneficial, pleasant-unpleasant, and likeable-dislikeable (Ajzen, 2001, p. 28). However, a person may have a positive attitude to a behaviour, such as recycling, but not undertake it because their current situation does not allow it, or the goal related to recycling is not active. Thus intention (as motivation) does not predict future action. Criticism of the theory of planned behaviour is based on consistent empirical evidence that demonstrates its weak predictive power, eventually leading to discussion on its theoretical usefulness (Sniehotta et al., 2014).

Later Cognitive Theories

Cognitive appraisal theory includes emotions in the study of goals. The valence and importance of a goal and expectations of achieving them (certainty) elicit anticipatory emotions, and hence motivational potential and value (Bagozzi et al., 1998; Braver et al., 2014). Appraisals are important in the understanding of cognition and emotion under the assumption that stimuli per se do not impact all individuals equally, let alone provoke in them the same emotions. Emotions act as a reward for goal achievement.

Goal Systems Theory suggests that human goals are dynamic and are represented mentally as cognitive constructs which are linked to subgoals and means of achievement (Kruglanski et al., 2002). In some ways, the theory is similar to means-end theory (Gutman, 1997; Pieters et al., 1995). A goal system is a set of hierarchically structured various subgoals and means associated with an end goal. A goal system is interconnected, characterised by the form and strength of links between goals and means within a given system. Theory of planned behaviour distinguishes between currently pursued focal goals of which goal status one is explicitly aware and background goals whose presence need not be consciously registered. Goals are subject to contextual framing effects.

Goals in Cognitive Science

Cognitive psychology provides a cohesive structure for discussing goals. However, like current thinking in other fields of psychology, cognitive psychology does not explain *why* one has a particular goal at a particular time. Goals are considered to be organised in schemata (similar to goal structures) which develop over time based on the personal experiences (Verschoor et al., 2010). Cognitive psychology does examine how goals are processed in the mind or brain (Custers & Aarts, 2010). It identifies differences in effort of goal achievement and the importance of executive control in achieving difficult goals. Thus, it distinguishes between routine and new behaviour. Cognitive psychology also examines how the chosen goals (goal framing) influence the filtering and processing of information in a particular situation, leading to adoption of a different perspective (Lindenberg & Steg, 2007), or how the context in which the brain evolved means that one may make biased decisions not optimal in a modern world (Korteling et al., 2018).

Descriptive decision theories such as the theory of planned behaviour attempt to describe what people do, not what they should do. The inconsistent, sometimes suboptimal, choices present less of a mystery to evolutionary psychologists, as the modular and adaptive brain (Sporns & Betzel, 2016; Steffen et al., 2022) has been sculpted by evolution to optimise reproduction and survival in the world that differed quite a bit from the one currently occupied. In the evolutionary past, you would never pass up the easy pickings of a jelly doughnut and you would not engage in exercise solely for the sake of burning off valuable calories. One is unaware that many of the decisions are made following simple efficient rules that were sculpted and hard coded by evolution, but still point to physiological and psychological limitations (Simon, 1956).

Disambiguating Goal Definitions

Conceptually, in psychology, goals are defined as 'mental representations of behaviours or behavioural outcomes that are desirable or rewarding to engage in or to attain' (Dijksterhuis & Aarts, 2010, p. 470), thus are imagined end states according to which one must deploy various means to make it happen. A more scholarly definition by Berkman states: 'a goal is desired future state (an end) coupled with a set of antecedent acts that promote the attainment of that end state' (2018, p. 2), by this emphasising the temporal dimension of human thinking and action. This indicates that goals are projective (envisioning the future). What they all have in common is to require a mechanism of the brain that translates them into action (Satpute et al., 2011). The nature of the 'mental representations' is generally not discussed.

More specifically, in cognitive psychology, a goal is treated as something that one wishes to achieve that has not happened yet and is not likely to happen on its own (Berkman, 2018). This definition means that going out with friends to celebrate someone's birthday is a goal, despite most people not thinking of going to a party that way. Similarly, driving a car to take a child to school is a goal although much of the goal planning is not conscious. However, both examples 'describe activities and outcomes with an abstractness and complexity far greater than what nonhuman animals probably would ever conceive' (Satpute et al., 2011, p. 49).

Goals differ in abstractedness (being a good parent, fetching a glass of water) and duration (losing weight, going to the supermarket), but despite these differences, they are always linked to action and behaviour. Behaviour driven by goals (or goal-directed behaviour) involves the initiation of the specific processes of goal setting, attention, motor control, response inhibition, and progress monitoring (Berkman & Lieberman, 2010).

Cognitive psychology identifies several goal-related schemata that guide goal-directed behaviour. For example, the life-span schema concerns developmental goals, defined as 'a knowledge scaffold that people can use to process and organise developmentally relevant information' (Ebner et al., 2009). Thus, a schema may be seen as a means of achieving a goal or subgoal.

Effortful Goals and Executive Control

Cognitive psychologists contrast goal-oriented (action outcome decisions) and habitual actions (stimulus response decisions) (Huang et al., 2020). Neuroscience research confirms this distinction as resulting from both processes involving the workings of two overlapping yet distinct brain regions (Huang et al., 2020; Watson et al., 2018). The distinction between goal-oriented and habitual behaviour is graded. One may control habitual behaviour through exerting cognitive control. Cognitive control provides the interface through which goals influence behaviour. First, once stated, goals influence implementation or prohibition of certain behaviours (Su et al., 2022). Additionally, goal-oriented behaviours require processes that enable people to maintain the focus on the information that is relevant to achieving that goal and ignore or inhibit irrelevant information; monitor progress towards the goal; and shift flexibly from one subgoal to another in a coordinated way. A goal-oriented action is based on previous experience of a causal relationship between the action and a particular effect, the representation of this effect as a goal, and the assessment of an expected reward (Miller & Wallis, 2009; Su et al., 2022). Goal-oriented behaviour requires the selection, retrieval, and manipulation of task relevant information. Working memory holds transient representations of task-relevant information. Working memory requires the interaction of the prefrontal cortex with other brain regions (Eichenbaum & Cohen, 2004).

Executive functions include the processes involved in the generation of plans and the direction of responding relative to goals (Cohen et al., 1993). In other words,

> executive function refers to a suite of higher-level cognitive skills and capacities that generally promote successful human functioning. Attention, task switching, working memory, and inhibitory control are usually described as executive functions. (Berkman, 2018, p. 4)

Executive function specialises in tasks, goals, and behaviour, combining acquisition of representations, planning, and execution functions (performed by the prefrontal cortex) (Barto et al., 2013; Kim-Spoon et al., 2019).

Executive function feels hard for a reason. It is a serial process (Berkman, 2018; Cohen et al., 1993) involving cognitive control by inhibition of responses and attentional orientation (Kim-Spoon et al., 2019), so the sense of effort that accompanies executive function is a signal that working on a difficult task necessarily means losing out on other opportunities. In other words, effort reflects the opportunity cost of finite cognitive resources (Kurzban et al., 2013). Executive functions can only be performed one at a time, so the most important ones should come first (Berkman, 2018). The factors that influence the capacity for executive function most directly are other concurrent cognitive operations and the relative importance of the task compared to other possibilities. Together, this suggests that it is optimal to carve out dedicated, distraction-free time to work on important novel tasks and challenges. The cognitive bandwidth, namely concerning limits posited on perception capacity, working memory capacity or storage capacity (Roldan, 2017), is precious and operates most efficiently in (mental) solitude (Berkman, 2018).

Goal Pursuit

Neuroscience research has focussed on executive function, involving attention, working memory, inhibitory control, and planning processes and the related neural systems and circuits. Goal pursuit is about coordinating neural and psychological/cognitive processes: goal setting, attention, motor control, response inhibition, and progress monitoring (Hommel, 2022). Goal pursuit requires goal representation, the mental process of maintaining a goal or desired end-state at least during a period of goal pursuit; goal intention, representing the link between goal representation and goal pursuit behaviour; and goal action. This consists of response inhibition: the mechanism critical to maintain goal pursuit by inhibiting goal-irrelevant responses (through self-control and self-regulation) and progress monitoring, the mechanism to monitor whether progress towards the goal is being made.

Neuroscience

The motivation to seek out and consume rewards has evolutionarily been driven by the urge to fulfil physiological needs. However, in a modern society dominated more by plenty than scarcity, one tends to think of motivation as fuelled by the search for pleasure. More recently, beyond hedonics, research has been exploring authenticity, meaning, or personal growth, introducing eudaimonia as another facet of motivation (Huta, 2015). Neuroscience identifies two separate but interconnected subcortical and unconscious processes that direct motivation: 'wanting' and 'liking'. Reward is fundamental to motivation-related effort, learning, wellbeing, and survival (Ryan et al., 2019). Environmental incentives are those environmental objects one learns to return to after experiencing their rewarding properties. Incentives have rewarding properties and promote approach-oriented behaviour because they send information through the five senses that reach the mesolimbic dopamine-based reward circuitry (Wise, 2002). 'Repeated experiences with objects and events allow people to form mental representations in which these environmental stimuli come to signal reward information in a predictive fashion' (Reeve & Lee, 2014, p. 367). In this way, past reward-related information helps establish an anticipatory motivational value of objects and events. This learnt reward-related information serves as the basis of future goals, values, and preferences (Berridge, 2004, p. 194).

> Motivation is conceptualised here as the strength of the desire to attain a particular outcome, irrespective of how pleasant or unpleasant the experience of actually attaining it is. This distinction between the motivational component of a reward – 'wanting' – and the hedonic component of consuming it – 'liking' – is maintained with remarkable

evolutionary consistency in the brains of both humans and animals. (Berkman, 2018, p. 20; Berridge & Robinson, 2003)

Wanting a reward is closely tied with activity of mesolimbic dopaminergic neurons, particularly within the ventral striatum and ventromedial prefrontal cortex which is sometimes also called the orbitofrontal cortex (Berkman, 2018).

Tourism Applications

Perhaps one of the most interesting questions for tourism researchers is 'why do we travel?' The discretionary nature of tourism focuses attention on goals. Essentially the question is an attempt to understand the motivation behind a person's goal (travel). In tourism literature, a 'goal' is considered 'the psychological representation of the individual's subjective assumption of her or his state of unfulfilled satisfaction and her or his activity outcomes' (Su et al., 2022, p. 941). And 'motivation' is defined as the energising of behaviour in pursuit of a goal, it is a fundamental element of one's interaction with the world and with each other (Cohen et al., 2013).

There is abundant research on motivation and goals in tourism (Beard & Ragheb, 1983; Crompton, 1979; Dann, 1977, 1981; Dunn Ross & Iso-Ahola, 1991; Iso-Ahola, 1982; Kruger et al., 2015; Oh et al., 2016; Plog, 1974; Su et al., 2022; Vada et al., 2019). The concept of motivation is used as a summative concept for how an individual's history and current state interact to modulate goal-directed activity. Goals are considered to affect tourism and tourist decision, activities, and destination choice (Pearce & Lee, 2005). When travelling, people make plans, consider engaging in diverse activities, or in behaviour that is goal-directed (Oh et al., 2016; Su et al., 2022). Oh et al. (2016) applied the theory of goal-driven consumer behaviour to explain slow tourism, describing it as going through a process comprising three main activities: goal setting, action planning, and goal attainment/failure.

Kruger et al. (2015) discuss tourist life satisfaction through tourism goal valence principle. The assumption is that satisfaction with tourism contributes significantly to the tourist's life satisfaction when important, enjoyable, and challenging goals are set. Achievement of tourism goals is seen to contribute and enhance positive affect towards diverse life domains. In general, this study's results indicate that tourist life satisfaction increases when they set intrinsic goals, growth-based goals, and goals related to flow activities. Vada et al. (2019) applies bottom-up and top-down theories of wellbeing from positive psychology to examine how tourism contributes to wellbeing. According to the bottom-up theory, tourism as a deliberate activity, is an important context for experiencing wellbeing; and according to the top-down theory, it is the tourist's goals, as internal dispositions towards a desired end state, and their achievement during the travel experience that impact on memorability and consequently on wellbeing. Goal systems theory has been used to study sustainability (Wong et al., 2021).

More recently, Su et al. (2022) tested a model of tourists' goal-directed behaviour based on construal level and goal-directed behaviour theories. Construal level theory introduces temporal distance in the analysis of goal-directed behaviours. In turn, goal-directed theory considers goal-directed behaviour an action strategy dependent on cognitive processes that lead to connecting actions to goals, and posits that goal-directed behaviours are influenced by external stimuli (incentives found in the environment), as well as by the individual's psychological needs (Kim et al., 2021; Su et al., 2022).

Particular Goals or Motivation for a Trip

Motivation receives a great deal of attention from tourism academics, given its importance in marketing decisions, such as segmentation, product development, advertising, and positioning (Bieger & Laesser, 2002; Su et al., 2022). Motivation is described in tourism studies as 'psychological/biological needs and wants, including integral forces that arouse, direct and integrate a person's behaviour and activity' (Yoon & Uysal, 2005, p. 46).

Many attempts have been made to identify reasons for tourism. The first relevant theory is attributed to Hull (1943) (Deci & Ryan, 2000), called 'drive theory'. It considers that innate physiological needs (for food, water, sex) that are based in non-nervous system tissue deficits give rise to drive states, push the organism into action, and must be satisfied for the organism to remain healthy. The drive states, when reduced, produce learning by linking drive stimulations to the responses that lead to drive reduction. Drive states and the stimulus-response associations were used to predict subsequent behaviour (Deci & Ryan, 2000). While successful in some situations, drive theory could not predict 'spontaneous' behaviours, which brings into discussion the recognition of intrinsic motivation.

Several tourism scholars seek to explain the motivation of travelers using some particular concept. Cohen considers that people travel because they desire novelty such as 'things, sights, customs and cultures different from his own'. Interestingly, desire for variety, novelty, and strangeness decrease as tourism becomes more institutionalised (1972, p. 165). Further, theories or models have been developed to explain motivation (Gnoth, 1997; Hsu et al., 2010) and early studies such as those of Plog (1974) 'allocentric psychocentric', Dann (1977) 'push pull', Pearce and Lee (2005) 'travel career ladder', and Dunn Ross and Iso-Ahola (1991) 'escape seeking' are instrumental. Pearce and Lee (2005) found that tourist push motivations are four-fold in nature (novelty seeking, escape/relaxation, kinship/relationship enhancement, and self-development). Gnoth (1997) specifically distinguishes between motives and motivations, arguing that the former is the tourist's lasting disposition, recurring with cyclical regularity (behaviourist approach), and the latter indicates object-specific preferences (cognitivist approach). However, it can be argued that tourist motivation is characterised neither by a behaviourist nor by a cognitivist approach but rather by a combination of both (McCabe, 2000). Accordingly, to date a theoretically robust conceptualisation of motivation remains elusive (White & Thompson, 2009) and researchers continue to treat the two concepts as one and the same.

The push–pull approach remains the most widely applied for explaining motivations, given its simplicity and intuitive approach (Klenosky, 2002). Tourists are pushed by their biogenic and emotional needs to travel and pulled by destination attributes (Yoon & Uysal, 2005). This process is moderated by such factors as involvement, imagery, and emotions (Goossens, 2000; White & Thompson, 2009). Iso-Ahola (1982) takes a psychological view. Psychologists/social psychologists generally agree that 'a motive is an internal factor that arouses, directs, and integrates a person's behaviour' (Murray, 1964, p. 7). This internal factor can be likened to 'an awareness of potential satisfaction in a future situation, meaning that motives are cognitive representations of future states' (Deci, 1975, p. 991).

Many of the earlier models or theories of tourist motivation are either conceptual or tested on small samples (Swarbrooke & Horner, 2004). Some studies (Pearce & Lee, 2005; Snepenger et al., 2006; Tran & Ralston, 2006) have sought to empirically test or extend the validity of such models and theories. They generally conclude that the original theories and models are still valid and applicable to pleasure-seeking tourists mainly. For example, Snepenger et al. (2006) test the validity of Iso-Ahola's (1982) motivation theory and conclude that the four-factor structure (personal seeking, personal escape, intrapersonal seeking, and intrapersonal escape) operates as a salient intrinsic motivation for tourism behaviour, but differences in motivation are notable between tourism and recreation experiences.

The relationship between motivation and other behavioural constructs, such as expectation and attitude, is surprisingly rarely studied (Hsu et al., 2010). Also, most studies focus on motivation per se, ignoring the formation of motivation (Gnoth, 1997). The relationship between motivation and emotions, moods, brand personality, and affective constructs, such as brand and destination attachment, remains unexplored. These are certainly worthwhile future areas of inquiry, along with the mechanism of how motivation stimulates actual behaviour.

MacCannell discusses tourism as part of a need for ritual and religion and hence a need for authenticity, noting that A, 'touristic consciousness is motivated by the desire for authentic experience' (1973, p. 597) and inauthenticity of life in the tourist's own society 'motivated the tourist for his quest (for authenticity)' (MacCannell, 1976, p. 188). On another front, Dann defines motivation as 'a meaningful state of mind which adequately disposes an actor or group of actors to travel, and which is subsequently interpretable by others as a valid explanation for such a decision' (1981, p. 205). Here state of mind refers to desires, needs at that particular point of time, and may be interpreted as a goal. Meaningful means that a person should be consciously aware of their goal. This is problematic as many decisions are taken and then subsequently justified. The word 'adequately' accounts for those mental situations involving a partial self-justification of behaviour. Again, this suggests a conscious weighing of decisions 'to travel or not to travel', which is often done post hoc. Disposition is considered as the pattern of thought (derived from social or individual experience) that may be used in a novel situation. It is a type of fallback mode of thinking when faced with a new situation. In cognitive psychology this may be interpreted as a schema. Interesting, Dann, does recognise the dynamic nature of the motivational process and how contexts affect behaviour. The approach used by Dann shares much in common with a cognitive psychology approach (1981, p. 206).

One of the results of such a long history of study on such a fundamental topic is that many terms in the business and tourism literature appear in some way related to goals. These include expectations/expectancies, involvement, needs, and so on. According to Deci and Ryan, 'Most [contemporary] theories of motivation assume that people initiate and persist at behaviour to the extent that they believe the behaviours will lead to desired outcomes or goals' (2000, p. 227).

Expectation refers to the individual's ability to anticipate, to form beliefs about and to predict future events and states (Maddux, 1999, pp. 17–39). Expectancy theory is a type of expectancy value theory. It is a cognitive approach that relates motivation to beliefs a person holds concerning expectancy (effort leads to behaviour/performance), instrumentality (performance leads to outcomes), and valence (outcomes are important or valued) (Grant & Shin, 2012). According to a study by Correia and Moital (2009), people are motivated by both personal expectancy and the value of the behaviour outcome. This theory presupposes that an individual is motivated to reduce a discrepancy between his present state (existing) and the desired (idealised) state represented by the goal. In the case of choosing a holiday destination, people will go to a destination that gives them the most value.

Although expectancy theory has several advantages, such as theoretical parsimony, it is limited by over relying on beliefs and calculation involved in the processes of decisionmaking, as calculation of the effort needed to perform in a particular way or calculation of the extent actions leads to the expected outcomes, neglecting the role played by intentions, subjective norms, social expectations, or affective components in these processes (Grant & Shin, 2012). These limitations make it difficult to use expectancy theory as a model to predict individual behaviour (Witt & Wright, 1992). Larsen (2007) highlights the importance of expectations in travel.

Involvement is said to impel the individual's physical, mental, emotional, social, or spiritual engagement in an experience, leading to increased satisfaction and memorability (Andrades & Dimanche, 2014). It is defined as 'a person's perceived relevance of the object based on inherent needs, values, and interests' (Zaichkowsky, 2012, p. 342) and as 'an unobservable state of motivation, arousal or interest towards a recreational activity or associated product' (Havitz & Dimanche, 1997, p. 246). Kim (2010) found that involvement, characterised by motivation and interest in experience, led to recollection and memory vividness. Kim, Brent Ritchie, and McCormick (2012) and Campos, Mendes, Oom do Valle, & Scott (2017) found that involvement, defined as personal relevance, contributes to memorability. It may be argued that involvement is similar to goal importance.

Needs are viewed primarily as acquired rather than innate (Deci & Ryan, 2000; Murray, 1938). In this approach, the concept of needs was very broadly construed as:

> a construct (a convenient fiction or hypothetical concept) that stands for a force (the physio-chemical nature of which is unknown) in the brain region, a force that organizes perception, apperception, intellection, conation and action in such a way as to transform in a certain direction an existing, unsatisfying situation. (Murray, 1938, pp. 123–124)

Later scholars sought to not only measure need strength but also need type. Essentially this research stream identified basic and universal psychological needs. For example, the early writing on push factors considered them as basic psychological needs. Dann (1977) identified two personality needs (anomia and ego-enhancement) as reasons for travel. Anomia is caused by inherent failures in society such as lawlessness (called *anomie*) which isolates people and in turn leads to attempts to reduce this isolation. Travel then is a means to 'transcend the feeling of isolation obtained in everyday life'. Alternatively, tourism may be a means to ego-enhancement, expressing the need to be recognised and achieve status (Dann, 1977, p. 187). Deci and Ryan focus on needs that 'specify necessary nutriments for healthy functioning' (2000, p. 228). The problem of this approach is that it assumes people's needs are directed towards certain goals or alternatively restrict the scope of their research to such outcomes.

The problem with the basic psychological needs approach is that researchers attribute universal needs to the traveller. For example, in self-determination theory (Deci & Ryan, 2000; Ryan & Deci, 2001), needs are innate rather than learnt and determine the motivational content to life. The theory has at its core the concept of basic, universal psychological needs for competence, autonomy, and relatedness. These needs must be satisfied for individuals to experience optimal psychological development, performance, and wellbeing within any domain and across cultural contexts. It is unclear if tourists seek optimal psychological development when they travel.

Attitudes construct is a mainstay for tourism research which is due to its functional role (Gnoth, 1997, p. 285; Hendee et al., 1968). Attitudes are seen to depend on a value system (Hendee et al., 1968). The reason individuals hold certain attitudes is that these serve psychological needs (Fodness, 1994), are aligned with motivation towards some goal (Hendee et al., 1968), and perform some type of function, either utilitarian or the social identity function (Katz, 1960). This is termed the functional theory of attitudes. As functions, attitudes are the mediators between needs and values as they arise within a subject. Thus, motivation for tourism can be determined by asking people attitude statements (Hendee et al., 1968; Sæþórsdóttir, 2010). In other words, the texture of motives is potentially richer than its function might indicate. The function of attitudes declares a causal relationship between a behaviour and a targeted object, but it does not reveal what both behaviour and object represent emotionally and to the tourist's self. The functional approach to attitudes relies on clear-cut relationships, whereas, in reality, motives underlying attitudes are often fuzzy.

CONCLUSION

The literature of motivation and goals in tourism is based on theories from psychology that have been superseded or severely criticised. The subconscious nature of many goals is often ignored (Aarts & Custers, 2012; Sitzmann & Bell, 2017) and, in some studies, a person's motivations and goals are considered rational, intentional, governable by consciousness or the will, and able to be verbalised (Aarts & Custers, 2012; Oh et al., 2016). In the psychology-based literature however, after decades of consistent evidence to the fact, it seems consensual that much of human functioning occurs outside the domain of consciousness (Murayama et al., 2012) and thus that 'behaviour is directed and motivated by goals outside of conscious awareness in the current situation or environment, even though we share the experience and belief that we consciously set and pursue goals' (Aarts & Custers, 2012, p. 233).

In addition, the concept of a goal overlaps with many other concepts (need, incentive, expectancy, priming, executive control, involvement), perhaps hinting at the different scientific disciplines and theories that have attempted to define it and explain goal-related mental processes and behaviour. To date, no unified theory exists that combines the behavioural, the cognitive, and the neurological facets of goals, and yet current research is increasingly joining the contributions of psychology and neuroscience (Berkman, 2018).

At the same time, a clear understanding of motivations and goals for behaviour has become important in helping to address important questions, such as how to motivate a transition to sustainability. Sustainability concerns are currently widely recognised in tourism development (Wu et al., 2021) and much discussion is devoted to discussing whether there should be limits to growth or not, or even if tourism degrowth is a viable option for the industry. Also, alignment of tourism development with societies and cultures and stakeholder wellbeing, especially following the COVID-19 pandemic, remain in the research agenda as critical topics in sustainability (Higgins-Debiolles, 2020). Sustainable tourism themes are today framed against the background of global sustainable development goals (SDGs). The UNSDGs are general statements concerning desired future states for societies and their people, widely accepted by the global community as important for the long-term future of humankind. Moreover, the underlying assumption of these goals is that they require an energetic, committed, and steady change of behaviour from social actors with impacts on the near and distant future. Accordingly, they present a high degree of abstractness and are set for the long run, to a great extent distant from the present time. These characteristics of the UNSDG are what make them accepted by almost anyone but difficult to attain. Psychological research has something to say about goal abstractedness, distance, and difficulty (Bandura, 1986; Bandura & Schunk, 1981; Gollwitzer & Oettingen, 2012; Heath et al., 1999; Schunk & Usher, 2012).

Firstly, goal abstractedness compromises the level of performance, as specific goals are perceived as more challenging and leading to improved performance. Therefore, UNSDGs may hinder social actors from being successful if sustainable goals are not translated into concrete intermediate goals and action. Second goal proximity (short-term goal) enhances motivation, learning, mastery, and efficacy and in turn, distance from overarching goals present difficulties in monitoring progress. As UNSDGs are visualised in the long term, social actors may find trouble in keeping high levels of motivation towards achievement. Lastly, goal difficulty enhances performance pushing it to higher levels; however, this involves higher effort and proficiency. UNSDGs are thus highly demanding also because they require high competence and commitment from a wide variety of factors that may not be as aligned in motivation and expertise as would be desirable. Taken together, these findings from the psychology of goals warn about the challenges ahead towards worldwide sustainability in tourism.

As mental representations, goals are fundamentally psychological, or cognitive, and per se distinct from extrinsic rewards (Heath et al., 1999). But they hold a close relation to motivation as they impact motivational processes related to affect, effort, persistence, and performance. One way to explain this influence is assuming they act as reference points, according to prospect theory (Heath et al., 1999; Kahneman & Tversky, 1979). Once reference point goals divide outcomes into regions of success (benefits) and failure (costs), they change the value of behavioural outcomes. Again, once outcomes' value changes, motivation, performance, and emotional reaction also change. This idea of goals acting as reference points suggests that this is why people motivate and experience positive or negative emotions towards outcomes. The question is whether intermediate goals affect effort and persistence of pursuit of overall sustainability goals if the former fall behind an accepted positive value, in turn eliciting negative emotions.

The affective significance of goals (Chiew & Braver, 2011) is noteworthy in sustainability issues and challenges as these link to perceptions of survival of the natural and human environments. Sustainability requires ongoing executive control and over a longer period. During this extended period other goals may become relevant and compete. Thus, the understanding of emotions relative to sustainability goals and how they influence executive control towards them remains a topic of interest.

It is known that goal-directed behaviour can be instigated in a bottom-up fashion by features of the environment that can be either consciously or unconsciously apprehended (Williams & Poehlman, 2017). Yet, while attentive to leads from industry professionals, such as when appealing to save electricity or water use, tourists are far from behaving sustainably or applying in tourism settings their sustainability principles as they do at home. This leads to examination of habitual behaviour and adoption of nudges, as small interventions to behavioural change are strategies seen to mitigate automatic undesirable habitual behaviour and effects (Dolnicar, 2020; MacInnes et al., 2022; Thaler & Sunstein, 2021).

One area for research is the 'vividness' of imagined futures. If a person is thinking about something that is very important to them then they can imagine it very clearly. The implication is that if a holiday can be imagined clearly then it is more likely to occur. A second area is the extent to which imagined events are personally relevant or self-concordant (Lehner & D'Argembeau, 2016). Self-concordant goals derive from autonomous motivation, personal choice, and endorsement. Accordingly self-concordant future events are associated with a stronger sense of 'realness' and of pre-experiencing the future, and are characterised by more positive and intense emotions (Ernst et al., 2018; Lehner & D'Argembeau, 2016). When goals are self-concordant, self-controlled behaviour is set in motion towards their achievement; the mediating factor linking goal setting and achievement is found in the increased subjective value of imagined events derived from self-concordance (Berkman et al., 2017). Expectedly, self-concordant goals are relevant to the utmost degree (Ernst et al., 2018), so one critical discussion concerning sustainability challenges in tourist behaviour change is how to work on non-concordant goals, including abstract, long term and difficult-to-achieve, and transforming them into concordant ones.

Research endeavours may address the application of research on foraging behaviour (Koy & Plotnick, 2007; Pyke et al., 1977) to tourism. Foraging describes the behaviour of acquiring and utilising sources of energy and nutrients to maintain the survival of the organism. This behaviour is carried out through calculation of the combination of energy spent and energy gained in a particular situation. Foraging involves 'the location and consumption of resources, as well as their retrieval and storage, within the context of the larger community' (Koy & Plotnick, 2007, p. 430). Evolution on foraging is determined by the organism's ability to improve decision-making by retrieving information from and during the processes of search, assessment, and exploitation of resources in the environment. Assessment of information collected from experience, in contrast with information derived from the present experience, is critical to the organism's decisionmaking process. Humans frequently face situations where they must choose between an option with a known value and one or more options of unknown value. The classic example of this is foraging: animals must make a decision about where to seek food and water, precious commodities that tend to occur in restricted locations. Foraging always requires decisions affecting future behaviour, such as 'do I continue to exploit the resource at hand or set out to explore in the hopes of finding a richer location?' Here the animal must calculate the value of the current option, resource richness of the overall environment, and the costs of exploration. The question is whether the same behaviour applies to a visit to the library or art gallery, a repeat visit to the destination, or exploring a different one.

Chapter 10

NOVELTY AND TOURISM

Liubov Skavronskaya
Griffith College, Australia

Abstract: Novelty detection in the environment is an important mental ability for evolutionary survival. It is also critical for understanding emotional arousal and its effect on memories in tourism. Tourism takes place, by definition, in an unusual environment and therefore intimately involves this concept. This chapter discusses novelty definition, processing, and effects on attention, goals, memories and emotions, and role in tourist experiences. Theoretical and practical implications have been discussed, including the potential utilisation of aspects of novelty in tourism experience design and how its use can improve experiences. Future research should focus on its antecedents and consequences in tourism experiences, as well as the effect of degree of novelty on emotional arousal. **Keywords**: Novelty; experience; consciousness; memories; emotions; arousal

INTRODUCTION

A useful place to start the discussion is the importance of the recognition of unusual or 'out of place' objects for survival. Senses are constantly receiving information about the environment and processing this data to determine how to act. Detection and response to novel events has played a significant role in the evolution of life on Earth. Evolutionary psychology explains that this concept is intricately connected to sensory processing of the state of an organism and its environment, which is critical for survival and adaptation (Bradley, 2009; Lisman & Grace, 2005). Rapid identification of new threats and opportunities in the environment has survival value. Thus, the brain has evolved to identify and respond to such novelty in the environment.

Cognitive Psychology and Tourism
Tourism Social Science Series, Volume 27, 115–126

Published under exclusive licence by Emerald Publishing Limited
ISSN: 1571-5043/doi:10.1108/S1571-504320240000027010

The 'folk-psychology' definition of novelty is the quality of being new, original, or unusual (Oxford languages). This recognises that objects such as new foods, new experiences, new social encounters, or new environments open opportunities or threats. However, this everyday definition considers it as an independent property of an object, stimulus, or environment. This may be true in some situations such as the development of a 'new to the world' invention, but for most situations, a novel object is something that a particular person has not experience before, or even an object previously experienced but of great important in the current situation (salient).

A personality perspective discusses it as a trait which may vary across the population. This trait is considered to be an inherent characteristic of a person and its degree is determined by genetic inheritance. The terms novelty seeking, risk taking, innovativeness, and the like are related terms. This differs from the cognitive psychology approach which views it because of an individual's assessment of a stimulus. One of the definitions of novelty in cognitive psychology is 'a motivationally salient learning signal that attracts attention, promotes memory encoding, and modifies goal-direct behaviour' (Bunzeck et al., 2012, p. 1310). Thus, it is a characteristic attributed to a stimulus that is involved with stimulus-driven attention, the detection of salient targets and the reorientation of attention (Gazzaniga et al., 2014).

Recognition of novel events in the environment from the information provided by senses may occur at low levels of processing. Thus, conscious recognition of some aspect of the environment as unfamiliar and contrasting with previous experiences is as a result of (say) visual detection triggering a saccade to focus the eye on a particular object, followed by further processing and the allocation of working memory to process it. Thus, novelty is recognised by comparing what is expected to be seen in the surrounding environment and what is observed. A discrepancy then triggers a cascade of processes, directed attention, arousal, and either low level responses (fight or flight – fast response) or conscious consideration (slow response). Table 10.1 shows the differences between these four paradigms.

Table 10.1. Novelty Definitions and Difference.

Discipline	Definition	Property of . . .	Personal Variation
Folk psychology	. . .is the quality of being new, original, or unusual (Oxford Dictionary).	Environment	Does not explain individual differences
Cognitive psychology	'. . .a motivationally salient learning signal that attracts attention, promotes	A person's appraisal of the environment (interaction)	Degree of unexpectedness

Table 10.1. *(Continued)*

Discipline	Definition	Property of . . .	Personal Variation
	memory encoding, and modifies goal-direct behaviour' (Bunzeck et al., 2012, p. 1310).		
Neuroscience	'. . .a fundamental dimension upon which the nervous system operates, significantly impacting an organism's internal states and altering external behavior' (Habib & Lepage, 2000, p. 275).	A person's appraisal of the environment	Associated with genetic variability in dopaminergic activity
Personality psychology	Novelty-seeking is a personal trait characterised by search for novel, varied, complex and intense feelings, and a readiness to take physical, social, legal, and financial responsibility for the sake of such experiences (Cloninger et al., 1994).	The person	Trait strength varies due to genetics and developmental processes

NOVELTY PROCESSING

Novelty can be processed in the brain at different levels. Dual processing theory states that its perception and processing can occur through two different processes. Identification of novel events subsequently influences attention, emotion, and memory processes and its identification is of critical importance for tourism experiences.

Dual-Process Approach

Dual processing theories highlight two distinct pathways for processing of environmental stimuli and other cognitive tasks: one (System 1) is fast, automatic, and non-conscious, and another (System 2) is slow, controlled, and conscious (Kahneman, 2011; Neys & Pennycook, 2019). The dual-process perspective has been widely adopted in studies on logic and reasoning (Neys, 2006), persuasion (Loken, 2006), decisionmaking (Kahneman, 2011), social cognition (Gawronski & Creighton, 2013), and other effortful thinking processes. An understanding of how novelty once detected is processed provides insights into how the brain accesses, transforms, stores, and uses new information.

Fast processing begins with an orienting response. Direction of attention towards novel stimuli in the environment is a fundamental biological mechanism necessary for survival (Lisman & Grace, 2005). An involuntary shift of attention as a response to stimuli in the environment is defined as an orienting response. 'Orienting is a rapid response to new (never experienced before), unexpected (out of context) or unpredictable stimuli' (Friedman et al., 2009, p. 356). Novel stimuli (food, sounds, odours, etc.) will trigger an orienting response by directing attention to the that stimulus. Next the novel stimulus is evaluated for its salience (Tapper & Molas, 2020). Detection of novel stimuli along with an evaluation of high salience can occur quickly in the amygdala – the brain region associated with the processing of emotional responses (Weierich et al., 2010). The amygdala is part of the brain's limbic system, the group of complex interconnected structures within the brain that are responsible for a person's emotional and behavioural responses (Blackford et al., 2010).

The amygdala may also activate a fight-or-flight response. This response can help people in immediate physical danger react quickly for their safety and security. For example, such fight-or-flight responses helped early humans respond to threats to avoid being injured or killed. The amygdala activates this fight-or-flight response without conscious awareness and may inhibit the front lobes. When the brain senses danger, it signals to produce stress hormones, preparing the body to either fight for survival or to flee to safety. This is sometimes called amygdala hijack (Robinson et al., 2014). Fast processing of external sensations by the amygdala lead to immediate emotional responses. The fast-thinking system (system 1) once activated produces emotions that can guide information processing, and activate goals (Henderson et al., 2015). If the danger is not immediate, the frontal lobes are used to evaluate what to do in response to the stress. For mild or moderate threats, the frontal lobes can re-revaluate the emotions the amygdala produces (see also Chapter 14 on coping).

In comparison, conscious thoughts are slow and involve the frontal lobes of the brain, two areas located behind the forehead. The front lobes are part of the brain's cerebral cortex. This area of the brain regulates voluntary actions like reasoning, thinking, movement, decisionmaking, and planning. It is responsible for 'slower' more 'rational' thinking. The frontal lobes are involved in the brain's executive control functions (Ardila, 2008). This intervention of the frontal lobe means that the evaluation of novelty to consciously experienced and part of the present moment thoughts (Mannell & Iso-Ahola, 1987). Slow processing also allows the response to novel stimuli to be affected by previous experience (memories and schema). Repeated encounters with low or medium threat novel stimuli mean that the unexpectedness of an event is reduced (Schomaker & Meeter, 2018). Therefore, from the perspective of stimulus salience assessment, high unexpectedness and high salience generates a strong approach-avoidance response. When a stimulus becomes familiar through repeated exposure, it may not be seen as threatening (Schomaker & Meeter, 2018).

Attentional Response to Novel Stimuli

There are numerous ways in which novelty (unexpectedness) affects human thinking and behaviour. The brain has limited capacity to process all sensory input present in the environment at any point in time and relies on attention to focus neural resources according to the contingencies of the moment (Katsuki & Constantinidis, 2014). When the brain detects novel stimuli in the environment, one's attention is directed towards it (Bradley, 2009). Attention can be directed towards external sensory stimuli (*bottom-up*) or influenced by a person's goals or intentions (top-down) (Corbetta & Shulman, 2002). The difference between bottom-up and top-down attention is based on the point of origin of the stimulus which is responsible for activation of the attentional response (Katsuki & Constantinidis, 2014). Therefore, attention can be defined as a complex cognitive function characterised by a focussed mental engagement when processing external sensations or internal memories and thoughts (Dayan et al., 2000). In other words, attention is a selection process for an external (sound, image, smell) or internal (thoughts, memories) event which is maintained at a certain level of awareness.

Bottom-up attention is externally driven and refers to automatic/involuntary allocation of attention with a stimulus in the environment external to the individual (Pinto et al., 2013). Bottom-up factor of attention refers to a 'novel object among usual, familiar ones; a rare, scarce or special item amongst those frequently observed; or vivid elements amongst unimpressive ones' (Campos et al., p. 3113). Thus, bottom-up attention can be understood as a response to novel stimuli and an automatised response to the environment, which occurs at the sensory-physiological and motor levels with low levels of awareness and therefore occurs subconsciously (Campos et al., 2020). When one's attention is directed towards unexpected novel stimulus, this occurs without conscious awareness. A stimulus that is sufficiently unexpected evokes the involuntary capture of attention enabling the event to enter consciousness (Friedman et al., 2001).

Effect on Attention and Consciousness

There is a debate regarding the role of attention for consciousness (Baars, 2002; Cohen & Dennett, 2011; Koch & Tsuchiya, 2007). Current cognitive and neuroscience studies suggest that bottom-up attention is more automatic, much faster, and less flexible than top-down attention. Top-down attention may be under conscious control, and thus not a cause, but an effect of consciousness.

Detecting novelty requires comparing real-time perceptions with the contents of memory, to determine if the stimulus has been previously experienced (Tulving & Kroll, 1995). Novel information is more likely to be retained and encoded in memory (Mather, 2013). Neuroscience and cognitive studies have confirmed that the first occurrence of a particular event that captures a person's attention leading to superior memory recollection (Heckler & Childers, 1992; Tulving & Kroll, 1995), which emphasises its critical importance for memory.

A schema is an association of existing knowledge stored in a person's memory that represents knowledge about a stimulus (a person, event, or object) (Gilboa & Marlatte, 2017; Kesteren et al., 2012). Schemata facilitate retrieval of processed information and guide human behaviour allowing an individual to act in a complex environment and interpret new information (Gwinner & Eaton, 1999; Reger et al., 1994). Recent cognitive and neuroscience studies confirm that incongruent information, inconsistent with a dominant schema, has the capacity to improve memory (Ghosh & Gilboa, 2014).

Schema theory identifies three levels of congruity between a stimulus object and an activated schema. Congruity occurs when the stimulus matches the schema it evokes (Clemente et al., 2014; Stumpf & Baum, 2016); incongruity or extreme incongruity occurs when an object and evoked schema are mismatched (Meyers-Levy & Tybout, 1989), and moderate incongruity occurs there is a partial match. Moderate or extreme incongruity are associated with surprise (Goolaup et al., 2018) which is an appraisal dimension leading to higher emotional arousal. Over time, one's memories of similar past events mean that the unexpectedness of an event is reduced. Thus, repeated encounters may lead us to expect the formerly novel stimuli and hence a less aroused response.

As previously mentioned in this book, any human experience is a mix of perceptions of external sensations, and memories (Chapter 11 on Memory), with the same brain mechanisms used for thinking about the present, past, or future (Skavronskaya et al., 2017). Memory is a mental process by which information is encoded, stored, and retrieved when needed (Atkinson & Shiffrin, 1968).

There are two major types of memory which distinguish one from another by its different capability and capacity. Short-term memory (working memory) can store information for a certain period but with limited capacity. Long-term memory has a larger capacity and contains memories of experiences and information that have accumulated over a lifetime (Landauer, 1986). Current literature distinguishes two types of long-term memory, semantic and episodic memory (Pearce & Packer, 2013). The former stores general knowledge about the world, whereas the latter stores memories about personal experiences and events (Braun-LaTour et al., 2006). If a stimulus has not been previously experienced, it will lead to superior memory recollection due to more significant mental effort the brain put into processing new information (Tulving & Kroll, 1995). Therefore, novel information is more likely to be retained and encoded in memory (Mather, 2013). Thus, improved memory for an event is one of the outcomes of novelty.

Cognitive Appraisal Theory of Emotions

Emotion is a complex psychological state. Cognitive psychology defines emotion as a

> mental state of readiness that arises from cognitive appraisals of events or thoughts; has a phenomenological tone; is accompanied by physiological processes; is often expressed physically (e.g., in gestures, posture, facial features, heart rate increases or pupil dilation) and may result in specific actions to affirm or cope with the emotion, depending on its nature and meaning for the person having it. (Bagozzi et al., 1999, p. 184)

Appraisal theorists maintain that emotions are the result of comparing an actual state with an expected or desired state on a small number of dimensions (Lazarus, 1991). One key dimension is the positive or negative fit (congruency) to a personal goal relevant to that situation (Massara et al., 2010; Scherer et al., 2001). Other dimensions are the degree of novelty, certainty, and agency (Ortony & Turner, 1990). Patterns of evaluations on these appraisal dimensions distinguish between emotions (Scherer, 1997b). For example, the set of appraisal dimensions, high goal congruence and importance, and high novelty are found to elicit the emotion of delight (Ellsworth & Smith, 1988a; Ma et al., 2013).

An appraisal of a stimulus as having high novelty elicits high levels of emotional arousal, and concomitant physiological responses (Barto et al., 2013). It is especially powerful when coincident with goal congruence, with a strong appraisal of a stimulus as novel eliciting delight and surprise with positive valence (Ma et al., 2013), which tourists associate with pleasure and enjoyment (Mitas & Bastiaansen, 2018). When associated with goal incongruence, it is considered to create strong negative emotions and a feeling of anger (Bradley, 2009). Novelty is, therefore, independent from emotional valence, rather it is the dimension of goal (in)congruency that determines the valence of elicited emotions. Consequently, it is a critical antecedent of both positive and negative emotions.

Personal psychology defines novelty-seeking behaviour as a search for novel, varied, complex and intense feelings, and a readiness to take physical, social, legal, and financial responsibility for the sake of such experiences (Cloninger et al., 1994). In personal psychology, it is one of the main factors included in the big five model of personality. The big five model identifies extroversion, agreeableness, openness, conscientiousness, and neuroticism. A desire for novelty is a part of openness along with curiosity, imagination, and creativity (Hirsh et al., 2009). Differences in trait novelty-seeking are associated with both genetic and developmental processes (Bidwell et al., 2015; Zuckerman, 2014).

Thus, personality psychology emphasises that novelty is not always a sign of a 'problem' in the environment. Research with laboratory animals indicates that novelty seeking is rewarding because it activates dopamine system (Rangel-Gomez & Meeter, 2016). Dopamine has been central to several theories of novelty processing (Düzel et al., 2010; Lisman & Grace, 2005), with recent studies confirming that variations in novelty-seeking behaviour are mediated by genetic variability in specific dopamine neurotransmitter activity in the brain (Cloninger et al., 1993). Therefore, high dopaminergic activity is associated with high novelty-seeking (Liang et al., 2017). Furthermore, dopamine plays a critical role in the studies on extrinsic and intrinsic goal-direct behaviour (Hansenne et al., 2002; Hirsh et al., 2009).

Tourism Applications

Novelty is commonly experienced in tourism, which according to UNWTO entails the movement of people to places outside their usual environment. Thus, this implies some degree of newness, as it is a critical component in tourists' motivation and behaviour. Since tourist experiences are usually associated with achievement of important goals, they are usually associated with positive emotions, such as pleasure, enjoyment, surprise, and delight (De Rojas & Camarero, 2008; Ma et al., 2013; Mitas & Bastiaansen, 2018; Pullman & Gross, 2004). This is in line with predictions from cognitive appraisal theory. The expansion of research involving this concept and the shifting nature of tourism discourse underlines the concept's growing significance in the field (Blomstervik & Olsen, 2022).

Multiple theoretical approaches are evident in studies that involve novelty, however studies report lack of coherent definition and structured comprehensive theory in tourism (Mitas & Bastiaansen, 2018; Skavronskaya et al., 2020). A recent review indicates three main clusters of studies in tourism, personal traits, experiences, and outcomes (Blomstervik & Olsen, 2022). The first considers preference for novelty as a personality trait through novelty-seeking. These studies conceptualise it as one of the influential factors for decisionmaking (Goossens, 2000; Lee & Crompton, 1992), exploration and motivation (Deci & Ryan, 1985), tourists' typology based on preferences (Cohen, 1972), and peak experiences (Cloninger et al., 1994). Thus, seeking novel experiences as a personality trait mostly discusses a motivational factor in tourism. The personal traits approach considers people in general travel for new experiences rather than for their own idiosyncratic goals (Baloglu & McCleary, 1999; Gartner, 1993).

The second group of studies are associated with experiences and use the appraisal process antecedent to emotions. Under this 'experiential' approach, prior experiences and current goals are a required condition for novelty to occur in tourism, which is unrelated to novelty seeking as a personality trait. Instead, it is considered as an appraisal dimension involved into emotions elicitation process which involves assessment of stimuli salience and a process of matching of current event with experiences from the past to determine the degree of newness and its outcome. This cognitive appraisal process determines the intensity of novelty and its outcome in the form of a discrete emotion and its valence. Appraisal of new wand goal congruency of the event responsible for elicitation of positive emotions, such as surprise and delight (Ma et al., 2017), and other emotions associated with pleasure (Mitas & Bastiaansen, 2018).

Adaptation of studies from cognitive and neuroscience to tourism sheds light on why and how novelty occurs in tourism, as well as adds value to the studies on important outcomes, such as memorable tourism experiences. Currently, this is one of the most studied topics in tourism research (Hosany et al., 2022; Hosseini et al., 2021; Tung & Ritchie, 2011) with the studies focussed on various aspects, such as satisfaction and loyalty (Zhong et al., 2017), co-creative experiences (Campos et al., 2018), and specifically the role of emotions for experiences. Emotions has been recognised as core buildings block of experiences (Bastiaansen et al., 2019) due to its capacity to enhances memorability. How experience is remembered and evaluated is determined by the succession of instantaneous emotional responses during an experience which occur in less than a second (Kahneman, 2011).

This evaluation mechanism can be further explained through the process of cognitive appraisal (Chapter 8 on emotions), with this chapter specifically emphasising the role of novelty for emotions and experiences memorability.

Thus, it increases the likelihood of experience being remembered (Skavronskaya et al., 2020). Furthermore, its importance for emotions elicitation determines its critical importance for MTEs both positive and negative, as novelty itself is neutral (Scherer, 2001). Overall, the experiential approach considers novelty as an antecedent to emotions, feelings, and memorable tourism experiences. Despite its critical importance for emotions and memories in tourism, research is at the early stage and associated with several problems.

Adapting cognitive theories of novelty from cognitive psychology to tourism context can provide new useful insights to existing approaches to understand tourism experiences. For example, Urry's gaze concept (1992) seeks to explain why people who have no prior experience with a place (tour groups) do not engage with a site and instead 'gaze' at it. This is interpreted from a sociological perspective as a leisure social activity and experiential consumption with predominately visual practice. From a cognitive psychology perspective, however, the gaze and lack of attention in a novel situation may be because there is no prior experience (cues), or goal related to viewing the scene. While the scene may be perceived as novel, it is not interpreted as highly arousing, and there is no orientation of attention to it.

Problems Associated With Research in Tourism

There are several problems associated with the research on novelty in tourism. First, current studies lack theory (Blomstervik & Olsen, 2022; Mitas & Bastiaansen, 2018; Skavronskaya et al., 2020). As a result, current studies on the concepts that underpin tourism experiences and implies novelty, experiencing difficulties in defining it and to distinguish antecedents with outcomes. For instance, novelty and novelty seeking are often used interchangeably in studies of tourism experiences but have different natures and therefore, meaning. Thus, it is often defined through 'folk-psychology', social science, or personality studies (Chapter 5 on disambiguation), ignoring advances in cognitive and neuroscience research on how the brain works.

Another problem of the research on novelty in tourism is associated with the incorrect assumption on emotions elicitation process. Tourism research demonstrates a tendency to rely on the implicit meaning of emotion in folk psychology, which is confounds the antecedent (eliciting condition) with the consequent (emotional response patterning, including feelings) (Scherer et al., 2001). Oversimplification of the appraisal mechanism to a general cognitive evaluation leads to the assumption that an external stimulus is a cause of emotions (Tuerlan et al., 2021). In turn, this leads to underestimation of the critical role of novelty for emotions elicitation and memorability formation. This assumption also ignores the importance of prior experiences and tourists' goals (Chapter 9 on goals). Under this assumption, tourists would experience novel events in exactly the same way. Instead, in cognitive appraisal process it is not a property of the environment but is an outcome of the appraisal process and antecedent to emotions, feelings, and tourism experience memorability.

Theoretical ambiguity and research inconsistency raise a methodology question in recent tourism studies on emotions and experiences that also involve novelty. Traditional methods used in tourism mainly adopt subjective self-report techniques and scales to measure different, often retrospective aspects of emotional and novel experiences (Pearson, 1970; Próchniak, 2014). Conversely, recent cognitive and neuroscience approaches utilise more objective psychophysiological methods that can provide real time data (Chapters 20–22).

CONCLUSION

This chapter contributes to tourism literature by explaining how novelty is processed in the brain at different levels and further explores its critical importance for tourism experience. This is achieved by utilising dual-process approach to perception of novel stimuli and its processing, and the exploration of its further connection with three main cognitive psychology concepts: emotions, memories, and attention that underpins tourism experience. By elucidating the key-related concepts and principles in cognitive psychology, this chapter discusses what role novelty plays in the processes such as attention and orienting response, cognitive appraisal of emotions elicitation and memorability, while also 'translated' them to tourism context.

Novelty is inherent to tourism when a tourist travels. They explore an unfamiliar environment that contrasts with their ordinary life. New experiences imply a high degree of newness, increased attention, and high emotional arousal, which improves memorability. Novel experiences in tourism are emotionally engaging with memories of these experiences accompanied by certain, usually pleasurable emotions. Emotions are what makes experiences memorable, with novelty being an antecedent condition to high emotional arousal. Therefore, the concept's importance for tourists' emotions and memorable experiences is evident with more studies required on its antecedents and consequences in tourism.

Future research can investigate the degree or intensity of novelty in memorable tourism experiences by further exploring its antecedent conditions, such as tourists' goals and prior experiences. Ongoing theoretical and methodological developments are required to improve theory and definitions of novelty towards establishing its typology in tourism context. Potential avenues for further studies include emotional valence with a specific focus on mixed-emotions and negative experiences elicited by novel events, its re-evaluation over time, and long-term outcomes. For instance, the global

COVID-19 pandemic is also a novel experience associated with mixed emotions. Agyeiwaah et al. argued that '... a positive outlook of COVID-19 can reduce the fear and uncertainty of travelling...and a positive outlook propels self-efficacy beliefs of overcoming challenging situations' (Agyeiwaah et al., 2021, p. 7). Therefore, theoretical advancements on novelty have a potential to better understand emotional aspects of COVID-19 impact on tourists over time.

Research on the outcomes of novel experiences has practical value. For instance, Volo emphasises the relevance of negative and mixed-emotional tourism experiences as they 'can be beneficially interpreted as a call to design educational elements of tourism' (2021, p. 6). Following Nawijn and Biran (2019) who confirm that negative emotions can lead to positive outcomes, such as pro-environmental and ethical choice formation. Future studies can focus on the role of novelty for elicitation of negative emotions in tourism experiences at different stages.

Understanding novelty also has significant practical value. For instance, it can improve understanding of high value tourists: (Australia's target audience), and improve experiences designed for them. This market has been defined as novelty-seekers. However, based on the findings from this chapter, novelty-seeking is a personality trait, while novelty is an appraisal dimension critical for elicitation of emotions and memorability of experiences. Therefore, to create experiences for high value tourists that are novel, memorable, and emotional, one needs to focus on understanding tourist goals and prior experiences.

Chapter 11

MEMORY

Noel Scott
Edith Cowan University, Australia

Biqiang Liu
Griffith University, Australia

Brent Moyle
Griffith University, Australia

Abstract: This chapter provides a holistic understanding of memory and the tourism-memory nexus. This chapter begins with an overview of what memory is and the history of research on it. Following this, the chapter outlines key memory-related themes in cognitive psychology. Next, the implications of the tourism-memory nexus for research on memorable tourism experiences are discussed. It provides a critical analysis of the research which examines tourism and memory from the viewpoint of cognitive psychology. The chapter concludes with an outline of key avenues for further research in order to delve into tourism-memory nexus. **Keywords**: Memory; collective memory; memorable tourism experiences; tourism-memory nexus; memory bias; schema

INTRODUCTION

Memory is a mental process that encodes, stores and retrieves information when required. Human memory is an evolved capability, which has been fine-tuned over generations through the process of natural selection (Nairne & Pandeirada, 2016). From the perspective of evolutionism, the function of

Cognitive Psychology and Tourism
Tourism Social Science Series, Volume 27, 127–142

Published under exclusive licence by Emerald Publishing Limited
ISSN: 1571-5043/doi:10.1108/S1571-504320240000027011

remembering (and forgetting) is to generate an evolved adaptation (or a set of adaptations) that ultimately enhances fitness (Klein et al., 2002; Sherry & Schacter, 1987). The implication of this evolutionary viewpoint for memory research is that memory processes should not be expected to be the product of a unified design. Instead, memory processes are the result of numerous incremental adaptations, one building upon another (Murray et al., 2017) in response to the needs of decision-making (Sherry & Schacter, 1987). Another implication is that what one remembers and how it is accessed and recalled is likely related to survival and related higher-level goals.

From a neuroscience perspective, memory processes occur in neuronal circuits. There have been significant advances in the study of encoding in such circuits and how short-term working memory is consolidated in long-term memory. From a cognitive psychology perspective, memories are individual phenomena, occurring in one's brain. Two people may share the same experience but have different memories of it. In comparison, the term collective memory, used in social psychology, refers to individual memories which share many similar or common characteristics concerning the identity of a community (Hirst et al., 2018).

Some 2,400 year ago, Aristotle wrote a philosophical treatise on memory, called in Greek *Peri mnêmês kai anamnêseôs* (On Memory and Recollection) in which many of the issues still under debate in cognitive science today were raised. Hermann Ebbinghaus was one of the first scholars to use experimental techniques to study memory. Ebbinghaus memorised lists of nonsense syllables and studied the ability to recall them. This led to characterisation of both the forgetting and remembering curves, and of sensory, short-term and long-term memory. In the early 20th century, Richard Semon identified that a memory was a specific web of neurons termed an engramme. In the 1930s, Frederick Bartlett used recall of stories to study memory storage. In the 1940s, neuropsychology began to explore the biological basis for encoding memories. Following this foundational work, the cognitive revolution in the 1960s led to many advances in memory research. For instance, Tulving (1972) proposed two distinct kinds of long-term memory, episodic and semantic.

MEMORY CONCEPTS

The traditional view of memory was of a 'unitary system' which 'stores' information and allows it to be accessed and reproduced. According to this view, memory is like a library. Information is encoded and stored to be retrieved later, like information written in a book is stored in a library to be read later (Salvaggio, 2018). This traditional view is no longer considered to be the case. Instead, reconstruction of memory is considered more like the process a detective uses reviewing available evidence to determine what most likely occurred. Much of what is remembered is not explicitly stored but is instead constructed or reconstructed when one attempts to recall it. In this reconstructive view (Salvaggio, 2018), when memories are retrieved, they may be altered and modified, resulting in inaccuracies and distortions. Memory recall is affected by the current context of the person doing the recalling as well as their background beliefs and other memories. Information from encoding and cues from retrieval, as well as generic information, are all exploited to construct a response (Bartlett, 1932). Today psychologists agree that there are several memory systems. Each memory system is a 'neurocognitive capacity to encode, store, and retrieve information' (Tulving, 2005, p. 36). These memory systems include sensory memory, working memory, declarative memory, semantic and episodic memory.

Sensory and Related Memories

Sensory memory involves the flow of information from sensory organs which is only stored for between 200 and 500 milliseconds. Such information underpins 'exteroception' (involved in touch, sight, sound, smell and taste) that enables interpretation of the external world, 'interoception' (involved in pain, hunger and other feelings) or body sensations and 'proprioception' (involved in body position, movement and acceleration) which enables motion. The sensory nerve system is responsible for detecting, deciphering and remembering such sensory data (Wan et al., 2020), and is related to the acquisition of motor skills. Through practice, motor learning is stored in the brain as a schema in memory. Consequently, performing skills such as riding a bike or driving a car are effortlessly and 'subconsciously' executed, even if someone had not performed these skills in a long period of time.

Short-term memory is related to attention (Oberauer, 2019). It can hold information temporarily in an accessible state. Short-term memory has strict capacity limits and decays over a short period (Cowan, 2008), but may be transferred to long-term memory through a process known as consolidation (McGaugh, 2000). Factors that can influence how long information endures in long-term memory include the situation when the memory was encoded and the number of times it is accessed. If a person is aware and alert during a personally important and relevant experience, then the memory of it will probably be more vivid. Empirical work demonstrates that emotional experiences also lead to stronger memories (Kensinger & Ford, 2020). Memories which are recalled often tend to become much stronger over time (Kim & Jang, 2016). Long-term memory was considered a unitary storehouse, but today the structure of long-term memory is considered as multiple systems. There is a fundamental division in long-term memory between declarative (explicit) and nondeclarative (implicit) systems (Fig. 11.1).

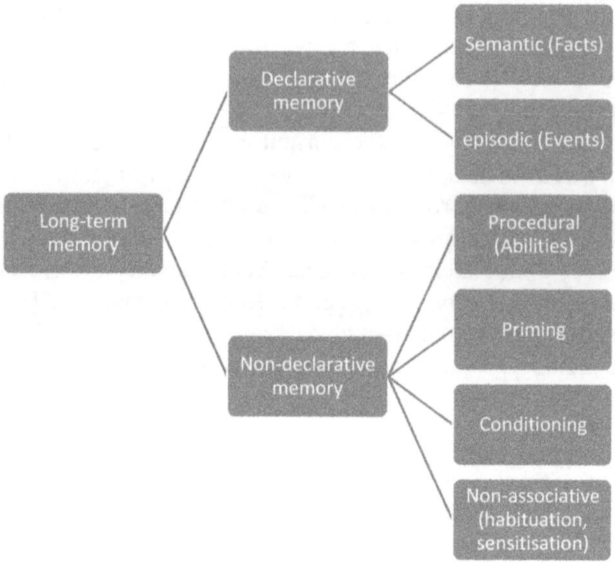

Fig. 11.1. Structure of Long-term Memory.

Declarative memory encompasses memories that are available to conscious awareness. Declarative memory is further divided into episodic memory (events that are encoded in terms of specific times and places of occurrence) and semantic memory (general knowledge of facts and concepts) (Jorgenson et al., 2019). Nondeclarative memory includes skill learning, priming, conditioning and habituation (Amodio & Berg, 2018). These different memory systems may function independently, in concert, or in competition, and are acquired, stored and expressed in different ways (Henke, 2010; Squire & Zola, 1996).

Episodic, Autobiographical and Semantic Memories

Episodic memory involves individuals' long-term storage of factual memories concerning personal experiences. This covers unique personal experiences in terms of their content (what), temporal occurrence (when) and location (where) (Tulving, 2005). The most impactful and easily recalled memories are those that are most emotional and meaningful, with studies demonstrating goal relevance is critical (Liu et al., 2022). This effect is attributed, in part, to increased attentional and sensory processing of emotional content (Palombo et al., 2021). These memory elements are highly accessible to cognitive operations and can be readily retrieved and integrated into current evaluative judgements (Clewett & Davachi, 2017; Gallo & Wheeler, 2013). Episodic memory also supports judgemental responses that are faster, more consistent across individuals and more predictive of people's thoughts when compared with non-affective, reason-based evaluations of stimuli (Pham et al., 2001).

Autobiographical memory is a form of episodic memory that integrates individual experiences of daily life to support cohesive identities (Fivush, 2011; Io, 2015). It encompasses events and experiences that occur to an individual remembering them (Jorgenson et al., 2019). These memories refer to representations of the self within a social and physical environment and have significance and strong emotional content. Autobiographical memories are constructed and reconstructed based on past experiences, context and information from outside sources (Tung et al., 2017). Studies examining the temporal distribution of autobiographical memories across the life span find an increased number when a person is 15–25 years, termed the 'reminiscence bump' (Tung et al., 2017).

Semantic memory encodes, stores and retrieves explicit information such as facts, ideas, concepts and knowledge of the world (Schwarz & Sudman, 2012; Tulving, 1972). The semantic aspect of autobiographical memory is responsible for processing personal, factual or self-referent semantic knowledge, for example, the name of the holiday destination previously visited (Meesters et al., 2000). Thus, semantic knowledge is self-connected generic knowledge as opposed to episodic components that are recollections of experiences that one had that were tied to a time and place (Piolino et al., 2002).

Encoding and Retrieval of Memories

Encoding and storage of episodic information in long-term memory varies with the kind of rehearsal given to information stored in short-term memory. Repeating a list of words is an example of maintenance rehearsal (Kim, 2018). Elaborative rehearsal is superior to maintenance rehearsal because it establishes links between the information held in short-term memory and the information already stored in long-term memory, for example, by forming visual images of the objects referred to by the words in a list (Roy & Pentland, 2002). Mnemonic techniques, such as the method of loci, are kinds of elaborative rehearsal. The method of loci involves imagining yourself placing new information around a room and then visualising going back and picking it up in the order you put it down (Roy & Pentland, 2002). The level of processing also affects learning success, with deep semantic processing supporting better memory than shallow sensory processing. Deeply encoded information is distinctive and easily retrieved in the future (Skavronskaya et al., 2017). Finally, the organisation and encoding of newly learnt information within existing schema is necessary for successful recognition and recall.

While encoding processes are important, they cannot be considered apart from retrieval processes. Retrieval refers to the process of reactivating information that has been stored in memory. The encoding specificity principle asserts that events are recognised or recalled only when retrieval cues at the time of test match the encoding cues at the time of learning (Frankland et al., 2019). Such retrieval cues allow one to activate the to-be-remembered episode and its context. From this perspective, forgetting represents a failure to access an episode because the retrieval cues are inadequate. Memory retrieval may be triggered not only when individuals actively recall a memory associated with a specific cue but also when individuals are passively exposed to cues which can spontaneously activate memory retrieval (Frankland et al., 2019).

Memories may be retrieved through several processes. Recall refers to the simple process of remembering something without any cues, and in the physical absence of that thing. Recognition is identifying the information about a previously known thing after seeing the thing or experiencing a memory again. Recollection refers to the rebuilding or piecing together of a memory. Ecphory is a process by which retrieval information provided by a cue is correlated with the information stored in an episodic memory trace, providing a baseline for the subjective experience of remembering and the corresponding memory performance (Steinvorth et al., 2006). Ecphory reflects the interplay of external sensory or internally generated cues and stored memory traces (or engrams) is a concept used in cognitive research and neurobiological research (Frankland et al., 2019).

Intensive emotions trigger the formation of strong memories, and weaker emotional experiences form weaker memories (Christianson, 1992). This is because strong emotional experiences trigger the release of chemical messengers, including monoamine neurotransmitters and hormones, which strengthen memory. Therefore, the memory for an emotional event is usually stronger than the memory for a non-emotional event (Buttafuoco et al., 2018). For example, when humans and animals are stressed, the brain secretes more of the neurotransmitter glutamate, which helps them remember the stressful event (Shields et al., 2017).

Flashbulb and Forgetfulness

A flashbulb memory is an exceptionally clear recollection of a salient event that carries with it strong emotional reactions (Brown & Kulik, 1977; Pillemer, 1984). These focus on personal recollections surrounding unexpected, momentous or emotionally charged events of public or personal significance, such as the terrorist attacks on New York's World Trade Center. As Pillemer noted, 'images of only a tiny subset of specific episodes – death of a loved one, landing a first job, getting married, hearing about public tragedies –

persist over a lifetime, with little subjectively experienced loss of clarity' (1984, p. 64). One explanation for why flashbulb memories are so well recalled is that they are highly distinctive events in long-term memory (McCloskey et al., 1988).

Forgetting is when the information encoded and stored in a person's memory is lost or modified. According to the seminal work of Ebbinghaus (1885), memory loss is explained by trace decay theory which identifies time as a main cause for memories fading (Brown, 1958). Connected to this, interference theory indicates forgetting occurs as a result of old and fresh memories interfering with each other (Postman, 1963), especially when it comes to similar events. Forgetfulness strength varies in terms of the effectiveness of encoding, which means ineffective encoding is more likely to lead to forgetfulness. In this case, the information people processed does not transfer into long-term memory (Nickerson & Adams, 1979). Furthermore, information received and encoded in the absence of memory-related cues has a less likelihood of being retrieved. This phenomenon is also known as cue-dependent forgetting or retrieval failure (Tulving, 1974).

Memory Structure

A number of different meta-cognitive structures relevant to memory have been noted (Ghosh & Gilboa, 2014). A schema is an associative network referring to a mental construct which consists of a cluster or collection of related concepts (Bartlett, 1932). There are many different types of schemata. When a schema about a person or object being observed is activated, the brain can make immediate assumptions about that person or object (Ghosh & Gilboa, 2014; Gilboa & Marlatte, 2017). Cognitive psychology defines schemas as, 'an active organization of past reactions, or of experiences, which must always be supposed to be operating in any well-adapted organic response' (Bartlett, 1932, p. 201). There are four necessary features that define a schema (Ghosh & Gilboa, 2014). First, a schema holds interconnected units where information can be stored, and meanings can be transformed. Second, a schema is based on numerous episodic events, and represents a cluster of similarities or commonalities. Third, a schema is flexible, implying that new events can generate a change in meaning. Fourth, when exposed to cluttered information, a schema is adaptable as humans use two ways to hold or alter their schemas, known as assimilation and accommodation (Ghosh & Gilboa, 2014). Unchallenging information is easily absorbed by the existing schema, which remains unchanged, while complicated information is accommodated with greater effort, leading to a changed schema (Chapter 12).

There are several functions of schemas which are increasingly evident during information processing (Schneider, 2004). They are critical for directing attention, as it is easier for humans to recognise schema-relevant information, based on the event episodes preserved in schemas. Information that is inconsistent with expectation and schemas attracts a higher level of attention (Hilton et al., 1991; Stern et al., 1984; White & Carlston, 1983). In addition to attention arousal, schemas help to evaluate new information and to interpret ambiguous behaviour (smiles and hand gestures). The more consistent behaviour is with schemas, the easier an evaluation will be achieved (Koehler, 1993; Munro & Ditto, 1997). Schemas also aid the encoding and retrieval from memory. People can better understand and remember information which activates a salient schema (Dooling & Lachman, 1971), and are prone to remember information inconsistent with a schema better than that which is consistent (Hastie & Kumar, 1979). The final function of schema is facilitation of inference (Taylor & Crocker, 1981), whereby prior knowledge structure or schemas guide cognitive reasoning and logic. A relatively static schema may be considered as a frame, while a dynamic schema is described as a script. Script theory indicates that individual behaviour is guided by 'scripts' (Tomkins, 1978). Scripts are considered to be the glue which links recollection, storytelling and consequent behaviour (Kawasaki & Uehara, 2020), as people are prone to behave in a way written by a script as a guideline for action (Hudson et al., 1992).

Tourism and Memory

Memory is an important concept in disciplines that study humans, including in the tourism field (Braasch, 2008). Pearce and Packer's (2013) thematic review suggests that tourism researchers should engage in a deep and meaningful way with contemporary developments in memory and cognition from mainstream psychology. In recent years, theory from cognitive psychology and neuroscience is increasingly being transferred to the social sciences, with discourse evident in tourism and cognate fields, such as marketing and management (Scott, 2020). Memory in particular has been studied from a cognitive perspective, to advance the conceptual understanding of why particular types of memories are recalled (Skavronskaya et al., 2017), how sensory stimulus information (sounds, tastes, touch) may trigger memory (Agapito et al., 2017) and how schema many be changed (Zhang et al., 2018).

The role of memory indeed permeates the entire process of a trip, from anticipation to recollection. During the pre-trip, a memory of a past trip may inspire destination selection, as tourists may visit the places associated with past experiences, or revisit the places connected with positive memories of an earlier journey. During a trip, memory plays a critical role in influencing the actual tourist experience at a destination. A seasoned tourist may recall earlier travel to similar destinations elsewhere and make comparisons with the present experience. After a trip, during the recollection stage of an experience, memories can be recalled and savoured (Filep et al., 2013; Marschall, 2012; Sthapit et al., 2021).

Memorable Tourism Experiences. Tourism is a context for experiences which are subsequently remembered and recalled (Ritchie et al., 2011). Tourism experiences are inherently personal and differ from an 'objective real experience', primarily due to a tourists' personal cognitive processes (Sthapit & Coudounaris, 2018). Accordingly, the study of memorable tourism experiences (MTEs) has been widely adopted by academia and the tourism industry, predominantly to assess the quality of experiences (Hayes & MacLeod, 2007; Loureiro, 2014). Numerous scholars have referenced the formation of memory in understanding MTEs (Wei et al., 2019). MTEs are acknowledged as an autobiographical memory of personal experiences (Chung et al., 2017; Kim, 2014; Larsen, 2007; Schwarz & Sudman, 2012; Servidio & Ruffolo, 2016). However, the application of theories and concepts from cognitive psychology to the study of MTEs, while emerging is still in its infancy.

MTEs are acknowledged as memorability-based construct where external stimuli are encoded, stored and retrieved (Skavronskaya et al., 2017). In tourism, several factors, such as involvement, have been found to influence how experiential information is processed (Chen & Rahman, 2018). In cognitive psychology, involvement could be interpreted as goal importance and congruence. MTEs have been noted to affect future travel (Chandralal & Valenzuela, 2013; Wirtz et al., 2003) and reflection (Servidio & Ruffolo, 2016, p. 153). Autobiographical memory of tourist experiences has been identified to positively affect revisit intention (Kim & Youn, 2017), word of mouth publicity (Kim, 2018), tourist satisfaction and quality of life (Chung et al., 2017).

Although research on MTEs is critical for the tourism-memory nexus, there remains a disconnect between the tourism field and higher-level debates in cognitive psychology. For instance, a large stream of research focuses on external objects as antecedents of MTEs, encapsulated by stimulus-organism-response theory derived from environmental psychology with an emphasis on individual behaviour, without focusing on cognitive processing. Bigne et al. (2020) identify three issues in using the theory: some external objects (elements) could be both objective stimuli and interpreted (authenticity and image); the linear sequence of stimulus-organism-response may blind researchers to important phenomena and dynamic cognitive processing; studies focus on behavioural outcomes. Hosany and Björk (2022) have systematically reviewed the existing body of knowledge on MTEs, proposing a framework for summarising the context, antecedents, core phenomena and outcomes of MTEs. Similarly, Hosseini et al. (2021) conducted a systematic review of 52 articles on the theory, utilising a qualitative approach, which identified nine main categories of topics.

Measurement Methods of Memorable Tourism Experiences. Kim et al. (2012) argue that existing tourism experiences measures are not qualified to provide deeper knowledge on experiential factors which influence future behavioural intentions. Indeed, tourism scholars found that experience memorability can be derived from multiple factors, such as novelty, involvement and familiarity (Skavronskaya et al., 2020). Despite this however, most MTEs measurement tools focus on aspects of the experiential context such as hedonism, refreshment, local culture, meaningfulness, knowledge, involvement and novelty (Kim et al., 2012), without recognising the subjective nature of experiences. Further, the effect of important variables such as goal importance and novelty on these subjective evaluations is yet to be recognised in existing measurement approaches. The application of concepts from cognitive psychology which determine a particular person's responses on the several dimensions noted by Kim et al. (2012) has substantive potential to increase the usefulness of existing measurement approaches.

In Kim et al. (2012), hedonism refers to pleasurable feelings that excite oneself. However, as discussed in Chapter 8 on emotions, whether a person evaluates an experience as hedonistic depends on personal goals at the time. Similarly, personal goals are also critical for dimensions such as restoration and meaningfulness. Novelty refers to an evaluation of the uniqueness of an experience based on personal memories of the past (previous experience). Subsequently, evaluation of local culture, activities and special events, hospitality, infrastructure, accessibility, quality of service, physiography and place attachment may lead to strong memorability for individuals; however, this is dependent on both current goals and previous experiences.

The memorability of an experience has been measured using frequency of recollection and vividness of memory (Kim, 2014; Wei et al., 2019). Recollection of MTEs refers to whether the former events or experiences can be entirely recalled, while the vividness of MTEs describes if specific items or features can be remembered (Ahn et al., 2017). Often, only an overall sense of a destination is recalled in existing studies (Chung et al., 2017). Although, upon prompting with specific cues, more detailed information is recalled (Wei et al., 2019). Sometimes a specific incident or feeling experienced during a vacation might be vividly remembered, which might be considered a 'flash bulb' memory (He et al., 2022).

Despite the remarkable number of studies assessing MTEs, previous measurement scales are predominantly based on a behaviourist paradigm. There remains a lack of measurement tools which draw explicitly on cognitive psychology. For instance, previous tourism studies have noticed the critical role of recall in measuring MTEs; however, they failed to capture its situational and personal nature. Similarly, the degree of personal involvement (goal strength) plays a pivotal role in understanding individuals' recall of tourism experiences (Kim et al., 2022a), yet is not often captured in existing measurement approaches.

Cognitive Psychology on Aspects of Memory in Tourism

Reconstructive Memory. The reconstructive properties of memory present an opportunity for retroactive management of tourist experiences (Braun-LaTour et al., 2006). For example, Braun et al. (2002) reported that when participants were shown a Disney ad that contained false information (picture of Bugs Bunny), 16% of participants reported meeting Bugs Bunny at Disneyland. Effectively this could not have happened, as Bugs Bunny is a Warner Bros copyrighted character and not at Disneyland. Compounding this, the reconstructive nature of memory has been further considered as a dynamic process that alters how consumers remember their previous experience because they create false post-experience 'information' (Sthapit, 2019).

Sensory Memory in Tourism. Empirical work demonstrates that the tastes, smells and sounds of a destination become sensory markers (cues) that initiate recollection. Subsequently, tourism planners could apply attributes in the sensory landscape of a destination in marketing to increase recall (Agapito et al., 2017; Kirillova et al., 2014; Ooi, 2005). The vital role of the sensory dimension of tourist experiences is stressed by current research and supported by multidisciplinary focus on the role of the five external senses in human perception, memory and behaviour. Agapito et al.'s (2017) two-step exploratory study conducted six months after tourists visited Southwest Portugal concluded that experiences which activate the senses play a significant role in the formation of long-term memories of travel experiences, and can encourage repeat visitation and positive behavioural intentions. Subsequently, Agapito et al. (2017) offers insights on the contribution of sensory diversity as perceived by tourists to memorable destination experiences and to explore the connection between long-term memory of sensory impressions and destination loyalty.

Forgetfulness in Tourism. The concept of forgetfulness in tourism has been largely overlooked and remains unexplored and conceptually underdeveloped. Ageing or physical injury often triggers forgetfulness; however, 'forgetting' is most often attributed to psychological causes (Farmaki, 2021). Farmaki (2021) has discussed forgetfulness in the context of crises in tourism, noting the likelihood of tourists forgetting about a crisis and continuing a journey is more profound when the effect of a crisis on the tourist is indirect.

Goossens (2000) conceptualised mental imagery as a mode of processing that includes perceptual or sensory representations encoded and stored in working memory (Cardoso et al., 2019; Kim et al., 2014; Walters et al., 2012). Subsequently, the concept of imagery sits at the boundary of consciousness and memory studies. In the tourism context, mental imagery has been identified as a multidimensional construct, consisting of two dimensions: elaboration and quality (Bogicevic et al., 2019).

The dimension of elaboration describes the volume of mental pictures created and the level of a person's engagement with fantasy imagery (Yoo & Kim, 2014). The quality of imagery captures 'how vibrant, intense, clear, and sharp mental images are, and is similar to the notion of "vividness"' (Bogicevic et al., 2019, p. 56). Recently, mental imagery has been considered to represent the extent to which individuals take action to psychologically process external stimuli, such as the psychological process engaged in by listeners during exposure to advertisements (Jia et al., 2021). In doing so, mental imagery is associated with cognitive psychology in tourism by its relevance to information cognition, learning process, reasoning and spatial ability (Bogicevic et al., 2019; Zheng et al., 2022).

Mental imagery plays an important role in out-of-the-ordinary experiences, and various visual stimuli (tourism ads and virtual environments) trigger imagery by employing a variety of imagery-eliciting strategies (Kim et al., 2014). In a hospitality context, research demonstrates a hotel advertisement has the capacity to inspire consumers to visualise the cosy bed or delicious meals found in a stimulus (Josiassen et al., 2016). Walters et al. (2007) demonstrated a consumer's purchasing decisions can be driven by his/her mental imagery of a tourism product. In virtual environments, space, though physically represented, is still perceived at a conceptual level which is why the human mind employs cognitive processes such as mental imagery to understand better the virtual environment (Bogicevic et al., 2019). Compared to traditional stimuli, digital technologies seek to stimulate tourists' multiple senses and engage them in the virtual world (Zheng et al., 2022). For example, during a VR hotel preview, users are led to envision themselves in many scenarios within the hotel suite and the mental imagery processing enhances their sense of presence (Bogicevic et al., 2019). Thus, further research is necessary to systematically examine the immersive experience of virtual tourism from the perspective of mental imagery processing.

Semantic and Autobiographical Memory in Tourism. Tourist memories of a destination are a form of semantic memory, whereas tourist memories of personal experiences are a form of autobiographical memory (Kim, 2014; Yin et al., 2017). Dream destinations are more strongly connected to semantic memory (with abstract or general facts), while a favourite destination is mostly based on episodic memory (personal and specific facts)

(Cardoso et al., 2019). Autobiographical memory is a type of long-term memory which relates to the study of tourist experiences (Sthapit et al., 2017). Cohen and Areni (1991) indicate that consumption emotions leave strong affective traces or 'markers' in episodic memory. The phenomenology of autobiographical memory allows researchers to understand major components of memory on individuals' recall of experienced events (Ballantyne et al., 2011; Kim et al., 2022b). Human autobiographical memory, containing information on the self and its sociocultural surroundings, is a unique reservoir of knowledge, valuable in rural social research as well as in several other fields of study (Charatsari, 2014).

As global competition intensifies, destinations increasingly aim to deliver MTEs and scholars accordingly explore the influential factors. Kim's (2010) analysis reveals that the experiential factors of involvement, hedonic activity and local culture positively affected the autobiographical memory of recollection and vividness of past experiences. To ensure that tourists remember past tourism experiences, Kim and Jang (2016) examine the effects of the personality trait of openness to a different culture and retrieval cues (auditory, olfactory and memorabilia) on memory retrieval. Recent studies focus on alternative stimuli (travel photographs) and found a significant positive impact on autobiographical memory by increasing positive emotions (Zhang et al., 2021).

Yin et al. (2017) examined the effects of autobiographical memory on behavioural responses. A conceptual model was developed to examine the causal relationships of product-destination congruity, travel satisfaction, autobiographical memory, evoked pleasure, destination attachment and purchase intention, with autobiographical memory found to lead to pleasure and a positive mood, higher levels of destination attachment, and greater purchase intentions for the product available in the tourist's hometown. The effect of senses on memory has been discussed (Agapito et al., 2017). Sensations are at the beginning of the process of perception – a dynamic process between experience and memory.

In tourism, several measures have been developed to identify the characteristics of a destination which are most likely to be remembered. For example, Johnson et al. (1988) initially developed the memory characteristic questionnaire to measure the phenomenology of memory based on six key dimensions. Following this, the tourism memory characteristics scale was developed and validated by Kim et al. (2022b) to measure the phenomenological characteristics of tourism memory in terms of a seven-dimension scale consisting of accessibility, trip details, vividness, sensory details, valence, emotional intensity and sharing (Kim et al., 2022b). Similarly, Tung and Ritchie (2011) interviewed 208 adults about MTEs, finding four major elements which enable a tourist experience to be memorable: affect (positive emotions and feelings), expectations (fulfiling and exceeding tourists' expectations), consequentiality (personal meanings and development) and recollection (sharing and re-living the memories) (Gross & Huber, 2020). Following this rationale, Jorgenson et al. (2019) tested the validity of a new tourism autobiographical memory scale that measures tourist experience through personal memory.

Schema and Schema Change. In the tourism literature schema, change is often termed 'transformation' which is considered an inner-oriented and eudemonic change leading to well-being and happiness (Tasci & Godovykh, 2021). Others consider that the transformative potential of tourism is particularly valuable as modern society transitions to the third-generation experience economy (3.0) (Kirillova et al., 2017a). However, in most cases, the tourism literature does not discuss the underlying cognitive mechanisms which underpin transformation (Pung et al., 2020). One 'mechanism' identifies 11 stages or phases of transformation, whereby a person makes meaning of new experiences and integrates new learning into a more inclusive and discriminating worldview (Reisinger, 2013a). Sheldon (2020) examines how outer journeys can transform the inner consciousness, leading to awakening or enlightenment. In this case, peak transformational moments can be designed to give glimpses or create persistent shifts in consciousness. Recent studies, such as Tasci and Godovykh (2021), attempt to develop a comprehensive model with antecedents, outcomes and moderators of transformation, which paves the way for empirical analysis on specific dimensions of transformation and its underlying process. Such research may benefit from examining schema change processes discussed in cognitive science (Zhang et al., 2018).

Script theory has been applied by tourism scholars to better describe the bonding role of memory in the information processing (Manthiou et al., 2017; Su et al., 2020). Scripts (a cluster of related schemata) have the potential to add further depth of knowledge into tourists' expectations as well as situational decision-making processes, which have implications for behaviour (Erasmus et al., 2002). Script theory has emerged in the tourism field as an approach developing a knowledge structure of emotional experience to facilitate the interpretation and formation of direct future behaviour when scripts are activated (Su et al., 2020). This theory can be used to interpret the formation of tourist's environmentally responsible behaviours, which means a positive experience will form a cognitive script to guide environmentally responsible behaviours (Su et al., 2020). Manthiou et al. (2017) have integrated cognitive appraisal theory and script theory to examine the relationships among cognitive appraisal, emotions, script and repurchase intention.

Script theory has a wide range of explanatory power (Wen et al., 2022). Tourism stakeholders (provider and tourist) are expected to behave according to a specific cognitive script that has been developed based on previous experience (Järvi et al., 2020). For example, drawing upon script theory, Miao et al. (2011) examined emotional responses (felt emotions and displayed emotions) to behaviours of other consumers in service encounters. By extending this theory into the theme park setting, Manthiou et al. (2016) developed an understanding of the impact of theme park visitors' experiences on their recollection-based memories, satisfaction and loyalty. Hsu and Chiang (2011) found that service encounter scripts can be used in the fast-food industry to elicit favourable impressions from tourists.

Some research highlights that hospitality practitioners can make script deviations to win guests' full attention (Magnini & Dallinger, 2018). Script deviations occur when something happens (positive or negative) that tourists did not expect. A variety of triggers can cause a deviation from a script during a trip. Magnini and Dallinger (2018) indicated that most tourists only pay full attention when there is a script deviation, whether positive or negative. Most research has focused on negative script deviations (service failure and recovery scenarios). Therefore, future research should examine positive script deviations.

CONCLUSION

This chapter has identified that while MTEs research is increasing, it remains fragmented. Its discussion offers and synthesises useful entry points for future tourism research to delve into the tourism-memory nexus. It proposes future research areas to advance knowledge.

There is a lack of conceptual engagement with key developments in cognitive science across MTEs-related studies (Hosany et al., 2022). Instead, subdisciplines such as behavioural psychology, positive psychology, environmental psychology and social psychology dominate existing discourse. Cognitive psychologists have suggested the use of a cognitive theoretical lens to develop new explanations and an in-depth understanding of experiences (Okhuysen & Bonardi, 2011). It is critical to note that experience memorability is dependent on an individual's cognitive processes. Further, it is vital to accept that memories of individual experiences are not 'caused' or 'determined' by external stimuli. Instead, they are the result of internal cognitive processes. Therefore, theories such as the stimulus-organism-response model, based on early environmental theory must be seriously questioned, and recalibrated according to a cognitive approach (Chapter 4).

Despite widespread usage, there is a mounting criticism regarding the continued application of Kim et al.'s (2012) MTEs scale (Hosany et al., 2022). Despite being validated in cross-cultural settings, the original seven dimensions (hedonism, novelty, local culture, refreshment, meaningfulness, involvement and knowledge) are not generalisable as tourism scholars have had difficulties replicating the scale. In addition, the measurement method could use cognitive psychological methods, since MTEs are based on tourists' autobiographical memory (Kim et al., 2022a). In response to such criticisms, measurement scales related to memorable experiences in tourism have been published in recent years, such as tourism autobiographical memory scale (Jorgenson et al., 2019), and tourism memory characteristics scale (Kim et al., 2022b). Future studies should be cautious when applying Kim et al.'s (2012) measurement, and instead apply other MTEs measures from the perspective of cognitive psychology.

MTEs have been overwhelmingly measured by using self-report surveys, which are limited due to subjective bias. This is not only because self-report measures rely on retrospective evaluation, leading to inevitable memory reconstruction and recall bias (Li et al., 2022), but also due to delays between experiencing and reporting which can trigger random and systematic retrospective biases (Liu & Li, 2021). In addition, Hosany et al.'s (2022) review found a large proportion of studies were cross-sectional, conducted post-visit and failed to represent the characteristics of personal or autobiographical memory which is inherently selective and influenced by multiple factors, such as situational cues, individual motivations and travel companions. Future research should apply a variety of data collection techniques which inspire tourists to accurately recall of their experiences by adopting a day-to-day reconstruction method, a reflective diary or journal, modifying sampling techniques and utilising self-report in combination with physiological and experimental design. Particular emphasis should be placed on combining self-report research methods with other neuroscience techniques, such as using eye-tracking technology, galvanic skin response, functional magnetic resonance imaging and event-related potential, to offer more accurate reflection of the cognitive processing activities of an individual (Bainbridge et al., 2017).

Chapter 12

MEMORY SCHEMA

Rui Zhang
Hainan Tropical Ocean University, China

Noel Scott
Edith Cowan University, Australia

Abstract: A schema is a concept used to describe the organisation of memory about an object. Schema are perceptual prototypes abstracted from numerous episodic events based on their similarities or commonalities. In tourism, a schema may be termed a network, mental model, social schema, stereotype, social role, script, worldview, heuristics or archetype. A schema is flexible. New information may be either assimilated or accommodated based on its degree of congruity. Schema incongruent information requires more mental effort to process. The concept is used in research on destination brand image, storytelling, social relationships, attitude change and transformation during travel. The concept is useful in discussion of the change to a sustainable worldview. **Keywords**: Congruity; image schema; attributes; spreading activation theory; episodic events; storytelling

INTRODUCTION

The term 'schema' is used to discuss the representation of knowledge in the human mind. It originated from the ancient Greek word σχήμα and philosophers (Socrates, Plato, and Aristotle) viewed it as the essence of things (Conway & Pleydell-Pearce, 2000; Ghosh & Gilboa, 2014; Halkias, 2015). These classical scholars defined schema as the inherent relationship between personal perception and understanding. Early work on schema was based on text comprehension (Bartlett, 1932). Researchers found that simple

Cognitive Psychology and Tourism
Tourism Social Science Series, Volume 27, 143–152

Published under exclusive licence by Emerald Publishing Limited
ISSN: 1571-5043/doi:10.1108/S1571-504320240000027012

associative models of memory were inadequate, and people use prior knowledge to understand narrative prose and to solve complex problems (Fiske & Linville, 1980).

The schema concept has been used in different branches of psychology (Rumelhart, 1984; Steffensen & Joag-Dev, 1984). Cognitive psychology defines a schema as 'an active organization of past reactions, or of experiences, which must always be supposed to be operating in any well-adapted organic response' (Bartlett, 1932, p. 201). In personal and mathematical psychology, schemata are mental knowledge mappings used for problem-solving and analogical reasoning (Halkias, 2015; Thorndyke & Hayes-Roth, 1979). Some social psychologists consider that schemata pre-exist in minds as stereotypes that may lead to biased understanding (Schneider, 2004). Ghosh and Gilboa (2014) provide a history of the development of the concept indicating that the term has evolved over the past century. Bartlett (1932) provided an early definition of schema as organised settings that are based on a person's experience and are used to infer meaning during information retrieval Bartlett (1932, p. 201).

SCHEMA

The schema concept involves processes such as abstracting information from experiences, indexing it, interpreting new information to allow inferences about it and invoking procedures or actions (Fiske & Linville, 1980). Examples of schemata include mental models, social schemata, stereotypes, social roles, scripts, worldviews, heuristics and archetypes. The schema concept is used in many models of cognitive processes. Schemata are critical factors which can explain how old knowledge interacts with new knowledge in perception, language, thought and memory. They are accumulated from infancy and are stored in humans' long-term memories (Brandon, 2016). There are associative and non-associative theories of schema (Iran-Nejad, 1989). For example, spread of activation theory (Collins & Loftus, 1975) involves associative networks, neural networks or semantic networks. Spread of activation is a method for searching these networks based on mechanisms of human memory operations. In both, schemata are based on experience generated from personal sensory, cognition, emotion and memory, as well as interpersonal relationships. People attempt to a maintain balance between their external environment (surrounding information) and their inner worlds (existing schemata) by seeking cognitive consistency (Levin et al., 1996).

Features and Functions of Schemata

There are four necessary features that define a schema (Ghosh & Gilboa, 2014). Its first feature is associability. A schema holds interconnected units where information can be stored, and meanings can be transformed. Next, a schema is based on numerous episodic events, and represents their similarities or commonalities. Schemata are flexible, implying that new events can be incorporated and can generate changes in meaning. Finally, adaptability is an important feature of schema. When exposed to cluttered information, people have two ways to hold or alter their schema: assimilation and accommodation. Unchallenging information is easily incorporated into an existing schema, which remains unchanged, while complicated information is accommodated with more effort, leading to a changed schema.

Four functions of schemata are evident during information processing (Schneider, 2004). First, a schema directs attention. It is easier for humans to recognise schema-relevant information, based on the characteristics of event episodes in a schema. Information that is inconsistent with expectations and a schema attracts more attention. Secondly, in addition to arousing attention, a schema can help to evaluate new information and interpret ambiguous behaviour (smiles and hand gestures). The more consistent behaviour is with a schema, the easier an evaluation will be achieved (Koehler, 1993). Third a schema aids the encoding and retrieval from memory. Consumers can better understand information which activates a salient schema (Dooling & Lachman, 1971). The final function of schemata is facilitation of inference (Taylor & Crocker, 1981), whereby a prior knowledge structure or schema guides people's reasoning. Processing information using schemata can generate superior memories (Craik & Lockhart, 1972).

Types of Schemata

There are a variety of types of schemata that have been discussed, for example, a schema for a particular person including information about their appearance, behaviour, personality and preferences. A self-schema concerns the self. Event schemata are focused on behaviour that should be followed for certain events. It is a script for what to do, how to act and what to say in a particular context.

In cognitive linguistics, an image schema is a type of schema considered an embodied prelinguistic structure of experience that motivates conceptual metaphor mappings (Rohrer, 2005). An image is 'derived mainly from the sensorimotor experience of motion and spatial relations in the process of one's interaction with the environment' (Merchant, 2016, p. 380). In psycholinguistics, this type of schemata is used to understand metaphors such as closed-lips reflecting a silence (Goossens et al., 1995).

More generally, the value of an image schema in cognitive science lies in its role as a perceptual prototype used in deriving meanings. For example, a particular type of sensorimotor experience can activate a schema which can then be transformed to other domains via analogical reasoning (Mandler, 1992). In tourism, image schemata have been adopted to study metaphors used by tourists to discuss their travel experiences (Laing & Crouch, 2009). For example, the use of the term 'dark tourism' is seen as a result of 'psychological experience grounded in the body' (Lv et al., 2022, p. 2). Haptic stimuli have been found to influence mental imagery although the use of the schema concept was not included (Lv et al., 2020).

Schema Congruity Theory

The foundation of schema congruity theory is the theory proposed by Bartlett (1932). Bartlett (1932) found that non-schema information either cannot be recalled or is adapted to fit previous schemata. New information can be recalled well when it is relevant to old knowledge. Thus, humans process information based on how new information can fit to their existing schema. Mandler (1982) developed schema congruity theory which indicates that the degree of matching between an object and its activated existing schema influences information processing and evaluation of the object. There are three levels of this matching: congruity, moderate incongruity and extreme incongruity. Congruity is where there is a match between the object and its activated schema. Moderate incongruity occurs where the object and its activated schema are partially unmatched, and extreme incongruity where there is a thorough mismatched relationship between the object and its activated schema.

The levels of congruity determine how further information processing occurs. Schema congruity makes information processing easier and more predictable. With congruity, no significant arousal occurs, and messages are easy to understand, resulting in moderate responses (moderate liking or familiarity). Under congruity, heuristic information processing is used rather than systematic information processing (Sujan et al., 1986). In comparison, extreme incongruity often results in negative outcomes (frustration and helplessness) and low evaluations, due to the mental effort (higher cognitive elaboration) needed to resolve mismatches. However, moderate incongruity only needs minor schema changes, and generates more arousal resulting in positive responses (interesting, satisfaction and preference) (Fiske & Neuberg, 1990) For example, moderately incongruent cultural information can enhance memorability (Hunzaker, 2016).

Schema and Attitudes

An attitude is a summary evaluation about an object or a person, and may be familiar or unfamiliar, positive or negative, friendly or unfriendly, and favourable or unfavourable (Eagly et al., 1978). Pratt and Sparks define an attitude as 'a global and relatively enduring evaluation of an object, issue, or person' (2014, p. 447). Consumers evaluate objects using schema (Halkias & Kokkinaki, 2014). There is a relationship between the level of schema congruity and its corresponding attitude (Harmon-Kizer, 2017). Fiske and Linville (1980) provide an assessment of the concept of scheme for social psychologists.

Applications

The concept of schema has been used in fields such as business and tourism, although often as a general concept to explain empirical findings. For example, Nishida (1999) found that schemata guide behaviour better in familiar situations (skilled drivers). Several definitions of schema from the business literature are listed in Table 12.1. Schemata are also called data structures, cognitive structures, mental structures, knowledge-units and a framework of cognitive knowledge. Definitions consider that stored knowledge in memory may be about entities (objects, events, actions and prototypes) or abstracts (concepts, situations, expectations and beliefs) (Cheong & Kim, 2011; Nelson, 2014).

Table 12.1. Definitions of Schema.

Authors (Years)	Definitions
Rumelhart and Ortony (1977, p. 101)	Schemata are data structures for representing the generic concepts stored in memory.
Nishida (1999, p. 775)	Schemata are generalised collections of knowledge of past experiences that are organised into related knowledge groups and are used to guide our behaviours in familiar situations.
Bellman and Rossiter (2004, p. 38)	Schemata are cognitive structures that guide information processing.
Aggarwal and McGill (2007, p. 470)	...a schema is a stored framework of cognitive knowledge that represents information about a topic, a concept, or a particular stimulus, including its attributes and the relations among the attributes.
Cheong and Kim (2011, p. 57)	... mental structures that organise expectations, beliefs, prototypes, and affect regarding a given domain.
Ghosh and Gilboa (2014, p. 113)	A schema is a memory structure capable of representing extremely complex constructs employing this information to influence encoding and retrieval of episodic memory, and guide elaborate, context-specific patterns of behaviour.

Schema Based on Commonalities. Cognitive psychology research findings concerning memory and schemata are applicable in a variety of fields of study, including tourism. For example, the study of social networks in tourism is based on collection of recalled data (Baggio et al., 2010). However, a study of the self-reported interactions among dyads and cliques is significantly different to actual behaviour (Bernard et al., 1979). This indicates that memory for such interactions is poor such that network data are entirely unreliable. Later research however found that reported networks are based on typical patterns of behaviour (Freeman et al., 1987). This suggests that people recall regular patterns of interaction (social structure) based on schema. The implication is that research provides useful information about a person's mental organisation of social structure rather than reliable individual interactions. This fits with schema in this case being a summary of interactions rather than a series of specific ones.

Image Schema. A stream of tourism literature discusses how destination information in advertisements, word-of-mouth or other sources is used to construct a perceptual prototype, described as a destination image schema. It allows filtering of masses of information about a place, and influences destination choice. The process of destination image formation involves schema (Kim & Chen, 2016). Prior to a trip, destination information processing is driven by a tourist's schema. These schemata are mental representations underpinning the travel decision-making of tourists (Dellaert et al., 2014).

According to Arentze et al. (2008), tourism schemata involve four components, attributes (A), benefits (B), situations (S) and causal relations. (Dellaert et al., 2014) propose that the structure of this schema involves five causal relationships among components: decision alternatives and attributes; situations and attributes; situations and benefits; attributes and benefits; and attributes and attributes. Decision alternatives and attributes relations are the image schema relationships that influence destination choice; for instance, the desired attributes of a city destination are different from those of a rural destination. This normative approach is useful but ignores the complex and individual network of relationships that a person may hold.

Spreading activation theory of memory has also been used in tourism studies. Here attribute–attribute relationship networks define a destination image schema (Li & Stepchenkova, 2012). In a brand attribute network, these cognitive units are named nodes associated with a brand, the activation of which originates from source nodes (brand name, logo or product) to adjacent nodes (high quality, famous or safety) (Teichert & Schöntag, 2010). During information processing, the important nodes in this attribute network influence consumers' brand evaluations (Keller, 1993). These related attributes form the cognitive/mental maps or image schemata for a destination.

A concept map is regarded as an effective tool to identify the most important attributes and their relationships for brands. Li and Stepchenkova (2012) developed associative-network maps representing stereotypical, affective and unique images of America for Chinese travelers. Atadil et al. (2017) generated mental maps for the image of Antalya, showing attribute connections based on the three most general themes – sun, sea and beach. These studies stress the important role that image schemata play in destination perception.

The general process for destination image formation then is perception of information about a destination which is abstracted to an image schema and then to destination image (Mandler, 1982). Consumers process a flood of destination information, and an image schema helps them select and organise the most important attributes to form a desired destination image. Thus, a destination image is the result of schema about a destination. Tourists not only use image schemata to form their own destination images during information processing but also to evaluate destinations for future travel choice based on the congruity of their existing image schema (what they already believe and have a favourable/unfavourable attitude towards) and new information. Thus, schema congruity theory assists in explaining how this evaluation occurs.

Storytelling. Stein and Albro suggest that stories are comprised 'episode schemata', or 'goal-directed action outcome sequences' (1997, p. 12). A successful narrative is structured in a similar way to the readers schema (Schank & Abelson, 1995). Therefore, the mental imagery process involved in storytelling involves schemata (Lee & Gretzel, 2012).

Schema Congruity Theory. Schema congruity theory has been applied in the social sciences, to examine topics such as perception, self-concept, political information processing, attitudes and stereotyping (Fiske & Linville, 1980). In consumer psychology, schemata impact the processing of self-related information from social experiences (Ghosh & Gilboa, 2014). In the study of marketing communication (brand, product, advertisement, events, spokesperson, salesperson, word of mouth), schemata help to understand how information is perceived and learnt (Sentis & Markus, 1986). Use of appropriate stimuli triggers interpretative schemata to enhance persuasive communication (Lopez-Guimera et al., 2010; Sánchez-Carracedo, & Fauquet, 2010).

In the domain of tourism, research using schema congruity theory is limited (Florek & Insch, 2011), apart from studies of destination advertising. Singh and Hu (2012) found that destination information incongruence impacts the persuasiveness of advertisements, consumers' attitudes and their visit intentions, but they did not measure the extent of incongruence. Singh and Formica (2007) ascribe the failure of destination marketing communication to a low level of positive image congruence. The higher level of affective image congruence between a destination and its promotional slogan facilitates correct identification of a destination (Lehto et al., 2014).

Schema congruity theory has been used in the context of event tourism. Events make significant effects on hosting destination images (Kim et al., 2014). The image match between a special event and hosting destination can enhance consumers' visit intentions (Cheung et al., 2016). Brand extension from a destination to its products (e.g. events) relies on their image congruence, particularly affective image congruence, since this congruous situation easily transfers the existing destination image to the extended products (Lim & Weaver, 2014). In an event sponsorship study, consumers' attitude transfer is strengthened when schema congruity occurs (Grohs & Reisinger, 2005). This enables an unknown city to gain positive attitude by using event-city congruent images; for example, in a movie festival (Kim & Jun, 2016). In tourism, image schema changes can be caused by the resolution of incongruity because of the visit resulting in a change of attitude towards a destination.

In marketing research, researchers use schema congruity theory in studies of product evaluation, brand alliances and extension, as well as the effectiveness of advertising. The effect of moderate incongruence between a brand and its related products has attracted much attention, particularly regarding sponsorship and endorsement events, as well as country-of-origin studies. Moderate incongruity between a brand and corporate social responsibility-related evoked schema in sponsored events has been found to lead to positive attitudes (Uhrich et al., 2014). Celebrity endorsement also influences brand confidence when sponsor-product congruity is relatively low (Clemente et al., 2014).

Product-country incongruence leads to negative product evaluation (Dekhili & Achabou, 2014). Consumer attitude for a brand with a congruent country-of-origin is better than one with moderate incongruence (Spielmann, 2016), and the lack of congruence reduces evaluations for new products (Johnson et al., 2016). Consumers have more favourable attitudes of a brand when the rewards are congruent with brand image (Stumpf & Baum, 2016). In a similar vein, product-endorser congruity leads to a more favourable attitude towards promotional advertisements (McCormick, 2016). Similar effects have been found in sports brand evaluation (Lee et al., 2016), hotel brand evaluation (Lee & Jeong, 2014) and green restaurant evaluation (Line et al., 2016).

CONCLUSION

The concept of schema informs several topics of interest to tourism scholars. These include transformation from travel experiences, use of self-congruity, social network data collection and importantly attitude change. Transformation from travel experiences has become a popular topic in the tourism literature. Transformation from travel is seen as a growth-enhancing and irreversible change that requires new knowledge and a fundamental break with current practices (Reisinger, 2013b). Transformative tourist experiences involve changes in understanding the self, revision of one's belief system and alterations in one's behaviour and lifestyle (Kirillova et al., 2017b).

The transformative literature would benefit from the use of the schema concept as a basis for the concept of belief systems and of schema congruity theory in explaining why transformation is such a rare event. The issue of transformation may be thought of as an extreme form of schema change. In this way research findings from cognitive psychology may be applied to issues such as changing attitudes to environmental protection and waste recycling. These are major issues and may for some conservatives represent an example of extreme incongruity (Purcell, 1986). In contrast, for others, the change of attitudes needed to address environmental issues may require only minor schema changes due to moderate incongruity (Loken, 2006).

Few scholars have explored how incongruity, especially extreme incongruity, can be resolved to obtain positive responses. Several postulate that information about tangible product attributes can assist in resolving incongruity. Carvalho et al. (2011) demonstrate that schema incongruity between country of brand origin and country of manufacture can be resolved through tangible product attributes, the strength of which also determines consumers' brand attitudes. If incongruity cannot be resolved, attitudes will be quite negative (Mandler, 1982). The concept of self-congruity concerns congruence of external perceptions with their self-schema. Self-congruity is the psychological process and outcome that a consumer engages in given that the consumer compares his or her perception of a brand image (more specifically, brand personality or brand-user image) with his or her own self-concept (actual self-image) (Sirgy et al., 2016). It would be useful to compare the findings of schema congruity theory with those of the study of self-congruity.

The basis for any investigation of the cognitive representation of social information is an understanding of how one's ability to develop and maintain relationships with others depends at a basic level on the ability to remember those others and their past behaviour. Brashears and Quintane (2015) provide evidence that humans store information about social relations using schemata. Schemata simplify and compress information about social networks allowing it to be stored in memory more easily and efficiently. Research has found that recollections of human network interaction are not consistent with observed behaviour at dyadic level (Bernard & Killworth, 1977). This suggests that self-report network data are entirely unreliable. Later studies found that individual recall tended to be an inaccurate indicator of any specific encounter (attendance at a colloquium talk), but a reliable measure of typical patterns of behaviour (typical attendance at colloquia).

Moreover, human recall for sub-groups in a social community has been shown to match well with observed patterns of association (Freeman et al., 1988). This is an illustration of the generality of social schema and the accuracy of recall of the regular patterns of interaction that form social structure. This suggests that the methodology of collection of social network data needs attention. The concept of schema may be useful in examining the persuasive effect of storytelling. Storytelling is considered a powerful marketing strategy for communicating with potential consumers (Moscardo, 2020). However, how stories fit into cognitive schema has not been discussed. In summary, there appear to be several areas where the tourism literature would benefit from the use of the schema concept and related theories.

Chapter 13

PROSPECTION AND TOURISM

Dung Le
VinUniversity, Vietnam

Noel Scott
Edith Cowan University, Australia

Abstract: When making decisions about their future dream holidays, tourists often spend significant time thinking about future scenarios. This is an example of 'mental time travel', a process called prospection. The concept of prospection refers to human ability of 'pre-experience' the future and hence how that future 'feels', which is why it is sometimes called affective forecasting. There is a large volume of psychological research on prospection, but in applied fields, the use of various terms makes the current literature complex, disconnected and hard to navigate. This chapter provides an introduction to prospection as an underlying mechanism explaining some of the effects of experiential marketing. It discusses tourism prospection-related terms, errors of prospection and mental biases and ways to better manage tourist experiences. **Keywords**: Prospection; mental imagery; mental time travel; simulations; emotions; rosy illusion

INTRODUCTION

Humans have a unique ability to mentally travel in time (Suddendorf & Corballis, 2007). This enables them to predict the hedonic consequences of future events based on their past experiences and knowledge (Gilbert & Wilson, 2007). If asked to imagine a holiday on a tropical island in the sun, a potential tourist might conjure up images of palm trees, white sandy beaches, turquoise water lapping up the shore, sun lounges and perhaps the odd

Cognitive Psychology and Tourism
Tourism Social Science Series, Volume 27, 153–161

Published under exclusive licence by Emerald Publishing Limited
ISSN: 1571-5043/doi:10.1108/S1571-504320240000027013

cocktail. During this exercise, the tourist would unconsciously combine incoming information (a request to imagine a future event) with stored information from memory (past experiences and or knowledge about tropical holidays) to build a representation of a future external world event. The tourist also may have experienced positive affect during this imagined holiday.

This process of imagining a future travel experience – conceptualised as prospection – is likely to occur for many tourists and be the source of innumerable travel decisions; from selecting a destination and/or places to visit (trip planning) to making specific choices of hotels/restaurants/events (tourism products/services) (Le et al., 2019). Prospection, sometimes termed as future-oriented thinking, mental imagery/simulation, telepresence, imagination or consumption vision – is particularly meaningful in tourism because it drives tourist behaviour. The hedonic nature of imagined future tourism experiences is reflected in these emotional responses (Kim & Hall, 2019; Miao et al., 2014).

However, prospection is fraught with error and biases which may result in inaccurate decisions about the future (Gilbert & Wilson, 2007). Therefore, understanding the cognitive mental processes involved in prospection, its effects and errors, are important for tourism marketers and managers in designing and managing tourist experiences. This chapter examines the mental processes involved in prospection and how they work. It examines prospection errors evident in travel decision-making and the application of prospection for marketing and experience design.

PROSPECTION FROM A MENTAL PROCESSING PERSPECTIVE

Humans have the ability to shift their attention to thinking about past, present or future events (Buckner & Carroll, 2007). Looking forward in time, or considering the future, is known as prospection (Gilbert, 2006). The ability to think about the future develops around the age of three or four in humans (Atance & O'Neill, 2005). The concept 'prospection' is the opposite of retrospection, which refers to humans' reconstruction of past events (Gilbert & Wilson, 2007). However, both prospection and retrospection are forms of mental representations, or internal models of the external world (Buckner & Carroll, 2007). These two temporal directions of mental time travel (forward versus backward in time) rely on similar cognitive structures and processes, such as mental imagery, self vs other referential processing, which are constructive operations that integrate various mental details together into a coherent mental representation (Seligman et al., 2013)

Neuroimaging studies support the hypothesis of mental time travel by demonstrating that a core brain network – the default network – is engaged in both prospection and retrospection, including medial temporal and frontal lobes, posterior cingulate and retrosplenial cortex and lateral parietal and temporal areas (Schacter et al., 2007, 2012). Buckner and Carroll (2007) argue that three areas of the brain are likely to be crucial to the process of prospection: the frontal lobe which is responsible for planning and structuring events in temporal sequence, shifting one's perspective beyond the present and mental simulation; the medial temporal lobe including the hippocampus and medial and lateral parietal regions, the former is important when accessing memories (damage to this area can lead to amnesia); and the frontopolar cortex and anterior midline frontal regions which have an important role in regulating shifts in perspective.

While retrospection is constrained by the recall of past events (even though some details may be modified or reconstructed), prospection is more volatile, involving the simulation of 'if-then' events that enables ones to 'pre-experience' different scenarios and anticipate one's responses to them (Rasmussen & Berntsen, 2013). It is generally understood that prospection promotes adaptation to the environment; the ability to imagine one's future allows avoiding pain and promotes pleasure (Buckner & Carroll, 2007; Eacott & Easton, 2012). In addition to this, Gilbert (2006) argues that people think about the future to exercise control over their lives.

Episodic memory and prospection underpin the capacity to appreciate the connection between one's past, present and future selves (McCarroll & Cosentino, 2020). Humans accumulate episodic memories of prior specific autobiographical events. This memory is rich in details, such as who was involved, what happened and how it was felt (Tulving, 1985). This episodic memory provides building blocks from which prospective simulation of future events are constructed (Szpunar & McDermott, 2008). These simulations in turn contribute to a process of human memory reconstruction (Lakshmanan & Krishnan, 2009). This means that memories continue to change and evolve throughout a person's life. In most cases, each time an event is recalled, stored details of this event are reconstructed a bit differently from the last time (Seligman et al., 2013). As people age, there is a tendency to remember the past and imagine the future with less episodic details. In other words, memory deficits often inhibit the ability to imagine the future due to a less effective process of extracting and recombining stored memories into a simulation of a future event (Schacter et al., 2013).

Prospection, defined in this book as mental simulations of future events (Gilbert & Wilson, 2007), is conceptually different from other forms of prospective cognition such as prediction, intention or planning (Szpunar et al., 2014). Prediction refers to the estimation of whether a future event or behaviour is likely to happen or not (Baumeister et al., 2016). Intention involves setting a specific vs non-specific/abstract goal in relation to a future event where simulation may be used to facilitate action to achieve that goal (Szpunar et al., 2014). Planning consists of a predetermination of a course of action aimed at achieving some goal (Hayes-Roth & Hayes-Roth, 1979).

Prospection has a close relationship with and often significantly influences prediction, intention or planning. For example, repeated prospective simulation of emotional interpersonal future events increases the perceived plausibility of the imagined events. In other words, repeated simulation alters one's prediction on the likelihood of a future event (Szpunar & Schacter, 2013). Inducing consumer prospection has been found as an effective strategy to influence consumer behavioural intentions (Bogicevic et al., 2019; Cowan & Ketron, 2019; Elder & Krishna, 2022; Park & Yoo, 2020; Yim et al., 2018) and trip planning (Ghosh & Sarkar, 2016; Le et al., 2021; Wattanacharoensil & La-ornual, 2019).

Different Forms of Prospection

There is a dense, complex, but disconnected body of research on prospection, which may easily confuse new readers. Table 13.1 below provides an overview of these related and interchangeable concepts to help researchers navigate the literature and have more comprehensive understanding of prospection. Various concepts related to prospection have been proposed in the current literature, which could be classified from low to high imagery elaboration (the richness and concrete details of mental presentations): mental images, mental imagery (Ha et al., 2019; Jia et al., 2021; Park & Yoo, 2020), mental simulation (Nardini & Lutz, 2018; Xie et al., 2022; Yim et al., 2021), consumption vision (Tercia et al., 2020; Yim et al., 2018; Yim & Yoo, 2020), narrative transportation (Ryu et al., 2019; Seo et al., 2018; van Laer et al., 2019), telepresence (Cowan & Ketron, 2019; Hyun & O'Keefe, 2012; Kang, 2020; Orús et al., 2021), imagination (Ghosh & Sarkar, 2016; Le et al., 2021; Philips, 2017) and fantasy (Chen, 2015; Williams, 2006).

Table 13.1. Overview of Prospection-Related Concepts.

High to low imagery elaboration	Imagination/fantasy (an internal representation that produces the experience of perception in the absence of the appropriate sensory input) (Philips, 2017)
	(Tele) presence (the subjective experience of being in an environment created or facilitated by augmented or virtual reality, whilst physically being situated in another (Yung et al., 2021)
	(Narrative) transportation (the phenomenon in which consumers mentally enter a world that a story evokes) (Van Laer et al., 2014)

Table 13.1. *(Continued)*

	Consumption vison (visual images of certain product related-behaviours and their consequences) (Walker & Olson, 1997)
	Mental imagery/simulation (the imitative mental representation of some event or series of events) (Escalas, 2004)
	Mental images (mental representation of object, person, subject in one's mind) (MacInnis & Price, 1987)
Referencing focus	Episodic prospection (self-imagery) (McCarroll & Cosentino, 2020) Others-prospection (others-imagery) (Escalas, 2006; Hur et al., 2020)
Content focus	Outcome-focused prospection (Escalas & Luce, 2004) Process-focused prospection (Zhao et al., 2009)
Sensory elements involved	Multi-sensory prospection (Elder & Krishna, 2022; Simmonds et al., 2020) Visual prospection (first-person vs third-person perspective) (Escalas, 2004) Haptic prospection (de Vries et al., 2018; Peck et al., 2013) Olfactory prospection (Krishna et al., 2013)

Prospection consists of a mental capacity necessary to navigate the complexities of life, and to evaluate how best to think, feel and act in diverse settings (Christian et al., 2013). When people engage in prospection, they can 'experience' future events/scenarios where they are involved personally in the scenes – called episodic prospection (McCarroll & Cosentino, 2020) or others are involved (Escalas, 2006; Hur et al., 2020). This allows an individual to anticipate their own behaviour and also others' responses and behaviour during future events. People tend to focus on the outcomes of future events and rely on their affective forecasting of those outcomes to make decisions.

On the other hand, during planning, people are more likely to focus on the process of how things would happen. This difference in the focus of prospection leads to different impacts on behavioural intentions and planning (Zhao et al., 2009). Most prospection involves multi-sensory elements which help people to have more 'real' pre-experiences of future events. Research has also explored prospection using singular senses, such as visual prospection (first-person vs third-person perspective) (Escalas, 2004), haptic prospection (de Vries et al., 2018; Peck et al., 2013) or olfactory prospection (Krishna et al., 2013).

Errors of Prospection

Despite the critical role of prospection in human survival and development, there are possible errors occurring in prospection that people are often not aware of (Gilbert & Wilson, 2007). Prospection and retrospection share the same neural pathways (Spreng et al., 2008), which may jeopardise the ability to accurately predict the future. First, mental representations of future events, which are formed based on personal memories and knowledge, are subject to the 'unrepresentative problem'. Because unusual events and recent events are more memorable (the peak and end effects) (Fredrickson, 2000), these unrepresentative memories are likely to be retrieved and used as a basis for prospection, causing prospection bias.

Second, prospection is essentialised and abbreviated. Due to limited capacity of working memory, people can only focus on essential details and/or a few selected moments when thinking about future events. Therefore, 'unessential' features are more likely to be omitted when the psychological distance of future events increases. According to construal level theory (Liberman et al., 2007; Trope & Liberman, 2010), mental representations of future events increase in abstraction in accordance with temporal, physical and sociocultural distance. Therefore, prospection tends to overrepresent future events and amplify the affective consequences of these events. This prospection bias consequently leads to inaccurate anticipation, evaluation and choice.

Third, prospection is decontextualised as people are unable to acknowledge the possible differences between the contextual factors at the point of reference (here and now) and future events. For example, people tend to rely on their current states (hunger or thirst) to form mental representations of how much they would like eating/drinking something in the future. As people fail to consider changes in contextual factors (their hunger or thirst tomorrow), prospection becomes biased. Only when people are encouraged to consider contextual factors of future events, does prospection become more accurate (Wilson et al., 2000).

Finally, prospection is associated with the uncorrected positive illusions. Compared to retrospection, which is constrained by the reality of things that have happened, prospection is more emotionally positive and idyllic. When people are asked to engage in both retrospection and prospection, prospection is reported as more extreme (either positive or negative). This gap between retrospection and prospection is larger for emotionally positive events (Rasmussen & Berntsen, 2013). Also, prospection is subject to fading-affect bias, whereby negative prospection fades more quickly than positive prospection, and this results in a tendency to remember a rosy imagined future (Szpunar et al., 2012).

Prospection in Tourism

Research on prospection underlies academic support for experiential marketing in tourism (Le et al., 2019). Due to the hedonic nature of tourism, experiential marketing has become an effective strategy to promote destinations and tourism products/services by evoking prospective and emotional responses (Rather, 2020; Skandalis et al., 2019), which in turn have both direct and indirect effects on tourist behaviour. Some authors believe that the mere act of thinking about an action could increase the intentions of performing this action (James, 1890). This belief underlies most tourism research on how to design and use sensory and symbolic elements to stimulate tourists' pragmatic prospection, defined as thinking about the future in ways that will have practical utility such as buying or consumption actions (Baumeister et al., 2016). Traditional marketing tools such as pictorial advertisements (Roose et al., 2019; Royo-Vela & Black, 2020) or websites (Dlodlo, 2018; Lee, 2018) are still widely used to trigger tourist prospection. However, there is a remarkable shift to more advanced information and communication technologies, including Facebook fanpage (Perez-Vega et al., 2018); social networking services communication (Ha et al., 2019); eWOM (Seo et al., 2018); augmented reality (He et al., 2018) and virtual reality (Bogicevic et al., 2019; Kim & Hall, 2019; Orús et al., 2021; Spielmann & Mantonakis, 2018; Tussyadiah et al., 2018; Wei et al., 2019; Yung et al., 2021) in prospection-generating marketing.

Interestingly, prospection is closely associated with pre-travel affective responses. People rely on prospection to 'pre-view' future events and use their affective responses to prospection as basis for affective forecasting (Wilson & Gilbert, 2005). Some authors even propose that without affective consequences, prospection may not influence tourist behaviour (Lang, 1977, 1979). Hence, affective consequences (positive versus negative) are included in prospection research, particularly mental imagery measurement scales (Miller et al., 2000; Weiler et al., 2016). Given the hedonic nature of tourism experiences, most tourism-related prospection is found to be biased towards positive emotions (Le et al., 2019). Positive prospection could, thus, improve tourist cognitive responses including attitude towards an advertisement (Lee, 2012; Lien & Chen, 2013) and attitude towards a brand (Chang, 2013; Fennis et al., 2012; Peck et al., 2013), confidence in attitude, attitude resistance (Lee & Gretzel, 2012), perceived usefulness (Lee, 2012), perceived value (Fiore et al., 2005; Overmars & Poels, 2015) and behavioural intentions (Ha et al., 2019; Park & Yoo, 2020).

However, recent evidence has emphasised the double-edge sword of prospection, demonstrating that in some cases prospection can reduce tourists' evaluations of tourism product (Jia et al., 2021), intentions to visit places (museum) (Deng et al., 2019) or intentions to join tourism events (Le et al., 2021). Cowan and Ketron (2019) review virtual reality research in tourism and suggest that product involvement plays an important role in differentiating mental imagery processing. Highly involved tourists are likely to engage in elaborated imagery processing, experience telepresence, imagination while low involvement tourists engage in less cognitively elaborated processing, only experiencing interactivity with virtual reality content. Tourists have different goals and motives when going on vacations (Bieger & Laesser, 2002; Chen et al., 2014; Park & Yoon, 2009), consuming hospitality services (Guttentag et al., 2018; Prayag et al., 2015) or participating in entertaining events (Smith & Costello, 2009; Tkaczynski & Rundle-Thiele, 2011). Thus, groups of tourists with different travel goals and needs find different destination/service/event marketing messages as interesting and relevant for prospection (Le et al., 2020; Ye & Tussyadiah, 2011). Therefore, marketers should apply appropriate virtual reality strategies in using virtual reality to target different consumer segments (Cowan & Ketron, 2019).

Errors of Prospection and Tourist Experience Biases

Errors of prospection, as found in psychology research, have been explored in tourism research and supported by evidence of tourist experience bias. For example, tourists who engage in prospection of future trips often overestimate their emotional responses compared to their real experiences, in accordance with the uncorrected positive illusions hypothesis (Mitchell et al., 1997; Pollai et al., 2010; Wirtz et al., 2003). This affective misforecasting phenomenon can be explained by errors of pre-trip prospection (unrepresentative, essentialised and abbreviated, decontextualised) that have been discussed in previous section. Another explanation is that after engaging in pre-trip prospection, tourists process actual experiences in a less detailed manner, resulting in a negative gap between experienced emotions and prospection-related emotions (Nardini & Lutz, 2018). To address this, marketers may want to maintain some mystique by encouraging tourists to 'come see for themselves'.

Continuous exposure to misleading marketing information before and after a trip – that stimulates tourist prospection and forms prospective memory (Hannon & Daneman, 2007) – may lead to memory distortion. Thus, tourists may believe that they experienced something they did not (Braun et al., 2002, 2006). The reconstruction process of tourist memories based on prospective memory may explain why tourists tend to remember and evaluate their trip based on pre-trip prospection and emotions rather than what they experienced (Wirtz et al., 2003). Prospective memory and memory reconstruction may be the underlying mechanism of tourists' 'rosy view'. Even when a trip proved disappointing, tourists' recollections of their trip are more positive than their actual experiences (Mitchell et al., 1997). The salience of emotions aroused in false memories during recollection may be different as they could be based on fabricated prospective events that do not represent autobiographical experiences as encoded (Tung et al., 2017).

CONCLUSION

By reviewing relevant psychology and tourism research on prospection, this chapter sheds lights on a mechanism underlying tourist experience biases and provides a useful guide to understand experiential marketing effects. While a rational marketing approach has been found to have little influence on tourist perception (Govers et al., 2007), experiential marketing with a focus on prospection-generating and emotion-evoking stimuli can be more influential.

Despite academic efforts in exploring prospection, this area remains in its infancy. Various research gaps and questions need to be further investigated to advance the knowledge of this essential mental mechanism. First, insertion of episodic memories during prospection, that is, memories of personally experienced events (Suddendorf & Corballis, 2007) is particularly problematic given memory is imperfect and is generally accepted as a constructive rather than a reproductive process. Further exploration of episodic memories in prospection may help to reduce errors of prospection and allow tourists to form more accurate pre-trip prospection and affective forecasts.

Second, semantic memory (general knowledge of facts not personally experienced) and conceptual knowledge could also play a role in prospection (Demblon & D'Argembeau, 2014). However, research on semantic memory and prospection remains scarce, with only initial efforts in examining tourists' prior knowledge of tourism events having been conducted by Le et al. (2021). Third, little is known about prospective memory, how it changes over time or what type of prospection is more memorable and influential in altering tourist experiences. One pioneering research study in this area shows that the retention of positive, negative and neutral prospection over delays of 10 minute and one day is subject to fading-affect bias, whereby negative reactions fade more quickly than positive reactions (Szpunar et al., 2012). Further research on such a rosy prospection bias could be beneficial for managing tourist experiences and tourists' 'rosy view'.

Chapter 14

COPING STRATEGY AND NEGATIVE EMOTIONS

Yawei Jiang
Griffith University, Australia

Zhiming Deng
The Hong Kong Polytechnic University, Hong Kong

Abstract: This chapter uses cognitive psychology to understand how individuals deal with negative emotions using coping strategies. It discusses the coping process and reviews three current debates about cognitive appraisal theory, coping and emotion regulation. The use of coping strategies in tourism, services and business management is presented to understand different stakeholders' responses to negative emotions. The literature discusses various elements (antecedents, coping strategies, coping consequence) for emotion coping process and the chapter summarises a synthesised conceptual framework for emotional coping based on the cognitive appraisal theory. The framework is used to examine coping in the context of COVID-19. The chapter concludes with future research opportunities in this subject area. **Keywords**: Negative emotions; anger; emotion coping strategy; cognitive psychology; cognitive appraisal; COVID-19

INTRODUCTION

There has been a proliferation of studies of emotion in the tourism literature and more recently emotional coping has also been used to investigate stress in various tourism situations. Coping is an integral part of the emotional process that allows one to deal with negative emotions and to manage stressors (Folkman, 2013; Lazarus & Folkman, 1984). Various coping strategies are

Cognitive Psychology and Tourism
Tourism Social Science Series, Volume 27, 163–176

ISSN: 1571-5043/doi:10.1108/S1571-504320240000027014

used for dealing with negative emotions. Coping is multidimensional and varied, as it can manifest as thinking or action (Taylor & Stanton, 2007). This chapter focuses on coping as discussed by cognitive appraisal theory (CAT) (Lazarus & Folkman, 1984). Stress, appraisal and coping are central to psychological well-being, and it is surprising that there have been few studies applying knowledge of the mechanism of coping in managing negative emotions.

This chapter discusses how individuals use coping strategies to deal with negative emotions, and to provide an integrated framework of the coping process based on CAT. Coping plays an essential role in the management of negative emotion in the business management, service and tourism sectors (Gabbott et al., 2011; Jordan et al., 2015). A second aim of this chapter is to discuss key dimensions and debates about the coping concept applying in these different fields. Although there is a sizeable amount of appraisal research using coping strategies, their application to the study of service and tourism relevant issues remains unclear or has been inadequately discussed. Moreover, the emotion regulation process and strategies are widely discussed in dealing with negative emotions in tourism-related studies, but lack of clear positioning in the existing coping process drawn from CAT. Therefore, a synthesised multidimensional framework of the emotional coping process based on 'antecedents – coping – consequences' (Lazarus, 1991) is developed and used to discuss practical crisis/disaster-related coping strategies for issues, such as service failure and crisis communication.

NEGATIVE EMOTION COPING STRATEGIES

A significant literature has developed concerning negative emotions (Arimitsu & Hofmann, 2017). Negative emotions are pervasive in an individual's everyday life. As Worthington and Scherer (2004) point out, an individual's ability to cope with negative emotions derived from stressful situations are connected to coping strategies (Fisher et al., 2007). Cognitive psychologists recognise coping as an individuals' cognitive and behavioural responses to stressful or unpleasant situations (Somerfield & McCrae, 2000). In their classic text on appraisal psychology, Lazarus and Folkman describe coping as 'constantly changing cognitive and behavioral efforts to manage specific external/internal demands that are appraised as taxing or exceeding the resources of the person' (1984, p. 141). In the stress and coping literature, CAT provides a theoretical foundation for coping. Scholars view cognitive appraisal as a mental process, whereby the individuals appraise a situation and assess coping strategies as one's cognitive and behavioural responses (Aldwin, 2007). The mental process consists of three sequential components: primary appraisal, secondary appraisal and the outcomes of appraisals – coping strategies (Dewe, 1991).

In primary appraisal, individuals evaluate a situation by first determining its importance and whether it is congruent with their goals (Moors et al., 2013). An appraisal of valence is based on a person's evaluation of goal congruence of an event. The degree of arousal is based on a person's evaluation of goal importance of an event. Thus, goal congruence and relevance are central dimensions of primary appraisal (Roseman, 2013). Prior research argues that a high level of goal importance and a low level of congruence can stimulate strong negative cognitive and emotional responses (Carpenter, 2016). In other words, if the situation is incongruent but important to one's goal, secondary appraisal of how to deal with strong negative emotions is elicited.

In secondary appraisal, individuals evaluate their available resources or options to determine how to cope with the negative emotions elicited and address the question 'what can I do?' (Moors et al., 2013). This is important to determine to what extent a stressful situation can be changed. Thiel et al. (2011) argue that secondary appraisal influences how negative emotions are displayed, based on appraisals on dimensions such as perceived control and/or perceived powerlessness. These appraisals then may trigger distinct negative emotions, such as anger, worry or fear.

In the final stage of CAT, individuals subconsciously choose a suitable coping strategy based on outcomes of primary and secondary appraisal process. The coping consequence is to mitigate negative emotions that an individual could experience from a stressful situation (Duhachek, 2005). Lazarus and Folkman (1984) provide a transactional framework of stress and coping, where an individual first cognitively evaluates what is at stake (primary appraisal) and then appraises resources to determine what can be done to cope with the situation (secondary appraisal). This cognitive appraisal process can evoke discrete emotions and determine suitable coping strategies to manage stressful situations (Carpenter, 2016). Therefore, executing a coping strategy does not directly reduce negative emotions raised by stressful situations, but rather provides a cognitive and behavioural response to manage specific external/internal demands through a cognitive appraisal process. Figure 14.1 summarises the cognitive appraisal and coping process model which will be discussed throughout this chapter.

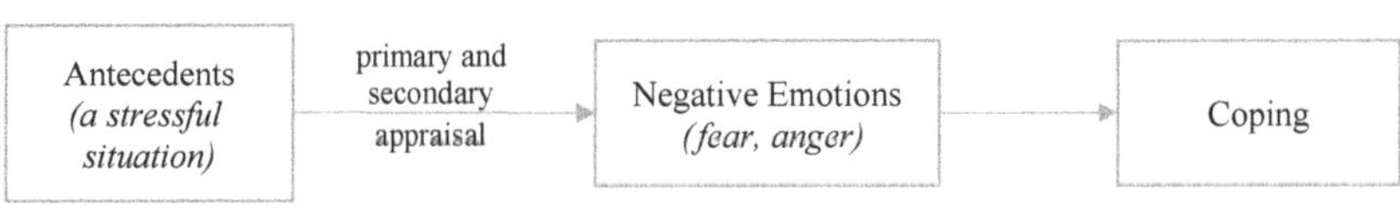

Figure 14.1. Cognitive Appraisal and Coping.

Coping Strategies and Several Debates

The selection of coping strategies depends on the cognitive appraisal process. Coping strategies act as a complex and dynamic adjustment mechanism to deal with appraisal emotions aroused by a stressful situation (Labrague et al., 2017). Prior research has identified an array of coping strategies (engagement, behavioural coping), which can be categorised into emotion-focused (inward) coping and problem-focused (outward) coping strategies (Folkman, 2013). Emotion-focused coping strategies (self-control, acceptance) attempt to manage emotional stress triggered by an event, whereas problem-focused coping strategies (planful problem solving) concern actions used to minimise or eliminate stress (Herman & Tetrick, 2009). Both types of coping strategies are adopted within a stressful situation (Hewett et al., 2018). Acknowledging the prior coping categories, Duhachek (2005) provides a comprehensive hierarchical coping structure based on cognitive precursors with three classifications: active coping (action coping, rational thinking and positive thinking); expressive support seeking (emotional venting, instrumental support and emotional support) and avoidance (denial).

These classifications of coping strategies have led to numerous debates. First, although such classifications of coping strategies are helpful, some specific coping strategies do not fit neatly into one dimension (Yi & Baumgartner, 2004). Individuals can adopt a variety of coping strategies. In the same vein, Stephenson and DeLongis (2020) argue that coping strategies cannot be examined alone, as coping strategies may interact and connect with one another. For example, students may control their anxiety (emotion-focused coping) to better prepare for exams (problem-focused coping). Therefore, emotion-focused and problem-focused coping strategies need to be considered simultaneously.

The second debate concerns coping strategy effectiveness (Carver & Connor-Smith, 2010; Kim & Duda, 2003), that is, which coping strategy works best to address a problem or to relieve negative emotions. Coping studies have discussed many factors affecting the efficacy of specific coping strategies, including the situation context (stimulus that is appraised) (Carver & Connor-Smith, 2010), individual characteristics (social-demographics, knowledge and emotional intelligence) (Strizhakova & Tsarenko, 2010) and time frame (Louvet et al., 2007). In reviewing the coping literature, scholars generally agree that problem-focused coping strategies are effective in reducing physical (injury) and mental harmfulness (anger) (Carver & Connor-Smith, 2010), while emotion-focused coping strategies are maladaptive and ineffective (Baker & Berenbaum, 2007).

However, this does not mean emotion-focused coping is not advantageous. Baker and Berenbaum argue that emotion-focused coping allows 'actively identifying, processing, and expressing one's emotions, thus providing information about one's goal status' (2007, p. 96). When the stressful situation is uncontrollable, emotion-focused coping is more likely to be effective (Troy et al., 2010). For example, a sense of humour can be used as an effective emotion-focused coping strategy to relieve negative emotions and enable positive interactions with others (Berk, 2015). Thus, the effectiveness of coping strategies varies between individuals and situational contexts (Nielsen & Knardahl, 2014).

Coping, Appraisal Process and Emotion Regulation

According to the cognitive appraisal process, emotions are manageable if measures of control are exerted before employing coping strategies. Emotion regulation is viewed as an umbrella term for these measures. Emotion regulation is defined as 'the process by which individuals influence which emotions they have, when they have them, and how they experience and express emotions' (Gross, 1998, p. 275). In his recent work, Gross (2015) views emotion regulation as activation of a goal to modify a person's emotional trajectory. On one hand, an emotion regulation goal can be targeted at changing the emotion felt. For example, parents may regulate their anger towards their naughty children. On the other hand, the goal can be aimed at achieving a meaningful outcome that goes beyond the emotion itself. For instance, an individual may regulate their fear of travelling to a terrorism-affected destination to obtain a memorable tourism experience.

The most recognised emotion regulation strategies are to regulate emotions either by manipulating input (antecedent-focused regulation) or modifying output (response-focused regulation) (English et al., 2017). Antecedent-focused regulation emphasises reappraisal of the situation before an emotion is evoked, whereas response-focused regulation focuses on altering appraisals after an emotion is generated (Kever et al., 2015). Each of these two broad classifications includes more specific approaches. For example, antecedent-focused regulation encompasses cognitive reappraisal, situation selection and situation modification (Livingstone & Isaacowitz, 2015). Response-focused regulation includes emotional suppression (Braunstein et al., 2017).

Emotion regulation is embedded in the cognitive appraisal process. Emotion regulation is a process that can redirect emotions. Antecedent-focused regulation is employed before an emotion is elicited by altering cognitive appraisal. Response-focused regulation acts after the emotion has been experienced and attempts to alter the expression of emotions. A scenario can illustrate emotion regulation in the cognitive appraisal process (Figure 14.2).

> A hotel frontline employee may encounter a customer complaining in an uncivil manner about their hotel experience (stressful situation). Negative emotions (such as worry and anger) are triggered through

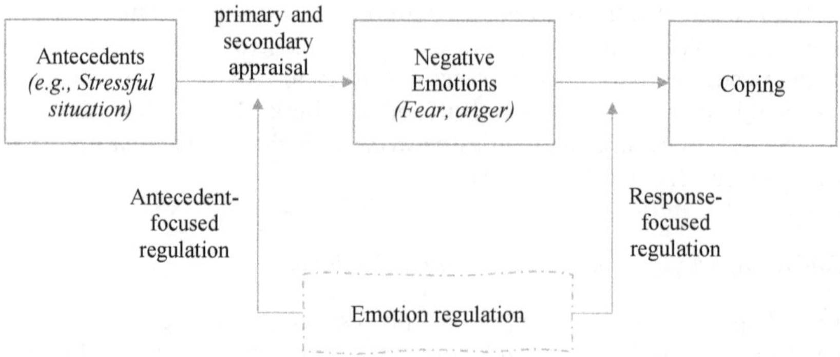

Figure 14.2. Emotion Regulation in the Cognitive Appraisal and Coping Process.

the appraisal process (appraisal). To regulate potential goal related personal consequences, the employee may try to empathize with the customer's situation and ask for a colleague's assistance (reappraisal and situation modification). In other situations, after the negative emotion has been generated, the employee may attempt to control these negative emotions by providing service with calmness (suppression).

Negative Emotion Coping in Different Disciplines

Negative emotions often are associated with a service failure and result from goal incongruent appraisal of a situation and subsequent coping behaviours (Mattila & Ro, 2008). For example, negative emotions may be experienced by employees coping with customer misbehaviors (Tummers et al., 2015). Understanding the emotional experiences and responses of different stakeholders from a cognitive psychology perspective is important to improve better organisational practices and improve industry performance.

The service and business management literature covers emotion coping from both demand and the supply side – specifically, customers coping in a service delay or failure situation and employees coping with job stressors or customer misbehaviors. Negative emotions experienced by customers have been demonstrated in a variety of service and business contexts including service delays (Kim & Park, 2016), service failures (Bonifield & Cole, 2007) and service recovery (Wang et al., 2020). A service delay or failure may lead to anger (Strizhakova et al., 2012), regret and frustration (Hoang, 2020) or worry and disappointment (Li et al., 2021; Yi & Baumgartner, 2004).

For example, Kim and Park (2016) note that perceived waiting time in a flight delay context can lead to anger and uncertainty, and anger may lead to negative word-of-mouth of customers as a common type of behaviour in such circumstances is to not use the company again. Similarly, service failure in a restaurant context may lead to anger, worry or regret. Customers feeling anger or regret are likely to respond with dissatisfied responses, such as direct complaining, negative word-of-mouth or switching intention (Mattila & Ro, 2008). There is little discussion in the literature of how coping strategies can be used as a means of minimising dissatisfied behavioural responses caused by negative emotions.

During a service failure, different regulation strategies, such as cognitive reappraisal and expressive suppression, may be used (Gross & John, 2003). Customers who adopt a suppression strategy and hide their emotions may engage in greater negative word-of-mouth intentions and retaliatory behaviour. In contrast, a reappraisal strategy allows customers to re-interpret the service failure situation with positive meanings which thereby mediate the relationship between negative emotions and repurchase intentions (Balaji et al., 2017). Service providers can also use their knowledge of coping to affect customer emotions. For example, after a service failure, an apology expressing remorse can positively affect customers' coping behaviours by increasing emotional empathy (Davis & Gold, 2011). This increased emotional empathy can result in forgiveness and thus reduce the frequency of coping behaviours such as revenge or avoidance (Wang et al., 2020). Therefore, acknowledging a mistake and providing a remorseful apology are service recovery strategies that can make customers feel that the service providers recognise their mistake or have little control over the incident.

Emotional intelligence in dealing with negative emotions has been examined as part of the appraisal and coping process. Emotional intelligence can involve both intrinsic (regulating one's own emotion) and extrinsic emotion regulation (regulating others' emotions) (Gross, 2015). Grandey and Gabriel (2015) posit emotional intelligence as the ability that enables perceiving, comprehending, controlling and communicating emotions and utilising emotions to support intellectual growth. Following Roseman et al. (1990), emotional intelligence involves four dimensions: appraisal and expression of emotions in the self; appraisal and expression of emotions in others; regulation of emotions in the self and utilisation of emotions to facilitate performance (Mayer & Salovey, 1997).

In service failure contexts, customers' emotional intelligence can be used in both emotion generation and the coping process. Strizhakova and Tsarenko (2010) find that customers with higher emotional intelligence can control and minimise the elicitation of negative emotions in stressful consumption contexts. They also find that emotional intelligence has strong positive effects on active and expressive coping, which can help customers actively seek issue-resolution and alleviate emotional stress by increasing engaging in expressive behaviour.

Emotional labour occurs when employees monitor and regulate their emotions and cope with stressful situations (Simillidou et al., 2020). Coping with occupational stressors requires employees to control or reduce demands created in stressful situations (Dewe, 2000). Coping strategies in workplaces are important factors in determining labour stress outcomes and employee well-being (Ben-Zur, 2009). In workplace settings, employees are usually required to follow standard rules for appropriate expression of emotions on the job to create customer satisfaction, even when they encounter uncivil customers (Diefendorff & Gosserand, 2003). However, the emotional labour of suppressing negative emotions can decrease employee well-being and affect their service quality performance (Hopp et al., 2012). Service organisations should thus avoid imposing demands to suppress negative emotions (anger) even they feel them, but rather encourage the expression of positive emotions (cheerfulness, sympathy) even when a worker might not feel them.

When dealing with customer aggression, employees usually adopt expressive coping behaviours through social support-seeking, negative emotion venting and avoidance coping behaviour through disengagement (Goussinsky, 2015). Employees with high self-efficacy are less likely to use emotional support and thereby reduce the negative impact of customer aggression on emotional exhaustion (Goussinsky, 2015). Service organisations and managers are thus encouraged to foster their employees' trust in their own ability to cope with customers' misbehaviour (Siu et al., 2007).

Tummers et al. (2015) classify coping strategies during stressful service delivery into moving towards clients, moving away from clients or moving against clients. These include various behavioural coping strategies that occur when frontline workers interact with clients to master, tolerate or reduce conflicts they face (Bartels, 2013), such as seeking social support from colleagues or complaining to managers after the service delivery (Van Yperen & Hagedoorn, 2003). Frontline workers commonly use cognitive coping through emotional detachment from clients during client–worker interactions or engaging in cognitive restructuring towards work after service delivery (Tummers et al., 2015). In most situations, workers have a tendency to provide good service to clients even under stressful conditions by adopting 'moving towards clients' strategies.

Emotional Coping in Tourism

In tourism/hospitality studies, a tourism experience is usually taken to lead to happiness and positive well-being and therefore negative emotions are seldom studied as (Gao & Kerstetter, 2018). Several tourism studies have adopted a cognitive appraisal perspective in understanding the antecedents of emotions and their impacts on decision-making (Breitsohl & Garrod, 2016; Li et al., 2021). Specifically, negative emotions in tourism studies have been studied under three main themes: tourists' negative emotions and coping strategies, emotional labour under stressful working environments and residents' emotional affection and appraisal.

First, CAT has been applied to examine tourists' emotions and coping strategies in the context of unethical destination incidents and service delays/failures. This research uses the 'cognition-emotion-coping' sequence in multiple tourism contexts. For example, Breitsohl and Garrod (2016) find the strength of different cognitive antecedents in eliciting tourists' holistic emotional responses to unethical destination incidents and the consequential effects on behaviours, such as negative word-of-mouth and avoidance. Studies have also investigated antecedents and coping strategies for discrete emotions. For example, Jiang et al. (2020) adopt CAT to examine the antecedents of tourists' worry and anger evoked by a flight delay and their respective effects on tourists' behavioural intentions. Worry and anger are found to have different effects on switching intentions, complaining behaviours and negative word-of-mouth. Tourists' emotion regulation strategies (reappraisal and suppression) in coping with negative emotions are found to moderate the process of elicitation of worry and anger and mediate the impacts on reduction of loyalty intentions (Li et al., 2021).

Second, emotional labour in stressful working environments is also an important topic in tourism studies. The utilisation of emotion regulation (surface-acting and deep-acting) as part of the cognitive appraisal and coping process has been discussed to manage negative emotions in coping with uncivil customer behaviour and job stress (Choi et al., 2019). Specifically, Teoh et al. (2019) identified four families of emotion coping strategies (opposition, rumination, emotional regulation and positive cognitive restructuring) to manage emotional labour, corresponding to surface acting, deep acting (Reyers & Matusitz, 2012) and genuine emotions (Goussinsky, 2015). The choice of different coping strategies depends on secondary appraisal of situations, such as individuals' perceived control of their stress and job tasks (Luchman & González-Morales, 2013), self-efficacy (belief in oneself) (Meier et al., 2008) and different motivations such as competence to perform a service, and autonomy of psychological freedom in their own actions (Skinner et al., 2003).

Third, tourism resident emotions have been discussed in relation to tourism development and planning. Past studies focus on the overall well-being of residents (Woo et al., 2015) rather than distinct emotional responses to tourism development (Zheng et al., 2020) and residents' emotional experiences at local dark tourism sites (Jordan & Prayag, 2022). Various appraisal dimensions have been examined for their relationships with residents' discrete emotions, such as outcome desirability, fairness, agency and certainty (Zheng et al., 2020). However, current studies have only discussed the appraisal-emotion model by identifying the relationship between key appraisal and emotions, and limited understanding has been revealed on residents' emotional coping and regulation strategies in managing different negative emotions at a community level.

Emotion Coping Framework for Negative Emotions

Drawing from CAT, an emotional coping framework has been developed based on an 'antecedent – coping approach – coping consequence' model. This model presents hierarchical relationships in the process of emotional coping. A multidimensional coping construct is further embedded in the framework to posit emotional or cognitive appraisal antecedents to coping (process 1), such as the discrete emotions of anger, sadness, threat or the cognitive appraisal of situational efficacy and control (Duhachek, 2005). Furthermore, the strategies used for coping also affect the nature of the subsequent emotional responses and consequences (process 2). Further research has expanded and identified additional cognitive appraisal elements, such as goal congruence/relevance, agency, certainty and fairness (Johnson & Stewart, 2005; Lazarus, 1991). It is interesting to note that in their stress coping framework, Jordan et al. (2015) discuss the role of personality in the emotional appraisal and coping process; however, this personality psychology view is not within the scope of the cognitive psychology theory used to discuss emotion and coping processes in this chapter.

Werner and Gross (2010) discuss a process model of emotion regulation developed from Gross and Thompson's five groups of emotion regulation strategies along the timeline of emotional appraisal and response – 'situation selection, situation modification, attentional deployment, cognitive change, and response modulation' (2007, p. 18). Gao and Kerstetter (2018) expand the process model to understand how individuals express emotional responses based on three phases of emotional regulation strategies (Gross, 1998) – Phase I Interpersonal (group-focused); Phase II Situational (situation-focused) and Phase III Intrapersonal (self-focused). Key emotional regulation corresponding strategies and sub-strategies are revealed and embedded in the framework, drawing from empirical study outcomes (Figure 14.3, adopted from Gao & Kerstetter, 2018).

In Phase I, situation selection can be built through interpersonal strategies, whereby people regulate emotions by soliciting social support (spending time together) from family members, friends, relatives and others. In Phase II, people consider alternative plans to prepare for problems that might occur in the future (situation modification) and use distraction and reveling in the moment strategy to move attention away from negative situations or direct attention to good feelings (attention deployment). In Phase III, self-focused strategies can be used to change their cognitive mindset or modulate their emotional responses. These include emotional reappraisal of the situation (Gross, 2014) and coping with

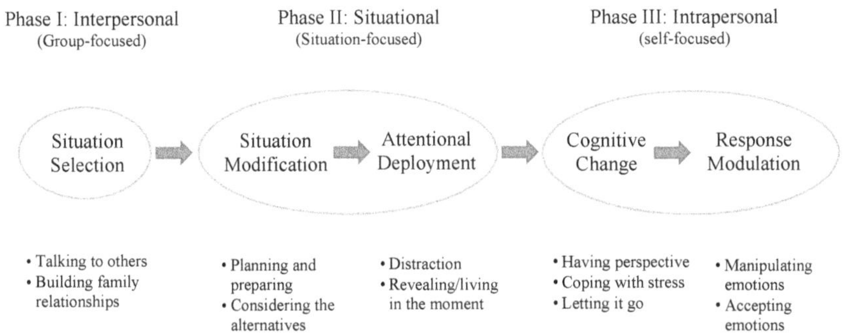

Figure 14.3. A Process Model of Emotion Regulation Strategies.
Source: Adopted from Gao and Kerstetter (2018).

stress (Iwasaki & Mannell, 2000) to decrease the generation of negative emotions, as well as holding in emotions or suppressions to control their expression of negative emotions (Gao & Kerstetter, 2018).

Furthermore, a dual process framework has been proposed to integrate both explicit and implicit forms of emotion regulation (Gyurak et al., 2011). Most studies conceptualise emotion regulation from an explicit view which requires constant monitoring of emotions and demands a certain level of insight and awareness. However, given the high demand for moment-to-moment emotion regulation in one's lives, the use of implicit emotion regulation (without awareness) can be important in the daily emotion coping process. This supports the view that emotion regulation goals can be activated outside conscious awareness (Custers & Aarts, 2010).

Explicit and implicit regulation processes are not mutually exclusive and have 'porous boundaries' between them (Gyurak et al., 2011). This involves an emotion change process which can range from more automatic (subconscious) to more controlled (conscious) forms (Braunstein et al., 2017). Four different classes of regulation can be generated based on implicit vs. explicit goals and automatic vs. controlled change processes. As an example, during cognitive reappraisal, the emotion regulation strategy alters one's emotional response through changing one's cognitive construal of an emotional situation (Gyurak et al., 2011). Reappraisal can be implemented explicitly (conscious) or implicitly (subconscious) (Shahane et al., 2019) and implicit cognitive reappraisal can be as effective as explicit reappraisal due to less cognitive engagement (Gyurak et al., 2011). A recent study finds that explicit reappraisal can receive better effects on emotion regulation over time if individuals have received a certain degree of reappraisal training and practice, especially for psychological distancing (Denny & Ochsner, 2014). Figure 14.4 presents a synthesised framework for an emotional coping process.

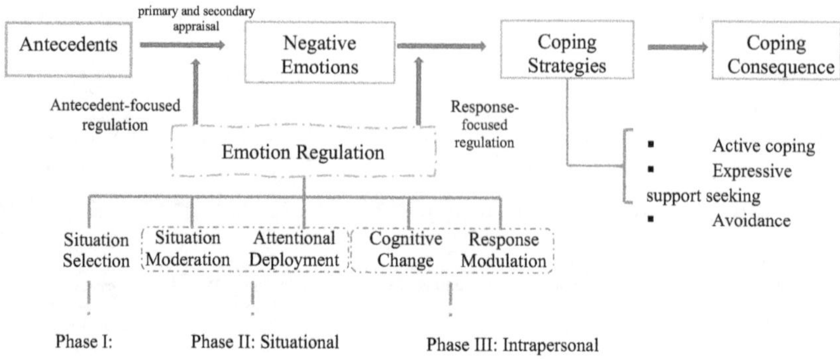

Figure 14.4. Synthesised Framework of Emotional Coping.

Illustration of Coping in the COVID-19 Context

Since December 2019, the novel coronavirus (COVID-19) has spread rapidly throughout the world and has greatly impacted the tourism and hospitality industry (Gössling et al., 2020). Given this uncertain environment, a growing number of studies have focused on negative emotions generated among tourists and tourism labour and the utilisation of different coping strategies. The most studied emotions in the COVID-19 context have been fear of infection (Kim et al., 2021b) and travel fear (Zheng et al., 2021). These are related to feeling anxious (Jungmann & Witthöft, 2020) and stressed (Xu et al., 2021) due to the perceived uncertainty of travel safety during the pandemic. However, CAT has not been fully utilised to understand the key appraisal dimensions in triggering these negative emotions at either primary or secondary level.

Importantly, a few studies have examined the emotion coping process in the pandemic environment following the 'cognition–emotion–coping–coping consequence' pattern in CAT. They aim to understand how to manage fear with different coping strategies (Ying et al., 2021) and what coping strategies can facilitate a positive consequence of encouraging travel intentions (Zheng et al., 2021). For example, in the hotel customer distrust context, perceived vulnerability, severity and cost were found to reflect the emotions of fear, helplessness (helpless) and anxiety (worry and uneasy) (Ying et al., 2021). The most salient coping behaviours were vigilance and dependence reduction (caused by fear), negotiation (caused by helplessness and anxiety), reference shifting (caused by anxiety) and problem disengagement (caused by helplessness). Furthermore, in the public's pandemic travel fear context, threat severity and threat susceptibility can trigger travel fear, which has led to protective travel behaviours, such as travel avoidance in the pandemic outbreak. Self-supported emotional coping and disengagement coping are the most effective coping strategies to increase people's psychological resilience and further evoke cautious travel in a post-pandemic world (Zheng et al., 2021).

Although Yu et al. (2021) find that subjective stress can provoke negative emotions in the COVID-19 environment, none of the primary nor secondary appraisal processes have been discussed. Notably, problem-/emotion-focused coping strategies can cope with stress. For example, business operators may adopt a 'situation modification' approach (increase income, reduce cost, prepare for recovery) to reduce their stress, or use 'attention deployment' (socialising, exercising, engaging in relaxing leisure) and 'cognitive change' approaches (cognitive framing, positive comparison) to improve their well-being and mental health (Xu et al., 2021). These reflect parts of the emotional coping framework presented in Figure 14.4. Other cognitive appraisals (positive appraisal, psychological distancing) and behavioural strategies (social connection) are also found to predict more stable well-being during the pandemic (Kim et al., 2021b).

CONCLUSION

Studies on tourism experience and emotions have seen a growth in the past decades and reflect the notion that emotions are critical to tourists' behaviour and their engagement in leisure seeking (Barnett, 2013). Although a tourism experience is often seen as a hedonic consumption context, negative emotions are common, and more studies are needed to reveal the emotional reappraisal process through effective coping and regulation (Gross & Barrett, 2011). This chapter has provided an understanding of coping in the cognitive psychology literature by introducing three sequential components of appraisals: primary appraisal, secondary appraisal and outcomes of appraisal–coping strategies (Dewe, 1991). Several classifications of coping strategies and construct debates are discussed, including the multidimensional view of coping strategies, the effectiveness of coping strategies and the positioning of emotion regulation in the cognitive appraisal process. An increasing number of tourism studies have applied CAT to understand the antecedents of discrete emotions for different stakeholders (tourists, employees and residents).

Emotional coping framework has been gradually developed based on an 'antecedent – emotion – coping – coping consequence' model. In recent years, studies have expanded the understanding of emotion coping through several aspects including various emotion regulation strategies in different phases (Gao & Kerstetter, 2018) and explicit vs. implicit regulation in the daily emotion coping process (Gyurak et al., 2011). However, these findings have not yet been transferred to tourism or general business disciplines, with current emotion studies still relying on the theory of behavioural psychology rather than cognitive psychology. Aiming to provide critical suggestions for future research opportunities, a synthesised framework is thus developed based on the CAT to summarise the emotional coping process.

Future research on emotions should develop in-depth understanding of how human brains deal with negative emotions through the cognitive process and what coping strategies can be used to cope with difficult situations. To achieve this goal, several future research agenda are proposed. First, theoretically speaking, constructive discussion and a disambiguation approach are needed to develop a clear distinction between emotion coping and emotion regulation in tourism research. This can help to enrich the existing understanding of the coping mechanism with the consideration of explicit vs. implicit processes (Gyurak et al., 2011) and controlled vs. automatic processes (Braunstein et al., 2017).

Second, it is important to identify the antecedents (relevant appraisals) leading to specific emotions towards negative tourism experience and subsequent coping consequences (Jiang et al., 2020; Kim et al., 2021a). These are missing in current COVID-19 emotional coping studies. This should also consider the diverse angle of research into the emotions of different stakeholders (tourists, residents, employees). Third, rather than studying negative emotions as a holistic concept, future studies need to further investigate different discrete emotions in a service/experience failure setting and discuss the most effective coping strategies in various situations. These can support service providers by providing a deeper understanding of customer/tourist emotional responses and the effects of different coping strategies in dealing with challenging situations.

SECTION 3
APPLICATIONS

Chapter 15

EXPERIENCE DESIGN

Noel Scott
Edith Cowan University, Australia

Brent Moyle
Griffith University, Australia

Jianyu Ma
Shanghai Normal University, PR China

Ana Cláudia Campos
University of Algarve, Portugal

Lynn I-Ling Chen
National Pingtung University, Taiwan

Dung Le
VinUniversity, Vietnam

Liubov Skavronskaya
Griffith College, Australia

Shanshi Li
Xiamen University, China

Cognitive Psychology and Tourism
Tourism Social Science Series, Volume 27, 179–195

Published under exclusive licence by Emerald Publishing Limited
ISSN: 1571-5043/doi:10.1108/S1571-504320240000027015

Rui Zhang
Hainan Tropical Ocean University, China

Shan Jiang
Capital Normal University, China

Lihua Gao
Soochow University, China

Arghavan Hadinejad
University of South Australia, Australia

Abstract: Effective experience design requires not only a knowledge of tourist goals, but an understanding of how these can be met in a particular tourism site. Research on experiences and experience design is supported by cognitive psychology concepts such as perception, attention, appraisal, emotion, consciousness, feelings and memory. However, these concepts are often used in a combination with others from sociology, social or environmental psychology in a manner that leads to confusion rather than clarity, without apparent understanding of the theoretical mechanisms by which these concepts are related. This chapter develops a series of propositions for potential application to tourism experience design. Future research should examine the efficacy of these propositions from cognitive psychology for tourism experience design. **Keywords**: Attention; appraisal; emotion; memory; experience design; consciousness

INTRODUCTION

The concept of 'experience design' is deeply entrenched in architectural design literature (Tussyadiah, 2014; Tussyadiah et al., 2008) and is based on meeting the desires and expectations of users (Gunn, 1972). Its process is based on the concept of pattern language (Tussyadiah et al., 2008) which is a set of interconnected patterns, each of which describes a problem and the core of a solution that can be used in many ways within a specific field of expertise (Alexander, 1977). Researchers interested in experience design using an architectural approach are overlooking a pre-existing body of theory and practice about the mental processes that occur before, during and after an experience. In essence they are using an architecture-inspired design process to solve a psychological problem, as experience is fundamentally a psychological concept.

Seminal work has identified the tourism experience as a psychological phenomenon (Cohen, 1979; Larsen, 2007; Mannell & Iso-Ahola, 1987; Pearce & Caltabiano, 1983). Often researchers use qualitative ethnographic methods (Tussyadiah, 2014; Tussyadiah et al., 2008), narrative analysis or plot analysis (Boje, 2001) to identify useful information, such as perceived tourist problems, solutions, sequences, episodes and schema (Alexander, 1977). However, this research is predominantly descriptive and based on a proposition that common problem-solution pairs can be identified from user narratives. The apparent lack of theory supporting the pattern language approach suggests that it may be better to adopt human-centred design. The goal of the design is to make 'a direct connection between users' internal state, design characteristics and the contexts of interactions between users and design' (Tussyadiah, 2014, p. 548). Thus, human-centred design moves from identifying patterns based on empirical observation to using prior concepts and theory to supplement the design process.

In tourism, the purpose of designing an experience is often ascribed to wellness (Buzinde, 2020), well-being (Vada et al., 2019) and the search for the optimal human experience (Csikszentmihalyi, 1990). Subsequently, theory from positive psychology, such as self-determination theory (Deci & Ryan, 2000) and flow theory, is becoming increasingly applied in the tourism field. Flow is an optimal human experience based on immersion and challenge (Csikszentmihalyi, 1990). In self-determination theory, people are considered to be goal-directed and that these goals address needs for competence, relatedness and autonomy (Deci & Ryan, 2000). It is argued here that the use of the psychology of mental processes and embodied actions associated with attention, satisfaction, schema and memory, goals, emotion and cognition are a way to address both optimal and ordinary experiences (Duerden et al., 2018; Skavronskaya et al., 2017).

The mental processes required to support tourism experiences are described in general by Kim and Fesenmaier (2017b) in their 'Physiological Foundations of the Tourism Experience'. Based on framework of experience creation adapted from Krishna (2012), this model focuses on perceptions of sensations and the interpretation of these through various filters as emotional and cognitive responses and subsequent attitudes, memories and behaviours. Subsequently, Kim and Fesenmaier (2017b) lend support to a cognitive psychological model of the tourism experience. Their model discusses cognitive appraisal theory and the centrality of goals in the elicitation of an emotion, termed as a psychological filter. There is some confusion in the distinction between cognition and consciousness (emotional appraisal is cognitive but subconscious), but experience is considered to be inherently emotional in nature. Further, emotion is seen as a source of information consistent with 'Feelings as Information Theory' (Schwarz, 2011), such that emotions subconsciously influence the level of attention.

Cognition has a critical role in experience design, yet is underrepresented in mainstream discourse on the tourist experience. It is the mental process of acquiring knowledge and understanding through thought, experience and the senses (Balota, 2000). Cognition is understood to operate through systems and processes, such as perception and attention, memory-related processes like mental schema and categorisation, emotion elicitation and states of consciousness including mindfulness, flow, retrospection and prospection (Lachman et al., 2015). These processes mediate how external sensations correlate with behaviour. They also mediate such internally generated states as cognition about the future (termed prospection like fantasy), or the past (termed retrospection, such as savouring).

However, the sensory abilities, stored memories and past experiences are different for each individual. For example, emotions are elicited by joint appraisal of the perceptions on dimensions including goal valance and importance, surprise and agency. These dimensions are the same for all people, but individuals may have different goals in the same situation. This allows the same mental processes assessing the same situation to elicit different emotions for everyone. Thus, a person's goals mediate between perception and emotion. Further the systems are interconnected. Goals both shape the emotional responses and additionally can direct one's attention to relevant stimuli, known as goal-driven, top-down attention.

Given the critical importance of cognitive processes, it is surprising that few scholars have applied cognitive psychology to examine tourism experiences. Most experience design papers adopt a sociological (Ek et al., 2008), behavioural psychology (Kim et al., 2010) or management perspective (Dias & Dias, 2019; Hwang & Seo, 2016; Moscardo, 2018; Pearce & Zare, 2017). There are a number of experience design papers that include emotion (Jiang et al., 2016) or various psychological factors (Wei et al., 2019) but without the underlying structure provided by detailed engagement with theory from cognitive psychology. Other papers have adopted a cognitive psychological view of particular aspects of experience, including emotion (Bastiaansen et al., 2019) and memory (Lin et al., 2018; Tung et al., 2017).

This chapter discusses how cognitive psychology can add value to the development of a psychologically supported experience design process. Accordingly, this discussion contributes to knowledge by establishing an interconnection between cognitive psychology and experience design through a critical analysis of relevant concepts and development of a framework of tourism experience design.

EXPERIENCE DESIGN

It is of fundamental importance to understand the relationship between cognitive psychology and other types of psychology that involve meaning. Cognitive psychology describes mental processes that enable the processing of meaning but do not involve it. A cognitive psychological process, such as a cognitive appraisal leading to an emotion, does not manipulate meanings or concepts. Instead, it may be considered as a matching process in the same way that an artificial intelligence programme identifies a face via an algorithm. While the exact process at the neuronal level is not yet available, the brain has a neural network that identifies a match between certain parts of working memory and (say) a neural pattern associated with a goal. The extent of this match determines the degree of goal relevance. Similar processes determine the degree of novelty, goal importance and other cognitive appraisal processes and these jointly determine the elicitation of a particular emotion (determined by valance and degree of arousal). The important point is that this process is not specified by a particular set of meanings or events and instead is general and used when any emotion is elicited. Thus, the matching process occurs at a lower level than meaning. However, one can infer from the strength and valance of the emotion elicited the importance of the goal and the degree of novelty that are involved. By considering the context, this may enable the actual goal and source of novelty to be identified. For example, if at the beginning of a conversation a person is starting to get angry, it can be inferred that some important goal is related to the discussion and that goal congruence is negative. However, it is difficult to determine exactly the person's goal, which may have to be confirmed separately.

A second point is that the brain is limited in its ability to process all information it receives and uses a variety of techniques to focus its capabilities on eliciting salient perceptions. To understand and respond to its surroundings, the brain must be able to identify salient events (parsing). When one perceives the surrounding world in its 'blooming, buzzing, confusion' (James, 1981/1890), the brain must

> make sense of that information stream and to act effectively in the world, we need to reduce it to useful chunks. One important kind of chunk is an integrated unit of space and time that has a beginning, middle, and end: having breakfast, attending class, seeing a movie. Events themselves are segmented into parts: buying a ticket, finding a seat, viewing the movie. A variety of cognitive and perceptual features converge at event boundaries; they are instants of relatively greater change in action, they are moments when goals and sub-goals are accomplished, they are times when predictability breaks down. Event units are convenient packets for perception, comprehension, inference,

prediction, imitation, learning, and planning (Tversky & Zacks, 2013, p. 83).

This idea of bounded events or experiential episodes (Bastiaansen et al., 2019) as the basis for making sense of the world may be useful for the study of patterns in architectural design. Thus, the 'micro-experiences.... which are "assembled" using a variety of cognitive and emotion-based processes to create meaning' (Kim & Fesenmaier, 2017b, p. 19) can be naturally segmented using the concept of boundaries (Tversky & Zacks, 2013).

While idea of a pattern described by Alexander (1977) is a practically useful analogy, application of event recognition or parsing concepts as discussed in cognitive psychology provide additional levels of information and links to theories and observations subject to empirical verification. Cognitive psychology describes how the brain imposes discipline to turn a stream of sensation into a 'stream of consciousness' and stream of actions. Thus, it describes what phenomenological experience is – 'consciousness' and some aspects of how perception can be transformed into action (Figure 15.1).

A third major point about the mental activity is that much of it (such as emotional appraisal) is subconscious. This does imply it is irrational, instead, emotions function to provide an action orientation towards pleasure and away from pain. Goal appraisals are fundamental to determination of the valance and intensity of an elicited emotion (Dijksterhuis & Aarts, 2010). The goals of an individual may be dysfunctional as in addiction or functional as in pro-survival, but in both situations, the same goal-directed cognitive process operates. The concept of rationality is associated with logic and an absence of emotion. Rationality in a social science sense means actions or

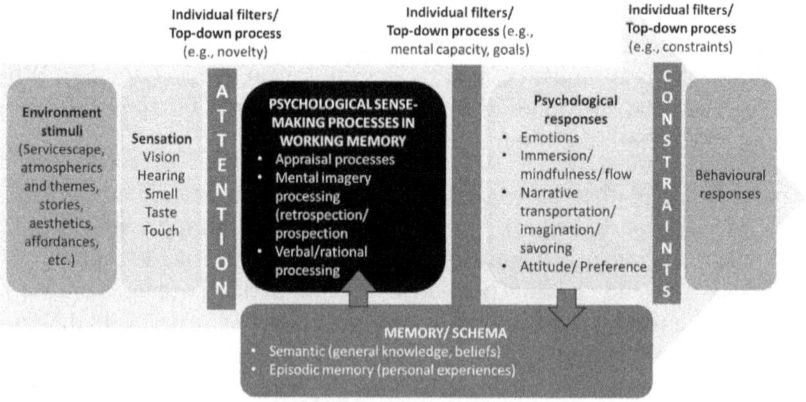

Figure 15.1. Framework of Experience Co-Creation From a Cognitive Perspective.

Source: Adapted from Kim and Fesenmaier (2017b) and Krishna (2012).

thoughts that 'contribute to certain goals, where these goals may be the pleasure or satisfaction of an individual or the guarantee of food or shelter for the members of a society'. When awareness and intention are present, the function is usually called manifest; otherwise, it is a latent function (Simon, 1978, p. 3). If emotion-based decisions are rational, they should be goal directed and as shown below, goals are an important emotional appraisal dimension.

Much of the theory of cognitive psychology could be considered common sense. For example, one pays attention to loud and unexpected noises as discussed by bottom-up attention mechanisms. However, the advantage of cognitive psychology is that it provides a process for such attention as well as how such loud noises may also elicit emotional arousal (surprise is an appraisal dimension). Such explanatory power, however, only applies within the scope of cognitive psychological theory and its core concepts (sensation, perception, attention, emotion, memory, consciousness). Therefore, when discussing these concepts, it would appear logical and indeed necessary to at the very least be aware of, and if possible, adopt the associated paradigm, relevant theories and concepts.

Tourism Experience Design

Cognitive psychology does not discuss directly some concepts that are used in studies of tourism experience, such as, for instance, atmosphere and themes (Åstrøm, 2019; Laing & Crouch, 2009; Lwin et al., 2016; Mari & Poggesi, 2013; Pearce & Wu, 2014), stories (Moscardo, 2017a, 2018, 2020), affordances (Brown et al., 2015; Pucillo & Cascini, 2014), technology (Bec et al., 2019) and co-creation (Campos et al., 2018). While these specific concepts are not explicitly embedded in cognitive psychology, they are able to be informed or related to relevant concepts.

Atmospherics and themes: The concept of atmospherics is commonly used in the design of experiences (Åstrøm, 2017; Gao et al., 2016; Laing & Crouch, 2009; Pearce & Wu, 2014). Arguably, both atmosphere and theme (and symbols) are related in that they provide cues as to the nature of the environment, and therefore supply a guide to appropriate goals and behaviour (Lwin et al., 2016; Mari & Poggesi, 2013). Subsequently, it may be expected that atmospherics, themes and symbols are able to evoke related goal-relevant activity. Additionally, tourists will usually select a particular atmosphere voluntarily, as they provide a self-selecting sample where goal alignment may be assumed. This explains why tourism is mostly associated with pleasant emotions.

Stories. A number of studies indicate the importance of stories for shaping and enhancing experiences (Moscardo, 2010, 2017a, 2018). It appears that narratives are somehow structurally similar to autobiographical memory (Tung et al., 2017). Stories also create a state called narrative transportation, which could be considered equivalent to directed internal attention (Ryu et al., 2019) and emotion (Zhang et al., 2021). The links between narrative and autobiographical memory mean that stories and narratives are an important resource for communication and persuasion. Narrative processing has been shown to affect persuasion through transportation. In narrative transportation, affective responses influence persuasion rather than the systematic analysis of message strength (Escalas, 2007). A narrative may engender similar emotion, engagement and adsorption, and self-memories as an experience, leading to persuasion (Appel & Richter, 2010; Phillips & McQuarrie, 2010) and is noted to shape tourist experiences (Moscardo, 2010). Note here that concepts such as narrative transportation, developed in behavioural psychology must be translated into the relevant concepts and processes of cognitive psychology. This translation process can be useful in allowing additional insight into research topics and gaps.

Affordances. An affordance is a concept from ecological psychology (Brown et al., 2015) and design studies (Pucillo & Cascini, 2014). It is 'action possibilities offered from the environment to the animals that are capable of seizing these offering' (Pucillo & Cascini, 2014, p. 162). The concept is used in perception research to examine how aspects of the environment are used by animals or humans to achieve their goal. The goals themselves are considered basic needs such mating, social interaction and so on. From a cognitive psychology perspective, an affordance may be considered as an aspect of the environment that is aligned to the achievement of common human goals.

Aesthetics. Aesthetics are discussed by tourism design authors (Kirillova, 2019; Kirillova & Lehto, 2016). A cognitive psychology approach to beauty considers it a product of the mental processes of the beholder (Kirillova, 2019; Kirillova et al., 2014). Aesthetic assessment, thus, reflects human responses to a scene and involves one's prior knowledge, experiences, values, emotional state and desires. While what a person finds beautiful is idiosyncratic, there is a common orienting response or direction of attention to objects found beautiful (or ugly) that can be used as an indicator of preferences (Scott et al., 2020). Thus, the relevance or meaning of an object (its beauty) for a particular individual can be inferred from measurement of their attention to it.

Co-Creation. The concept of co-creation is a management or marketing concept and is discussed in tourism design studies as part of the design process and also as valuable for customers. In terms of cognitive psychology, co-creation of experiences may increase active attention and engagement of the customer both physically and mentally as well as stimulating interactions with other people. Co-creation may therefore increase the goal relevance and importance of an activity leading to increased emotion. Thus, a visit to an attraction might provide a setting that increases social bonging through the interaction with family members (Campos et al., 2018). Another effect of co-creation is to increase task attention which can lead to increased memorability (Campos et al., 2014).

Key Concepts in Cognitive Psychology

Attention. Attention refers to focused mental engagement while processing external sensations or internal memories and thoughts, and is involved in perception, memory and learning. The salience of an external stimuli may be determined by top-down or bottom-up processes. Attentional focus is considered bottom up when the senses are drawn to stimuli from the external environment, like colour and movement or top-down when it is affected by endogenous mental goals (Dijksterhuis & Aarts, 2010). One's present mental goals, whether conscious or subconscious, can 'guide behavior through attention' to particular parts of the flow of information (Dijksterhuis & Aarts, 2010, p. 467), alerting a person to goal-relevant information and hence influencing overt behaviour (Monsell & Driver, 2000). Attentional focus is directed to novel stimuli (Mather, 2013), stimuli considered beautiful (Lindell & Lindell, 2014), and to faces which are of biological and social significance (Ro et al., 2007).

In other situations, attention is directed internally as is found in the concepts of engagement (Brodie et al., 2011), immersion (Ellis et al., 2018) and narrative transportation (van Laer et al., 2014). Higgins considers 'Engagement is a state of being involved, occupied, fully absorbed, or engrossed in something – sustained attention' (2009, p. 102). Similarly, a high level of immersion is characterised by [in part] 'high focus of attention on a limited stimulus field' (Ellis et al., 2018, p. 101). Narrative transportation has been related to immersion (van Laer et al., 2014) and conceptualised as co-activation of attention, imagery processing and emotions (Appel & Richter, 2010). Similarly, flow is considered as a state of optimal experience characterised by focused attention, clear mind, mind and body unison, effortless concentration, complete control, loss of self-consciousness, distortion of time and intrinsic enjoyment (Csikszentmihalyi, 1990).

Ooi (2005) provided an early application of the concept of attention to tourism, by noting that tourists visit a place for a relatively short period of time, lack local knowledge, and therefore it is important to manage their attention through use of symbols and visual cues. Attention has been empirically related to experience memorability since those activities or objects to which one pays attention to are more likely to be remembered. Active participation, interaction and enhanced attention improve experience memorability (Brunner-Sperdin et al., 2012; Campos et al., 2016; Moscardo, 1996). Attention is also central to meditative mindfulness which may help tourists obtain eudemonic outcomes from experiences (Chen et al., 2017). Knowledge of attentional processes is useful for improving marketing stimulus material, wayfinding, as well as service and tourism experience design (Scott et al., 2017).

Similarly, as already noted, attention is found to relate to perceived beauty (Scott et al., 2020). Stimulus novelty has the capacity to attract, redirect and increase attention (Lisman & Grace, 2005). Attention given to new stimuli determines the amount information that becomes encoded and stored in memory (Fivush, 2011). The first occurrence of a particular event that captures a person's attention leads to superior memory of it during the process of recollection (Tulving & Kroll, 1995). In tourism contexts, such as museums, customer satisfaction and achievement of objectives are dependent on the degree of attention tourists pay to what they are experiencing (Antón et al., 2018a).

Experiences Based on Mental Imagery Processing

One of the important contributions of cognitive psychology is to provide an explanation of various biases and distortions involved in thinking about the future, present or past. An essential observation from cognitive psychology is that the brain mechanisms used for thinking about the present, past or future essentially utilise the same neurological pathways (Skavronskaya et al., 2017). As previously determined, the tourism experience is a 'continuous process which shapes and reshapes itself through interaction in time and space' Kim and Fesenmaier (2017b, p. 27). This definition is consistent with the workspace model of consciousness whereby perceptions of the world around one is held in 'working memory' (Cardoso et al., 2019; D'Esposito & Postle, 2015; Tung et al., 2017). Consciousness provides a mental 'representation' of the world that is subject to recursive appraisal by other cognitive processes, allowing 'continuous reshaping' and 're-evaluation' (Bec et al., 2019). Consciousness is constantly created, reshaped and reinforced, utilising a mix of perceptions of external sensations and memories.

Importantly, the same mental processes are used to engage in mental time travel, which involves thoughts about future events (prospection), current imagined events or past events (retrospection) (Suddendorf & Corballis, 2007). However, the process of imagining future events draws on past memories to shape perceptions of future experiences, and therefore may be less detailed or vivid than the current experience. This is why dream beach holiday destinations are associated with prototypical attributes such as a beach, blue sky, and sea and palm trees, perhaps with a hammock, sailboat and seagull (Kane et al., 2012). Such mental images are networks of interrelated concepts clustered by existing schema (Zhang et al., 2018). Understanding the schema of the holiday category into which a destination is usually categorised is of great importance for tourism marketing. Such a mental image is often described as a 'fantasy' and has been termed consumption vision (Walters et al., 2012). There is a significant literature on episodic future thought construction which presents promising directions for the tourism field (D'Argembeau & Mathy, 2011; Schacter et al., 2008).

Importantly, memories of the past or imagination of the future are constructed (Schacter & Addis, 2007) and emotions do not appear to be stored in memory. Just as memory in general is reconstructive (Bartlett, 1932), memories of past emotions are reconstructed based on subsequent reinterpretations of earlier events (Levine, 1997). For example, recollection of emotional responses to an event can be altered if the current goals used to appraise an event have changed from the time the event had occurred (Levine et al., 2001). Such a reconstructive process provides a mechanism for savouring holidays, and continually revaluating tourism experiences during recollection (Filep et al., 2013). This suggests from a cognitive psychological perspective, a current 'experience' is the state of a visitor's consciousness and the mental processes occurring at that time. Critically, the processing of future (imagined) or recalled (past) tourism experiences utilises the same mental processes.

Memory and Schema

Memory is the mental process by which information is encoded, stored and retrieved (Atkinson & Shiffrin, 1968). Tourism research on experience design will undoubtably benefit from cognitive psychology theory which facilitates a movement beyond identifying the components of the servicescape components critical for memorable experiences (Larsen, 2007; Wei et al., 2019) towards an advanced conceptual understanding of why they are memorable (Campos et al., 2017). According to Kim, 'specifically, the causal connections between the experiential factors and one's ability to recollect past experiences and vividly retrieve information are unclear' (2011, p. 781).

A schema represents inherent relationships between personal perception and understanding (Piaget, 1926). An early definition of schema is as 'an active organization of past reactions, or of experiences, which must always be supposed to be operating in any well-adapted organic response' (Bartlett, 1932, p. 201). In social psychology, a stereotype is a type of schema that may lead to biased understanding (Schneider, 2004). In cognitive psychology, a schema has an associative structure, is based on memories from multiple episodes, has some generality and is malleable or adaptable (Ghosh & Gilboa, 2014). Schemas are also called data structures, mental structure (Cheong & Kim, 2011) or stored framework (Aggarwal & McGill, 2007) and represent information about a topic, a concept or a particular stimulus, including its attributes and the relationships among the attributes.

Emotion

Appraisal theorists consider that emotions are subconscious mental events elicited by an appraisal of relevant stimuli on a limited number of dimensions. According to Desmet and Hekkert, emotional experiences

> ...arise from encounters with products that are appraised as having beneficial or harmful consequences for the individual's concerns, that is, his or her major goals, motives, wellbeing, or other sensitivities (2007, p. 61).

Similarly, for Volo

> ...knowledge of the emotional tourists' space is essential in designing services which enable providers to relate with tourists in personal and memorable ways creating moments of engagements [that]....amplify tourists' emotions. Indeed, it is argued that congruity with tourists' goals realization can elicit positive emotions while incongruity can generate negative emotions (2017, p. 38).

Otto and Brent Ritchie state that 'perhaps more than any other service industry, tourism holds the potential to elicit strong emotional... reactions' (1996, p. 168). Following this, contemporary work has examined emotion in detail (Hosany & Prayag, 2013; Nawijn et al., 2013; Volo, 2017); however, most have only sought to correlate environmental stimuli with emotional states (Brunner-Sperdin et al., 2012) or to examine the effects of emotions on behaviour (Song et al., 2014) without developing constructive discourse focused on the reasons why a specific stimulus elicits particular affective responses. Descriptive approaches used to study emotions in tourism, such as the consumption emotion scale (Richins, 1997), differential emotions scale, circular model of emotion, the circumplex, the pleasure-arousal-dominance and the positive affect negative affect schedule are not able to explain how a tourist's consumption experience leads to one specific emotional response rather than another (Hadinejad et al., 2019).

More recently, a number of studies have applied cognitive appraisal theory (Cai et al., 2018; Choi & Choi, 2019; Le et al., 2020; Zheng et al., 2019) to understand tourists' emotional experiences towards destinations (Hosany, 2012; Hosany & Gilbert, 2010), and some have applied appraisal dimensions (Ma et al., 2013; Manthiou et al., 2017). Emotional involvement has been found to increase recall of memorable tourism experiences (Servidio & Ruffolo, 2016). Several scholars use the term cognitive appraisal in a manner different from cognitive appraisal theory (Breitsohl & Garrod, 2016; Su & Hsu, 2013). CAT defines an emotion as a mental state that results from subconscious appraising of personally relevant information on a limited number of dimensions.

Within cognitive appraisal theory, it is critical to recognise that external stimuli do not convey affective meaning. Instead, it is the mental appraisal of the external stimuli in relation to current goals, combined with other dimensions determines which emotion is elicited. Consequently, understanding an emotion requires an appreciation of the subjective evaluation of events by a person. This is different from many previous studies whereby researchers consider people in general travel for novelty or escape rather than for their own idiosyncratic goals (Baloglu & McCleary, 1999; Gartner, 1993). In addition, accordingly to cognitive appraisal theory, such dimensions as novelty, when combined with concurrent goal congruent appraisal is intricately connected to the elicitation of pleasurable emotions (Ma et al., 2017; Mitas & Bastiaansen, 2018). Cognitive psychology and neuroscience propose that novelty is free from emotional valence (Scherer, 2001), and as a consequence, novelty in tourism experiences is accompanied by positive as well as negative emotions (Kim et al., 2012; Lin et al., 2014). However, as tourism is a voluntary goal-directed activity associated with novel stimuli, tourism experiences are usually pleasurable. As such, a designed experience perceived as novel and salient will increase tourist attention, emotional arousal and memory recall.

Goal Appraisals and Emotions

In cognitive appraisal theory, the unexpected achievement of an important goal will elicit an arousal of positive emotion such as excitement, joy or delight (Desmet & Hekkert, 2007, p. 63). The theory supports the central role of goals in determining the strength and valance of emotions (Deci & Ryan, 2000; Le et al., 2020). Several scholars note the central importance of goals in experience design:

> An experience designer is foremost an author of experience. Only after having outlined the desired emotional and cognitive content of an experience, the action involved, its context and temporal structure, [she] may start designing the 'product'. And then, each and every detail (content, functionality, presentation, interaction) has to be scrutinized according to its potential to create or destroy the desired experience. (Hassenzahl, 2010, p. 68).

It is actually the fulfilment (or frustration) of psychological needs that renders an experience positive (or negative) and personally significant, that is, meaningful (Hassenzahl et al., 2013, p. 22). Many authors have sought to identify common travel goals. For example, Kock et al. (2018) consider that motives (goals) for travel have an evolutionary basis – evading harm, exploration, affiliation, status and mating. Similarly, self-determination theory (Deci & Ryan, 2000) discusses a number of goals such as 'being competent', 'being related to others', 'being special'. As discussed previously, cognitive appraisal theory does not rely on a particular goal but instead relies on appraisal of goal strength and goal importance. Subsequently, the emotional state of a tourist plays a central role in determining the summative evaluation of an experience reported such as satisfied or delighted. Emotions are subconscious but rational mental processes, with rational referring to goal directed.

Due to the central role of emotion, the design of a memorable experience should be based on an understanding of an individual tourists' goal(s) and their importance and relevance. Tourist goals are important for experience design as they determine (along with other appraisal dimensions) the valence and intensity of elicited emotions from the tourism experience. Often self-selection bias may mean that many tourists share the same goals and hence experience similar emotions. Once goals are determined and experiences designed accordingly, the favourable appraisal of experiences (positive valance) by tourists will motivate future action and behaviour.

Feelings

Often the terms feelings and emotions are used interchangeably, however can be distinguished. Emotions are subconscious mental events elicited by an appraisal of relevant stimuli. Feeling is used to describe diverse phenomena such as thoughts (I feel like I am being misunderstood), a bodily sensation (I feel happy) associated with an elevation in heart rate caused by an activation of the autonomic nervous system or a wish (I feel like taking a holiday). Bodily sensations can emerge as the result of a perception of external physical stimuli (touch, taste, smell and vision), while others are closely related to emotions and are the result of an 'emotional body state' being signalled back to the brain, not only to the limbic system but also to the somatosensory system (Damasio, 1995). An emotion of happiness may result in a smile and thereafter change the state of the muscles around the mouth, releasing neurotransmitters, and these changes are subsequently reported back to the conscious brain reinforcing a feeling of happiness. This is why bodily feedback or unconscious priming, such as asking a person to smile, can successfully induce emotion (Briñol & Petty, 2003). More generally, emotions and feelings have a recursive relation, each making the other more likely

(Han et al., 2007). This leads to the idea that feelings provide information to the brain as is discussed in feelings as information theory.

Experiences can influence consumers through more personally relevant and symbolic dimensions than products. However, consumers' evaluations of experiences, as opposed to products, are often more multidimensional, uncertain and complex (Gallo, 2013). Experiences are evaluated on a greater number of dimensions than products and because of uncertainty associated with how to integrate the numerous dimensions to form a judgement about the experience (Carter & Gilovich, 2010; Urbany et al., 1989). This leads to increased use of 'feelings' in choice and evaluation processes. Despite this, there is a limited discussion of feelings in the tourism literature (Buda et al., 2014), with most studies focuses instead on perception of external physical stimuli (touch, taste, smell and vision), rather than perceptions of conscious or unconscious emotional signals (Agapito et al., 2014). A recent contribution discusses how the secretion of neurotransmitters elicits measurable bodily changes governed by the autonomic nervous system, such as heartbeat, blood pressure and increased sweating (Moyle et al., 2019). Feelings as summative evaluations of tourism experiences would appear to be a useful area for further research. They are the result of emotions and are caused by neurotransmitters acting on the autonomic nervous system. Feelings provide information to the brain for use in further cognition.

CONCLUSION

This chapter provided a critical analysis of the efficacy of concepts from cognitive psychology in tourism experience design. Specifically, it focused on consciousness, prospection and retrospection, attention and schema, memory, emotion, feelings, which are concepts central to cognitive psychology. There are many applications of cognitive psychology in tourism experience design, but key concepts, such as in the appraisals of novelty and goal importance and congruence, are important in a number of mental processes like attention, emotional appraisal and memory processes. The proposed framework of tourism experience co-creation process from a cognitive psychology perspective (Figure 15.1) can assist practitioners to design memorable tourism experiences. By definition, tourism involves 'life' outside the usual environment and therefore novelty, surprise and unexpectedness are to be expected in tourism experiences related to this unusual environment. Further, tourism that is voluntary is goal directed. For these reasons, its experiences are likely to lead to memorable experiences. Prior research in tourism has noted the importance of emotion but has primarily described what emotions are correlated with what stimuli, rather than how stimuli, are appraised through the process of cognitive appraisal. This is a serious issue as the approach is often taken to mean that the stimuli are the cause of the emotion, rather than the consequence of cognitive processes derived from one's individual goals and prior experiences.

Opportunities exist for tourism researchers to further explore recall of emotional episodes in long-term memories (Pearce & Packer, 2013). Cognitive psychology can help research on the customer experience to move beyond identifying the components of memorable experiences (Larsen, 2007) to seeking an understanding as to the role of emotions and attention on experience memorability. How the manipulation of attention (through immersion, interactivity and visualisation techniques) commonly used in tourism design, affects appraisals of novelty, goal congruence, agency and the emotions and behaviour is an avenue for future research (Yung & Khoo-Lattimore, 2019).

Cognitive psychology can also help to understand why holidays are beneficial to health and well-being (Moyle et al., 2019). While travelling, people are away from the normal environment, separated from family, friends and colleagues, so are likely to think, feel and behave in ways different from their usual environment (Kottler, 2002). There are some indications that creativity (Dietrich, 2004) may be stimulated by the activation of the default network during relaxation, a characteristic typical of a holiday (de Bloom et al., 2014). Similarly, travel as a transformative experience is accepted, but the psychological processes supporting transformation and well-being are only beginning to be studied (Chen et al., 2017). There has been much discussion of the negative impacts of tourism on communities, but not so much research on the positive benefits of travel for individuals.

Several other important research topics may benefit from the use of cognitive psychological theory. For example, understanding promotion of destinations creates the expectations against which appraisals of novelty are made (Hosany, 2012; Hosany & Gilbert, 2010). Similarly, hotel design can affect relevant cues and novelty appraisals and thus activation of relevant schema (Ro et al., 2013). Imagination and expectations about a future holiday are based on prospection, which is subject to certain biases, such as prototypical thinking. Marketing is part of the experience design process. This is why narrative, immersion and mental imagery are used to persuade customers to visit a particular destination (Appel & Richter, 2010; Phillips & McQuarrie, 2010), and can enhance and shape experiences (Moscardo, 2010). The effect of desired future-oriented psychological states (hence imagination of a perfect holiday) on tourism requires further research as it sets the frame for designed experiences (Pearce, 2009), emotional episodes and long-term travel memories (Pearce & Packer, 2013).

The role of novelty in the creation of memorable tourism experiences warrants further attention. Understanding nuanced aspects of novelty is critical for experience design and has potential to assist practitioners to engage tourists at a deep and profound level (Bec et al., 2019). For instance, cutting-edge technologies, such as augmented or virtual reality used in immersive digital tourism experiences, themselves constitute a form of novelty (Bec et al., 2019; Yung & Khoo-Lattimore, 2019). They also tend to influence attention and mental elaboration. Utilising core concepts from cognitive psychology will enhance the capacity to explore the connection between novelty and emotions in tourism experience memorability formation (Skavronskaya, Scott, et al., 2020). By implementing novel, unexpected and surprising components while designing tourism experiences and marketing campaigns, tourism practitioners can increase emotional engagement with the destination (Michalkó et al., 2015), and generate tourist's satisfaction (Chen & Rahman, 2018; Kim, 2018).

Tourism and service experiences share many similar characteristics including intangibility and therefore the concepts discussed in this review apply equally to both. It is unlikely, for example, that the limitations of human attention would not apply to service processes, or that the cognitive appraisal process works differently in eliciting emotion in services. Cognitive psychology concepts have been applied in the study of conceptual changes in knowledge representations (Zhang & Soergel, 2016), mental time travel (Manisaligil & Haksoz, 2016), consumers' affective expectations (Wirtz et al., 2000) and emotions in services (Obeidat et al., 2017). Further application of these concepts – both connecting to and drawing from conceptually related fields – is recommended to advance research on tourism and service experiences and shift from a cognitive to a behavioural paradigm. There is need for further discussion about the application of these concepts in the arts (Ponsignon et al., 2017) and landscape design (Kara, 2013), as both the arts and tourism are contexts for experience design.

Chapter 16

EXPERIENCE CO-CREATION
Attention and Involvement

Juan Yang
Griffith University, Australia

Ana Cláudia Campos
University of Algarve, Portugal

Biqiang Liu
Griffith University, Australia

Brent Moyle
Griffith University, Australia

Anna Kralj
Griffith University, Australia

Truc H. Le
Griffith University, Australia

Abstract: Co-creation experiences are psychologically complex phenomena. This chapter discusses how cognitive psychology can add value to the co-creation of experiences. There are different meanings of co-creation and here the focus is on the customer's mental experience. This chapter analyses the theoretical underpinnings of co-creation and discusses key dimensions of the concept from a cognitive perspective,

Cognitive Psychology and Tourism
Tourism Social Science Series, Volume 27, 197–209

Published under exclusive licence by Emerald Publishing Limited
ISSN: 1571-5043/doi:10.1108/S1571-504320240000027016

highlighting the importance of attention and active involvement. Furthermore, it discusses how technologies (augmented reality, robotics, intelligence systems) can prompt experience co-creation. Finally, this chapter examines the literature of co-creation and experiential learning overlap in knowledge building. Opportunities for future empirical research in this area are suggested. **Keywords**: Co-creation; experiences; cognitive psychology; attention; involvement; experiential learning

> Tell me and I will forget; show me and I may remember; involve me and I will understand. – Confucius, Chinese philosopher.

INTRODUCTION

Tourists increasingly are interested in being a producer of, and actor in, their own experiences, instead of a passive sightseer (Binkhorst & Den Dekker, 2009, 2013). Co-creation occurs when a tourist is involved in the design and production of tourism experiences. When doing so, a tourist applies their own personal (cultural, intellectual, physical) resources (Prebensen et al., 2013) to add value to the consumption experience (Neuhofer et al., 2013). The concept of co-creation emphasises the critical role of tourists in creating experiences and therefore it is especially important to understand the goals, thoughts, and emotions of individuals.

There has been an exponential growth in literature on co-creation in tourism over the past decade (Font et al., 2021; McCartney & Chen, 2020). The concept has been applied in tourism contexts, such as festival settings (Chen et al., 2020), dining experiences (Ji et al., 2018), the sharing economy (Zhang et al., 2018), adventure tourism (McLeay et al., 2019), animal-based tourism (Campos et al., 2016), heritage settings (Ross & Saxena, 2019), gambling (Noronha et al., 2021), spas (Hjalager & Konu, 2011), deserts (Michopoulou et al., 2021), and skiing (Chekalina et al., 2014). Co-creation has been used as an approach to study the interactions between tourists and travel specialists (Michopoulou et al., 2021), tourists with online travel-related content generators (Jiménez-Barreto & Campo-Martínez, 2018), and relationships with non-human actors, such as a tree house, or a dolphin (Bertella et al., 2019; Campos et al., 2017b; Phi & Dredge, 2019).

Active participation during consumption experiences requires both physical and psychological involvement (Bertella, 2014; Hsieh et al., 2018; Minkiewicz et al., 2014). Co-creation has been noted to involve interactivity, involvement, sensual stimulation, and authenticity, and is commonly regarded to elicit enduring memories (Campos et al., 2017b; Stasiak, 2019). Psychological involvement during its activities has been connected to learning, as well as the process of gathering and exchanging information (Yi & Gong, 2013). The cognitive benefits of co-creation in tourism experiences are purported to include new knowledge, skills, attitudes, beliefs, and visions (Mele et al., 2021; Payne et al., 2008). Co-creation has also been demonstrated to connect to positive emotions, such as joy, satisfaction, and contentment (Ind et al., 2020; Prebensen & Xie, 2017; Rachão et al., 2020). Consequently, it is considered to provide a path to memorable experiences as well as an opportunity for learning and personal growth.

However, as discussed in previous chapters, using the concepts from cognitive psychology to examine experience design and co-creation requires 'translation' of the terms usually used in these fields of study. This is a process of disambiguation whereby relevant cognitive psychology theories and concepts are matched with those used in the tourism literature. In this process, those parts of the tourism discussion that are relevant must be sorted from those parts that fall outside the scope of cognitive psychology. However, it is argued here that cognitive psychology provides a cohesive framework for understanding a consumer's mental processes. Notably, at least some concepts from cognitive psychology will be applicable and useful in better understanding the mental processes underpinning co-creation of experiences.

DISAMBIGUATION OF CO-CREATION

The concept of co-creation has two different meanings. First, it is a management or marketing concept applied in studies on experience design as part of a process to increase customer value. From this perspective, the priority of managers is value creation from experiential consumption (Campos et al., 2018; Vargo & Lusch, 2004). From a product and firm-centric viewpoint, the sense of value and the process of value development are improved by customised user interactions. Consumers who are well-informed, connected, motivated, and engaged can co-create value with businesses. In this view, the firm-consumer relationship is the core of value development and extraction.

Consumers' readiness to become active participants in production and consumption activities has led companies and brands to develop their offerings through co-creation (Cherif & Miled, 2013). This is an emerging trend in business that is leading companies to develop and adopt innovative organisational practices meant to integrate the consumer in the process of constructing a rewarding consumption experience (Payne et al., 2008). From this business perspective, co-creation defines a new perspective of business management which views consumer participation as a source of competitive advantage spreading among companies in a growing diversity of industries and services, such as education, health, or banking (Prahalad & Ramaswamy, 2004).

For tourism businesses, co-creation is considered as a process of interrelated interactions and activities that connects the tourist and other actors and experiences are the context in which those interactions and activities occur (Bertella, 2014; Campos et al., 2018; Mathisen, 2013; Mehmetoglu & Engen, 2011;

Mossberg, 2007; O'Dell & Billing, 2005; Volo, 2009). This perspective on co-creation is that found in the management, consumer behaviour, and marketing research literature and underpins concepts of the experience economy, the performance turn, and S-D Logic (Arnould & Price, 1993; Holbrook & Hirschman, 1982; Pine & Gilmore, 2013; Vargo & Lusch, 2004, 2008). This approach dominates the tourism literature of experiences (Neuhofer et al., 2020). For example, it has been argued that tour operators should increase interactions with customers to stimulate their contribution to design, production, and consumption of holidays (Zouni & Kouremenos, 2008). However, it does not explain why customers consider co-creation valuable or what psychological mechanisms support experience co-creation.

A second perspective on co-creation focusses on tourism experiences as enacted in situ from the perspective of the customer (Prebensen & Foss, 2011; Prebensen et al., 2016). Embedded within, co-creation is discussed as a tourist's active participation and interaction with others during a consumption experience (Campos et al., 2018). This interpretation is related to current ideas of the performance turn (Perkins & Thorns, 2001), tourist agency (Kreziak & Frochot, 2011), and more recently to creative tourism (Richards, 2011, 2020). This perspective is closely related to psychology, and thus to a psychological explanation of why tourists choose to practice tourism and express themselves through active participation in experiences.

Drawing on this perspective, Campos et al. (2016) discussed a tourist's subjective experience as a set of psychological processes and events that take place during performance-based activities and interaction with other people within the experience environment. This psychological perspective conceptualises experience as involving cognitive processes, which are connected to different stages of the experience (Larsen, 2007). The literature of co-creation identifies possible sources of value that can be delivered to customers. Schmitt (1999b) introduced five dimensions of customer experience that are broadly related to cognitive psychology: sensory experiences, such as sight, touch, and smell; affective experiences, such as moods and feelings; thinking or conscious processes; physical experiences, behaviours, and lifestyles (which may be 'translated' as benefits from achieving goals); and social identity experiences resulting from relating to a reference group or culture, which may be 'translated' as achieving social bonding goals. Gentile et al. (2007) and Hwang and Seo (2016) identified a similar set of dimensions while other scholars have extended and modified them. In this chapter, co-creation of a tourism experience is considered as the sum of the psychological events a tourist goes through when contributing actively through physical and/or mental participation in the event and interacting with other tourists in an experience environment (Campos et al., 2018).

Co-Creation of Experiences in Cognitive Psychology

The experience co-creation process has emerged as a context for the application of existing cognitive psychology concepts and theories. Here some concepts will be identified that may be emphasised by the above definition and the dimensions previously identified. In particular, these are increased active attention (to and through the senses), physical participation (activation of senses and perceptions), involvement of the customer (attention and goal activation leading to emotion), and stimulating interactions with other people (goal congruence from social interaction leading to emotion).

Co-creation has been found to increase attention, expressions of emotion and the memorability of an event (Campos et al., 2017a,b). Tourism experiences that are both mentally and physically engaging lead to focussed attention, encoding, and recall, and thus memorability (Campos et al., 2016, 2017b). Therefore, facilitating active participation, interaction, and attention are considered paths to improve experience memorability. Active participation, either in physical or mental terms (Bertella, 2014; Minkiewicz et al., 2014; Prebensen & Foss, 2011), has been found to focus attention on the experience, leading to higher levels of memorability (Campos et al., 2016) with emotional benefits. Managers are urged to analyse co-creation mechanisms based on an understanding of cognitive processes rather than the outcome of co-creation, such as memories (Campos et al., 2018; Grissemann & Stokburger-Sauer, 2012; Richards & Marques, 2012).

The emotional outcomes from co-created experiences include joy and pleasure (Dekhili & Hallem, 2020; Pera, 2017; Rihova et al., 2018). However, cognitive appraisal theory indicates that an emotion of joy is derived from an event congruent with an important goal and which is novel and emotionally arousing (Chapter 8). On the other hand, pleasure is a mild positive emotional state. Now, it may be expected that because tourists mostly voluntarily choses their experiences then they are very likely to be experienced as pleasurable, unless there is an unexpected service failure, or the experience is undesirable. However, those tourists who experience joy would be expected to have greater expectations of achieving an important goal and/or have not had the experience before (Hosany, 2012; Ma et al., 2013).

Campos et al. (2016) found that active participation in experiences and interactions with others are significant contributors to enhanced attention. Their research found that tourism experience outcomes were mediated by attention. In experiential contexts, such as co-creation environments, attention can be directed to guides, instructors, or trainers of different sorts who all play a role in facilitating involvement (Mossberg, 2008). Increased attention to the context is therefore an effect of co-creation which can increase memorability. In turn, memorability is perceived as an outcome of co-creation experiences (Bertella, 2014; Campos et al., 2017b). The memorability of the experience is also linked to meaningfulness (Minkiewicz et al., 2014), which may be interpreted as another way of saying that the experience was positively goal congruent. Thus, memorability is enhanced in once-in-a-lifetime experiences (Tung & Ritchie, 2011) which stimulates highly aroused emotions.

However, a person may not begin an event expecting that it is going to be meaningful or novel. However, in these cases, co-creation leads to activation of goals and hence increases goal relevance and importance of an activity. As already discussed, increased goal importance leads to arousal and hence stronger emotion. Experiences which allow tourists to *do* things rather than just *look at them* (Azevedo, 2010; Campos et al., 2018; Eraqi, 2011) and engage in activities for self-development, explore multisensory environments, and connect to other people are thus more likely to elicit more intense emotions (Binkhorst & Den Dekker, 2009; Ihamäki, 2012). Self-development and connecting to people may be goals for some visitors. Also, when tourists are directly involved in creating and choreographing their activities from moment to moment, then their interest may be activated (Bertella, 2014; Binkhorst & Den Dekker, 2009; Kreziak & Frochot, 2011; Prebensen & Foss, 2011). Involving tourists in activities that meet their interests and capture their attention is critical for co-creation (Andrades & Dimanche, 2018). The natural environment has been found to be a space where staging of exploration, play and knowledge may activate tourists' involvement with onsite experiences (Mathisen, 2013). Kyle and Chick (2004) suggest that such collaborative involvement is linked to social relevance and meaningfulness of others which in turn creates bonding, attachment, and memorability.

As an example of goal activation through co-creation leading to emotion and memorability consider the following example. An experienced tourist is watching a re-enactment of a traditional marriage ceremony along with a large group of others. The tourist has seen similar traditional cultural practices being demonstrated several times before and is not interested in such practices. However, the tourist is surprised by being picked out of the watching group to participate in the marriage ceremony. The male tourist is dressed in traditional clothes, then 'married' to a beautiful young woman. The ceremony ends with the couple then carried off in a traditional sedan chair for their honeymoon. Such a co-created experience produces strong emotion and vivid memories. These are very different from the 'normal' emotions derived from simply watching such a cultural re-enactment.

Third, stimulation of the senses and associated mental processing of these perceptions may lead to capture of attention (Ooi, 2005). The experience economy paradigm and the experiential marketing approach have already called attention to the importance of the senses in consumer experience (Campos et al., 2016). Agapito et al. (2013) have reviewed the literature on the senses, discussing the relevance of the theme in the context of tourism experiences and highlighting the relevance of sensescapes. This multi-sensoriality is thought to leave a permanent imprint on memory. Ballantyne, Packer, and Sutherland reported vivid memories based on multi-sensory impressions and that 'to be physically close to the animals (. . .) made the experience novel or remarkable' (2011, p. 773). The same is found in the *Dolphin Emotions Experience* (Campos et al., 2016). Nevertheless, further investigation of the effect of multi-sensorial stimuli on attentional behaviour or memorability of co-creation tourism experiences need not be animal- or nature-based. Many contemporary creative tourism activities require the tourist's use of their body and physical engagement (Richards, 2020). Artistic performances (dance, music), gastronomy or crafts, are experiential contexts that involve active participation, senses, attention, and memorability.

In addition to the effect of attention capture, sensorial stimulation can also produce attention diversion and distraction. The selective attentional processing determines individuals can only focus on one thing at any given moment, and multitasking could generate attention diversion and distraction, even lead to negative psychological outcomes (mental stress and self-presentation concerns) and subsequent negative evaluation of tourism experiences (Liu et al., 2022). Li et al. (2022) found that there is a 'vampire effect' at play in celebrity endorsement, which means celebrities could overshadow the advertised destination scenery by competing for tourists' limited attention. Ooi (2005) presents the basis for an attention structure framework and explain how competition for tourists' attention can arise, highlighting four interrelated characteristics of attention structures: the ability to pay attention to only one thing at any given moment; behaviour and experience are affected when one's attention shifts; human attention is its scarcity; there are many different reasons why attention is paid to one thing. This seminal framework considered that there are complementary internal-push (person's motivation) and external-pull factors (environment and surroundings) in the competition for attention.

Fourth, co-creation has been recognised as a means of using personal resources (knowledge, time, and creativity) in contextual, personalised, and interactive experiences (Binkhorst & Den Dekker, 2009; Tajzadeh-Namin, 2012). Some researchers, such as Michaelidou and Dibb (2008), have distinguished between enduring and situational involvement connecting the first with stable motivations and personal relevance towards objects or activities rooted in attitude structures and the second with personal relevance felt towards transitory, contextual stimuli. In this regard, the tourist's own resources and behaviours (skills and knowledge, physical, social, and cultural resources) act as influencers of tourists' involvement in the conception and design of their experiences. Mathisen (2013) found tourists with more interest in nature had more pleasurable experiences in a natural setting favourable to active participation. In Campos et al.'s study (2017a), interest in nature was raised through exploration, creating a stage to play and enact roles, and providing new knowledge through these activities.

The difference between levels of attentional effort has been highlighted by distinguishing focus (the action of directing the mind and senses to a particular stimulus and singling it out, ignoring by that reason competitor stimuli) from concentration, characterised by sustained executive control (Campos et al., 2017b; Posner et al., 2004). Focussed attention occurs for no more than a few seconds and could involve shallow cognitive processing, however it has also been found highly effective in the stimuli selection process (Liu & Li, 2021; Yantis & Johnston, 1990). For example, individuals' tendency to look first at the central option has been revealed as an initial central fixation bias. Compounding this, further investigation of the attention process found a central gaze cascade effect which means individuals progressively pay visual attention to the central option right prior to decision (Atalay et al., 2012). However, only the central gaze cascade effect was related to choice (Atalay et al., 2012), which means initial attention may indicate the fast stimuli selection process and involve shallow cognitive processing. Changes in attentional effort influence action and choice, and consequent outcomes. For instance, individuals actively engaged in activities are more attentive to the experience environment because active engagement requires strategic allocation of attention (Liu et al., 2022), and also demonstrate better recall (Campos et al., 2017b). Similarly, research in learning contexts has found that environments which stimulate active participation and cooperative interaction among individuals lead to better performance in terms of maintaining interest and attention (Cavanagh, 2011).

Co-creation may involve goal congruence from social interaction. This may be through interactions and activities that connect the tourist and other actors before travel (before consumption) (Andrades & Dimanche, 2018; Binkhorst & Den Dekker, 2009; Prebensen et al., 2013), during stay at the destination (during consumption) (Campos et al., 2018), and/or after the travel (after consumption) (Binkhorst & Den Dekker, 2009; Neuhofer et al., 2013). Goal congruence focusses more on the fit/match-up between an expected goal and the situation (Liu et al., 2022). Sheng et al. (2022) study proposes a framework of how goal congruence plays a pivotal role in the resource integration process between all actors in the joint sphere. Specifically, when the level of goal congruence between a customer and a service provider is high, it stimulates the exchange of valuable resources from all actors, which leads to the actual experience of co-creation resulting in value-in-use and improves service outcomes. In the same vein, a visit to an attraction might provide a setting that increases social bonding through interaction with family members (Campos et al., 2018).

Cognitive Technologies and Experience Co-Creation

One stream of literature explores the use of technology to co-create certain experiences. Recent calls suggest the need to investigate new forms of technology-based experience co-creation and provide insights for more valuable, contextual types of human-to-machine interactions (Mele et al., 2021). The Internet of Things and related technologies, such as augmented reality, robotics, and intelligence systems, are developing rapidly and increasingly creating synergies, further contributing to technological acceleration. A key consideration is the role of different technologies and how persuasive cognitive technologies might support experience co-creation. In their search to improve their viability, actors act together with technologies to develop new understanding and practices (Weinmann et al., 2016). To specify how cognitive technologies create predisposing factors and choice architectures that alter human decisionmaking, and thereby increase experience co-creation and prompt a re-institutionalising process, Mele et al. (2021) propose a framework of the role of cognitive technologies in actors' experience co-creation efforts (Figure 16.1 adapted from Mele et al., 2021). This framework offers a deeper conceptualisation of co-creation experience processes, delineating how cognitive technologies can enact resources in context and guide individual decisions. By expanding and unbinding the range of actors' possible choices, cognitive technologies contribute to activating actors' agency, thus boosting co-creation experiences (Mele et al., 2021).

People make decisions and take actions every day to improve their viability, and they increasingly turn to cognitive technologies (artificial intelligence) to assist with their decisionmaking (Tussyadiah, 2020). Cognitive technologies rely on computational components that deliver cognition as a service (which implies language, learning, and levels) and that empower and scale human expertise (Mele et al., 2021). These technologies augment human intelligence and capabilities across the spectra of sensory perception, deduction, reasoning, learning, and knowledge (Kelly, 2015). Such trends suggest the need to determine how

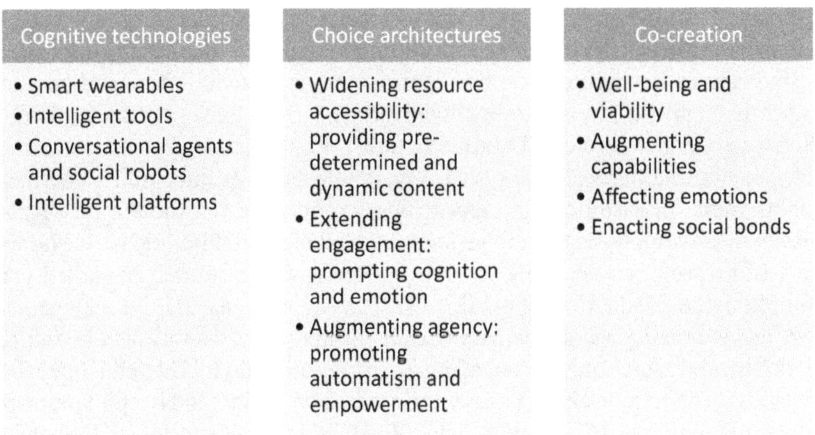

Figure 16.1. Cognitive Technologies, Choice Architectures, and Co-Creation.

Source: Adapted from Mele et al. (2021).

cognitive technologies affect experience co-creation. Several choice architectures driven by digital choice environments affect experience co-creation, by \ widening resource accessibility; extending engagement; or augmenting human actors' agency. Although cognitive technologies are unlikely to engender smart outcomes alone, they enable designs of conditions and contexts that promote smart behaviours, by amplifying capacities for self-understanding, control, and action (Mele et al., 2021).

Experiential Learning

Experiential learning involves 'hands-on' activities, out-of-classroom educational experiences, and experimentation (Breunig, 2017). The concept of co-creation overlaps with experiential learning. 'Tell me and I will forget; show me and I may remember; involve me and I will understand' (Chinese philosopher Confucius). This saying indicates the critical role of experiential learning in individual development. It means knowledge is more effectively acquired through active involvement rather than by simply being taught an abstract definition of a term, and better retained through active experiences. The process of knowledge building by accumulating experience is termed experiential learning (Antón et al., 2018b). The learning cycle consists of four steps, namely concrete experience, reflective observation, abstract conceptualisation, and active experimentation (Tanaka et al., 2016). The increasing popularity of experiential learning in management education raises several new opportunities and challenges for instructors, particularly with regard to shifting instructor roles and attention to expression of emotions during learning (Kisfalvi & Oliver, 2015). Pee (2020) proposes that online solution co-creation and solution sharing enhance cognitive learning (perceived knowledge about a subject topic), while decision co-creation enhances epistemic learning (perceived understanding of criteria of knowing).

Experience co-creation has been considered as a combination of body and mind through the physical work involved in exploration, play and role-play during experience activities (Mathisen, 2013). For example, tourists performed roles of hunting heroes or dog-sled drivers in a dog-sled race and a Northern Lights hunt. In a same vein, geocaching, comprising the dimension of play intertwined with physical exercise, studying and tactical thinking, is an example of a transformative 'learning by doing' experience embedded in skilled consumption and production (Ihamäki, 2012). As a result, such active engagement and involvement generates self-development and gains new skills (Tan et al., 2014). In adventure tourism, engagement in play and playful role-enactment can be seen in Arctic trekking (Gyimóthy & Mykletun, 2004), philanthropic adventure activities (Coghlan & Filo, 2013), or hiking (Nordbø & Prebensen, 2015). Such physical activities performed in adventure tourism expose the interconnectedness between the physical and mental dimensions of experiences

(Campos et al., 2017b). In the context of heritage tourism, the tourists' experiential learning lies in co-production, engagement, and personalisation in performance of heritage activities through allowing tailoring of the experience to personal needs (Minkiewicz et al., 2014).

According to Kolb's experiential learning theory which describes 'knowledge is created through the transformation of experience' (Kolb, 1984, p. 38), more recent research in tourism considering the experiential learning approach (Čavlek, 2015). It has been regarded as one of useful alternatives to classroom-based teaching for students of tourism who need the opportunity to visit destinations and be involved in co-creation experiences (Arcodia et al., 2020). By conducting a longitudinal approach, Arcodia et al. (2020) compared data collected at two points in time (immediately at the end of the experiential learning and two months afterwards), finding that three factors emerged as important for meeting expectations, social and professional connections, learning, and traditional yet engaging teaching.

In line with Mele et al.'s (2021) framework focussing on the role of cognitive technologies in actors' experience co-creation efforts, Moorhouse et al. (2017) argue augmented reality applications have the potential to engage students to develop skills for the future and enhance the overall learning experience. In the same vein, Bos et al. (2015) agree holidays can provide a context for experiential learning, and social tourism plays an important role in contextualising classroom-based teaching through relearning. Subsequently, its critical role in facilitating links between theory and practice drives the call for further research on the links between tourism and education.

From a managerial point of view, these findings may stimulate organisations to design experiences that involve learning opportunities in a leisure-based context so that tourists may feel they are acquiring new knowledge and developing new skills in a friendly, non-compromising manner. In fact, the results suggest managers need to help in creating the right blend of educational and entertainment experiences (Hertzman et al., 2008).

CONCLUSION

Previous tourism studies considered that onsite co-creation tourism experiences involve parties connected in different ways (emotional, physical, social) in proximity and intensively. However, the literature has paid limited attention to the issue underlying tourists cognitive processing (Campos et al., 2018). This chapter, adopting a cognitive psychological approach, indicates that it is of great importance to recognise subjective cognitive processes underlying co-creation experiences. This involves discussing the interplay between co-creation and other cognitive concepts, such as involvement and attention.

The psychological perspective on the co-creation experience adopted in this chapter examines internal, subjective phenomena which need to be understood and connected to active participation and interaction if experience management and marketing are to be effective and competitive. The discussion highlights the need to concentrate reflection and empirical investigation on the onsite co-creation experience, where strong emotions, learning, and meaningful memories emerge. Furthermore, there is a need for clarification of the definition of key constructs. This chapter focusses on how cognitive technologies can enable one to make better decisions and build experience co-creation. Finally, this chapter examined how co-creation and experiential learning overlap in knowledge building through accumulating experience.

An examination of the literature findings reveals opportunities for future research, especially empirical studies about onsite co-creation experiences and the psychological processes involved (Campos et al., 2016, 2018). More particularly, research is required on one; identification and description of psychological processes (perceptual, cognitive, and affective) involved in co-creation tourism experiences that may influence memorability; two, measurement of the influence of active participation and interaction, as dimensions of co-creation, on focussed attention and other psychological processes occurring during onsite experiences; three, analysis of the experience mediators' role and the extent to which they are important as attention leaders in co-creation experiences; four, examining if and how co-creation experiences contribute to a better management of top-down and bottom-up attentional processes during the consumption experience as a way to better reconcile the experience environment with the tourist's needs and goals; five, investigation of the impact of the sensory dimension of co-creation experiences on memorability, and also exploration of the multiple senses as attention capturers and maintainers during these experiences; and, six, examination of the degree of spontaneity and unexpectedness in co-creation experiences and relevance as memorability enhancers; seven, description and discussion of participants' interactions and social spheres in onsite co-creation experiences (Campos et al., 2016).

In addition, many scholars (Mele et al., 2021; Verleye, 2015) suggest that future research should investigate the impact of technologisation and connectivity on the co-creation experience in different situations. Cognitive technologies are shaping service ecosystems and enabling new practices, so a broader focus should be assigned to actors' decisionmaking, taking into consideration the larger service ecosystems. It is of paramount importance to analyse how human and non-human actors interact and make decisions within shaping digital ecosystems. Actors' decisions, intertwined with cognitive technologies, determine such shaping, through consequent actions that can reform and renew practices, as well as by following such practices and making them institutionalised. Finally, further research might benefit from generating a better understanding of co-creation preferences of customers in various co-creation situations, since their expectations in terms of co-creation benefits might vary across co-creation situations. Future research should also pay attention to the stability of customers expected co-creation bernefits over time.

Chapter 17

INSPIRING AWE IN IMMERSIVE NATURE TOURISM EXPERIENCES

Vikki Schaffer
University of the Sunshine Coast, Australia

Lee Kannis-Dymand
University of the Sunshine Coast, Australia

Abstract: Awe is a multidimensional emotional experience that has the potential to adjust one's beliefs and schema. Vastness is the core cognitive appraisal required to elicit awe. An appraisal of vastness can challenge an individual's present frame of reference. The cognitive process, need for accommodation, can be activated by the appraisal of vastness. Both an appraisal of vastness and the need for cognitive accommodation are essential to experiencing awe. This chapter presents an overview of the rapidly emerging research on awe through a cognitive appraisal lens. It highlights the importance of immersive experiences and nature contexts within tourism, illustrated by examining immersive, marine tourism activities on the Sunshine Coast, Australia. The case study discusses tourists' experiences of engaging scuba diving, swimming with whales, whale watching, and the elicitation and experience of awe and associated cognition. **Keywords**: Awe; vastness; need for accommodation; nature; pro-environmental; wellbeing

INTRODUCTION

Modern society is becoming separated from nature with various sectors purporting the importance for reconnecting with the natural world. The biophilia hypothesis (Wilson, 1984) suggests that humans have an innate

Cognitive Psychology and Tourism
Tourism Social Science Series, Volume 27, 211–226

Published under exclusive licence by Emerald Publishing Limited
ISSN: 1571-5043/doi:10.1108/S1571-504320240000027017

desire to connect with nature. The natural places and spaces, flora, and fauna are essential components of tourism experiences. The benefits of nature-based tourism experiences include positive emotions and improved mental health (Buckley, 2020; Kotera et al., 2021; Lackey et al., 2021), place attachment, transformation (Meidenbauer et al., 2020; Wolf et al., 2015), personal challenge, learning, skill development, involvement, and interaction (Qiu et al., 2021). In addition to the potential environmental benefits, Buckley and Cooper (2022) emphasise the social and economic opportunities and benefits of nature-based tourism for mental health.

Increasingly, there is an emphasis on linking nature-based experiences and the importance of sustainability (Hanna et al., 2019; Ives et al., 2018). Tourism researchers and industry practitioners are increasingly aware of the importance of operationalising sustainability throughout business and product offerings. Cave and Dredge highlight the growing concerns and need for action to address 'climate change, overtourism, declining employment and labour conditions and resource degradation' and the need 'for tourism to move beyond "business as usual"' towards not just sustainable, but regenerative tourism (2020, p. 504).

Regenerative tourism reiterates that human health is connected to, and reliant upon, the health of all natural systems which requires one to learn from the nature (Becken & Kaur, 2021). This links with the United Nations sustainable development goals that aim for tourism as a driving force towards economic growth, inclusive development, and environmental sustainability (UNWTO, 2022). Designing meaningful experiences that facilitate engagement with nature can influence pro-environmental, sustainability-related attitudes, and behaviours by connecting tourists with nature (Clark et al., 2019). This chapter draws on the literature of sustainable, regenerative tourism, and psychology to discuss immersive experiences in nature and the connection with a specific emotional response, awe.

IMMERSIVE TOURISM EXPERIENCE

In the late 1990s, tourism researchers began to actively investigate experiences in tourism and leisure as 'subjective mental states' (Otto & Ritchie, 1996, p. 166). Blumenthal emphasises that 'immersion is a part of the total visitor experience and is hence a subjective phenomenon experienced inside the mind of the individual (2020, p. 5)'. Research into the holistic nature of experiences emphasised the importance of understanding 'the customer's cognitive, affective, emotional, social and physical responses' (Verhoef et al., 2009, p. 32). Immersion has been described as feeling completely present, undistracted by thoughts (Weinstein et al., 2009), total and deep involvement (Hansen & Mossberg, 2017), or a process of becoming one with, or part of, the experience (Carù & Cova, 2007). A sense or feeling of being immersed is subjective and generally refers to spatio-temporal belonging, characterised by deep engagement in the present moment (Hansen & Mossberg, 2017).

Exposure to real or simulated natural environments can enhance or restore one's attention and increase positive emotions (including awe), reflection, and connectedness to nature (Mayer et al., 2009). Immersion in nature can influence an individual's environmental behaviours (Ahn et al., 2014) to inspire and initiate transformation in intent, attitudes, and behaviours. Tourists are increasingly seeking immersive experiences that foster personal meaning, self-development, and growth (Kirillova et al., 2017a; Knobloch et al., 2017; Smith & Diekmann, 2017). This is leading to an increase in transformative experiences and eudaemonia in tourism research (Smith & Diekmann, 2017). Immersive experiences can produce higher emotional intensity and higher levels of mastery, existential authenticity, human flourishing, self-development, self-fulfilment, and happiness (Løvoll, 2019; Seeler et al., 2021). Immersion may produce a fuller experience, eliciting a stronger reaction to natural or non-natural stimuli (Weinstein et al., 2009) to influence memorability, and increase inherent motivation and enjoyment (Makransky et al., 2019).

Drawing on the criteria for ecotourism, *eco-immersion* considers direct participation that initiates higher levels of mental and/or physical involvement for multifaceted engagement of tourists: physically, emotionally, socially, and/or intellectually, that create meaningful and respectful connectivity with the natural environment and communities within. Natural environments may elicit emotional responses, such as awe, that engender a sense of meaning, place attachment, connection, or purposefulness (Hidalgo & Hernandez, 2001; Williams & Vaske, 2003). Cognitive processes and their pivotal influence on the tourist experience is emerging and requires greater examination.

Cognitive Appraisal Theory

Cognitive appraisal theory is a theory of emotion which emphasises the role of the individual's interpretation (appraisal) in the emotion elicitation processes and explains how emotions are activated and differentiated based on the significance of a situation for each individual (Ellsworth & Scherer, 2003; Frijda, 1986; Lazarus, 1991; Roseman, 2001; Roseman et al., 1990; Watson & Spence, 2007). The theory is increasingly being employed to study emotions in tourism (Chapter 8). Notably, a number of appraisal dimensions have been identified: goal relevance or importance, goal congruence (also called motive consistency), certainty, coping potential (also called control and/or power), agency (also called cause) (Moors et al., 2013), surprise or uncertainty (in relationship to expected experience) (Ma et al., 2013), novelty (Skavronskaya et al., 2020), and norm/self-compatibility (Scherer, 1984).

Cognitive appraisal theory posits that each emotion has a unique appraisal structure. For example, the appraisal structure of anger in a social setting involves appraising an event as relevant to a goal; appraising the event as incongruent with the goal; judging a threat to one's social- or self-esteem; and blaming someone for the threat (Scherer, 2001). In contrast, sadness can be elicited when as an environmental feature has been destroyed and can be interpreted through appraisal as something that was desired, that has now been lost (goal incongruity) and cannot be recovered (Roseman & Smith, 2001). In the context of emotions with positive valence, an appraisal of high goal congruence and importance, and novelty are found to elicit the emotion of delight (Ellsworth & Smith, 1988a). The variation among individuals' subconscious assessment of an experience on the appraisal dimensions (due to different goals, etc.) explicate upon the differences between individuals and the diversity of emotions resulting from exposure to the same stimulus (Ma et al., 2017).

Awe

Emotional experiences are among the most important benefits realised by those engaging in the natural environment (Ulrich, 1983). Psychologists suggest emotions such as awe, surprise, anxiety, and thrill, may result from external stimuli associated with immersion in, or interaction with, natural (dark skies) or built environments (artificial reefs) and encounters with flora and fauna (whales). Emotions may result from cognitive processes undertaken to appraise an external stimulus or assess other internal triggers such as personal memories (from childhood, previous trips, or experiences) (Lerner & Keltner, 2000). Emotions are affective states involving complex psychophysiological constructs linked to episodic feelings associated with referent stimuli that prompt responsive behaviours (Cohen & Areni, 1991). Some emotions are simple and others more complex such as awe.

Awe is an aroused state of cognition whereby tourists marvel or wonder at, for example, the magnificence, enormity, uniqueness, beauty, or 'awesomeness' of their surroundings. Awe may lead to behavioural and attitudinal responses that challenge one's perception of the world (Shiota et al., 2007). Otto described awe as 'a harmony of contrasts', that 'humbles and at the same time exalts us, circumscribes, and extends us beyond ourselves' (1958, p. 41). Research asserts that experiencing awe is sensually and emotionally absorbing, spiritually uplifting, and allows time out from daily routines to enable contemplation and reflection (Hill et al., 2014). Awe enhances wellbeing, fosters prosocial behaviours, increases social and environmental connectedness, and is an important emotion in sustainable tourism (Lu et al., 2017). Given these benefits at individual, community, and environmental levels, awe is a valuable human emotion and instrumental experience that likely contributes to human and planetary health and well-being. Studies suggest tourism-related experiences that evoke awe may influence and encourage environmentally responsible behaviour (Wang & Lyu, 2019); helpful for promoting sustainable tourist destinations (Zhou et al., 2020).

Research examining the multifaceted nature of awe have determined two core and essential characteristics, an appraisal of vastness and a cognitive process termed as a need for accommodation (Dong & Ni, 2020; Keltner & Haidt, 2003; Piff et al., 2015; Stancato & Keltner, 2021; Zhao et al., 2019). While vastness and the need for accommodation are often categorised as appraisals within the awe literature, the need for accommodation is arguably a cognitive process. That is, the individual cognitive processes of an experience that involves accommodation into a person's existing schema. Schema are theorised as knowledge representations, typically developed in early life, comprising of explicit beliefs accessible to consciousness (one's awareness) and implicit knowledge, as well as behavioural-procedural and emotional information (Jacob & Arntz, 2013; Young et al., 2003).

Schema are enduring, broad patterns of involving information processing, thoughts, beliefs, emotions, memories (past experiences), and attention tendencies (Jacob & Arntz, 2013; Young et al., 2003); they inform or underpin an individual's responses. Potentially, the need for accommodation is comparable to the cognitive process that transpires in the cognitive model of posttraumatic stress disorder (Ehlers & Clark, 2000), wherein people struggle to cognitively integrate a traumatic experience into their existing knowledge base. Occasionally awe may have negative valence associated with threat (Taylor & Uchida, 2019); however, it is evident in posttraumatic stress disorder that the appraisal of threat is dominant and leads to anxiety rather than awe. This emphasises the essential role of appraisal and the need for cognitive accommodation in overwhelming emotional experiences (awe and traumatic distress).

Vastness can refer to size (large, immense) or to something powerful or overwhelming and perceived to be larger than life (Bonner & Friedman, 2011; Keltner & Haidt, 2003; Shiota et al., 2007) that challenges one's frame of reference (Keltner & Haidt, 2003; Shiota et al., 2007). Vastness can be elicited when encountering something beyond the current frame of reference, and therefore, requires perceptual accommodation or mental adjustment to 'take it all in' (Keltner & Haidt, 2003). When an individual's mental frameworks (past experiences, values, beliefs) are surpassed or challenged by unfamiliar or novel experiences, cognitive modification is required to 'accommodate' the new experience and all it entails (Keltner & Haidt, 2003; Guan et al., 2019; Joye & Dewitte, 2016). The experience of awe, including the activation of the need for accommodation, can temporarily or permanently influence our beliefs and knowledge about ourselves, others, the world, and one's spiritual views (Chirico et al., 2016). Thus, awe can have a transformative effect due to the potential to alter a person's views, beliefs, schema, and identity (Kitson et al., 2019).

Scholars have argued that both characteristics of awe, vastness, and the need for accommodation, are needed to evoke awe (Ke & Yoon, 2020; Keltner & Haidt, 2003). Keltner and Haidt point out that 'surprise' involves a need for accommodation but not vastness, whereas 'deference' involves vastness but not the need for accommodation (2003, p. 304). Appraisal theorists have noted that while most emotions are either pleasant or unpleasant in their composition, a limited set of emotions can have divergent appraisals, such with the emotion of awe (Briñol et al., 2018). Awe's unique, multifaceted nature may be influenced by several facets that can affect its valence and character (Keltner & Haidt, 2003; Yaden et al., 2019). For example, appraisals of threat, perceptions of beauty, or self-diminishment can affect awe's tone and valance (Yaden et al., 2019). That is, threat may result in awe with a negative valance, while beauty may be associated with positive valance. Self-diminishment can include an awareness of one's physical size but also a subjective sense of personal goals and sense of being (Piff et al., 2015; Yaden et al., 2019). Regarding cognitive appraisal theory and its embedded appraisal and reappraisal processes (accommodating the 'awe' experience), awe cultivates evaluation and modification of an individual's perspectives on life, goals, and purpose (Bonner & Friedman, 2011; Chirico, 2020; Schneider, 2009; Silvia, 2010).

As highlighted, awe is described as ambiguous (Lazarus, 1991, p. 83); a composite emotion with variable valence. That is, awe can elicit positive and negative responses and is considered 'foundational' to 'human experience that defines the human existence' (Schneider, 2011, p. 249). In addition to awe's impact on beliefs, schema, and wellbeing, awe's effect on people has been conceptualised as an expanding process 'from a physiological or neurological level, through to a psychological one, to an existential one' (Chirico & Gaggioli, 2021, p. 2) and may be related to a range of immediate or ongoing behavioural intentions or responses (Chirico & Gaggioli, 2021; Schaffer & Kannis-Dymand, 2022) (Figure 17.1).

Awe can stimulate nature–human relationships, spiritual connection, transformation, goal clarification, and a sense of humility (Powell, et al., 2012). Coghlan et al. (2012) offered three components relating to awe: physiological response (feeling goosebumps; feeling joy or fear); comparative uniqueness (unforgettable, amazing, novelty, memorable); and schema-changing for the future (inspirational, challenging, change making). Nonetheless, 'despite its potentially transformative and schema-changing effects' (Coghlan et al., 2012, p. 1711), awe 'is a little-studied emotion with a great transformative potential' (Chirico et al., 2018, p. 1; Pearce et al., 2017; Schneider, 2011; Silvia, 2010).

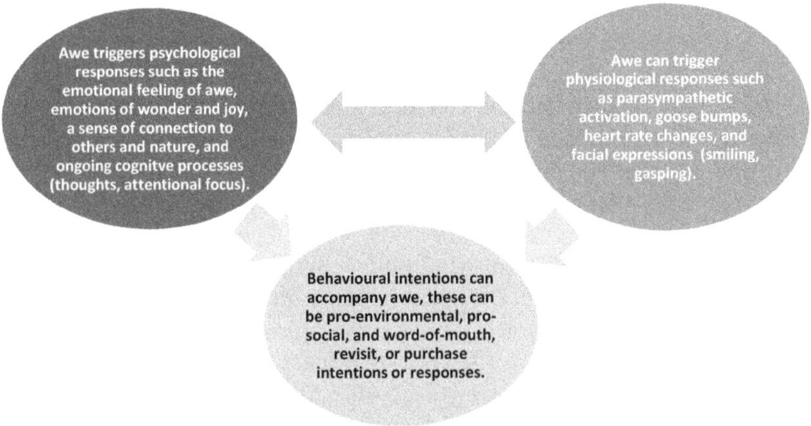

Figure 17.1. Effect of Awe.

Case Study

Real, immersive tourism experiences can generate long-lasting and significant memories (Ballantyne et al., 2011; Falk et al., 2012). Immersive experiences that elicit awe can facilitate a sense of connection that potentially influences intent, attitudes, and behaviours (Coghlan et al., 2012; Clark et al., 2019; Shiota et al., 2007). Three immersive marine tourism activities were investigated: scuba diving, swimming with whales, and watching humpback whales. These activities offer the potential for physical and psychological immersion. A deeper understanding of the tourists participating in these activities assists in designing experiences relevant to the consumer and that align with sustainable tourism management (Orams, 2000). Taking a trans-disciplinary approach drawing upon sustainable tourism management, environmental psychology, and cognition and emotion to investigate emotion, particularly awe, was undertaken in the context of marine tourism. Data was collected using online surveys from consenting participants aged 18 years or over, completing a commercial marine tourism activity in 2021.

Scuba Diving. The former Royal Australian Navy warship, Her Majesty's Australian Ship Brisbane, was scuttled 2.9 nautical miles off the shores of the Sunshine Coast, Australia, in 2005. The artificial reef that has formed on the wreck supports a growing reef ecosystem. The Ex-Her Majesty's Australian Ship Brisbane Conservation Park is considered one of Australia's premier wreck-dive destinations and is a restricted access area, with entry requiring a permit or written approval (https://parks.des.qld.gov.au/parks/ex-hmas-brisbane). Early estimates found that revenue derived from divers visiting the

Conservation Park was approximately US$3.2 million (AU$4.5 million) per year (Schaffer & Lawley, 2010).

The sample comprised qualified scuba divers (n = 84) visiting the Conservation Park with a commercial dive operator. Most divers were male (55%), aged between 25 to 34 (29%). Almost one third (32%) had completed a post graduate degree. Three-quarters had completed at least one previous scuba diving experience (n = 76; 90%). The main motivation for taking this scuba diving trip was for leisure (n = 39; 46%). The desire to pursue a personal interest (n = 8; 10%), to spend time with family/friends (n = 6; 7%), to connect with nature (n = 6; 7%), and to learn more (n = 5; 6%) were other identified trip motivators.

When asked immediately after the experience, participants indicated the scuba diving activity was engaging (av. 6.41/7; *StD* 0.89), enjoyable (av. 6.17/7; *StD* 0.97), and that it was exciting to be immersed in this environment (av. 5.83; *StD* 1.29). In addition, the activity stimulated reflection on new ideas (av. 6.18/7; *StD* 0.95) and resulted in some participants discussing what they had seen with companions/friends (av. 5.8/7; *StD* 1.24). Scuba diving at the ex-Her Majesty's Australian Ship Brisbane Conservation Park occasioned divers to experience something surprising or unexpected (av. 5.78; *StD* 1.05) and facilitated an emotional connection with the environment (av. 5.17/7; *StD* 1.19).

Scuba divers most often felt positive emotions including awe, wonderment, happiness, and enjoyment (Figure 17.2, higher frequency = larger text). Negative emotions were also expressed including worry and fear (n = 14).

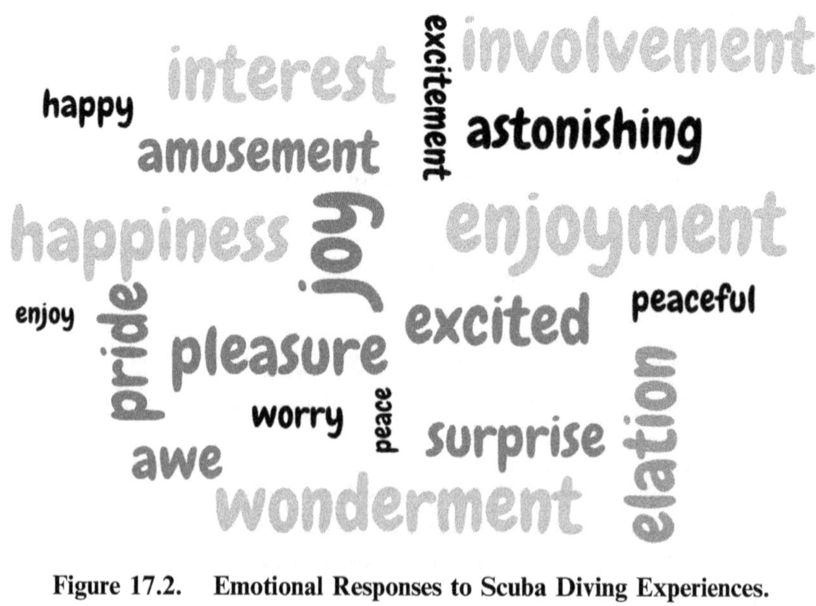

Figure 17.2. Emotional Responses to Scuba Diving Experiences.

Participating scuba divers indicated the strength of each emotion using a 5-point scale (1 = weakly felt emotion and 5 = intensely felt emotion). All positive emotions were moderately, strongly, or intensely felt (rated 3 or above). While less frequently stated negative emotions varied in strength, with worry and fear being strongly to intensely felt (rated 4 or 5). According to participating scuba divers, the diving itself stimulated enjoyment, concentration, involvement, and offered an opportunity to overcome fear.

Whale-Related Tourism. Southern hemisphere humpback whales (*Megaptera novaeangliae*) undertake northwards annual migration from Antarctic waters to northern Queensland, between April and November. Whale-related tourism has grown into a global marine tourism industry with significant environmental, educational, scientific, and socioeconomic impacts for coastal communities (Hoyt, 2001; O'Connor et al., 2009). Swim-with whale tourism is offered globally, with humpback whales the most targeted baleen whale species. It is an activity enabling tourists to swim, snorkel, or scuba dive with cetaceans or to observe cetaceans while in the water with them (GBRMPA, 2017). Swim-with whale creates increased immersion as participants physically enter the ocean wearing snorkel equipment. Globally, the number and type of its programmes continue to grow. Swim-with humpback whales experiences are offered in Australia, Dominican Republic, French Polynesia, Kingdom of Tonga, New Zealand, and Africa. Tours provide opportunities to swim-with minke whales (Australia), blue whales (Sri Lanka), sperm whales (Dominica), southern right whales (Patagonia), orcas (Norway and Canada), and even belugas (Canada).

Whale watching is a multi-billion-dollar. In 1991, an estimated four million tourists took tours offered in 31 countries, spending approximately US$317.9 million. By 2008, the whale-watch numbers had increased to some 13 million, with a total spend of approximately $2.1 billion (O'Connor et al., 2009) and were offered in 119 countries (O'Connor et al., 2009). Whale tourism annual tourist numbers in Australia and New Zealand increased from 3,400 to 873,000 in the 10 years to 2013 (Lundquist, 2014).

Most participants were female (n = 72; 66%). Age groups varied across all categories from 18 to 84 years, with over a quarter being aged between 35–44 years (n = 30; 28%). Most had gained a technical college/TAFE certificate (n = 29; 27%) closely followed by an undergraduate (bachelor's) degree (n = 26; 24%). Half of the participants had previously undertaken a whale-related tourism experience; for the remaining half, this was their first experience. Most participants were motivated for reasons of leisure (n = 25; 43%), to spend time with family/friends (n = 20; 34%), or as it seemed like fun (n = 13; 22%).

Emotions elicited from whale-tourism-related experiences were predominantly positive in nature. Over two-thirds (70%) of the stated emotions were awe, wonderment, enjoyment, happiness, and joy. Two negative valanced emotions were expressed by swim-with whales' participants: disappointment and relief and one negative emotion, disappointment, was expressed by whale watching tourists. To understand emotional elicitation further, participants stated the strongest emotions felt and what was happening at the time. The stimuli are clustered into five groups focussed on whale behaviours, seeing whales and other marine animals: proximity; participants past experiences; physical (change in heartbeat); and psychological responses (the need for accommodation because of the size of the female whale, and the vastness of the ocean) (Figure 17.3).

When asked, more than three-quarters of scuba divers (n = 63; 83%) and two-thirds of whale-tourism participants experienced awe (n = 73; 67%). For divers, awe was triggered by the sight of the wreck and marine life, especially large schools of fish. For whale watchers and swimmers, awe was stimulated by the size of the whales, the sound of the whales singing, and proximity to the whales. For the scuba divers who reported feeling awe, awe appraisals and ensuing states included being 'humbled and at peace', 'elated, relaxed. [An] absence of stress', feeling 'interested in my surroundings and a sense of achievement', 'like I was floating on clouds', 'grateful to be able to experience such beauty', feeling 'like being in a different world and travelling a long time', 'like a kid in a candy shop', 'relaxed and refreshed', 'without thoughts, weight', 'one with the surroundings', 'fascinated and absorbed', and 'at one with the ocean'.

Moreover, for the scuba divers who experienced awe, the overall impression they were left with, when reflecting on their diving experience, included 'humans are only a small part of our environment', 'beauty of the

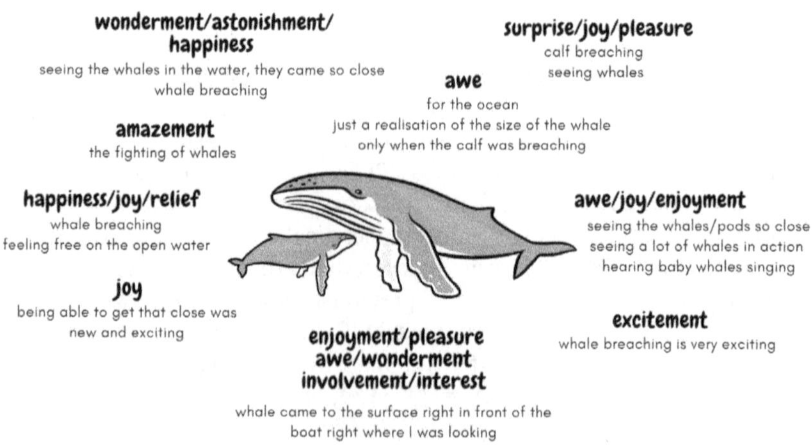

Figure 17.3. Emotional Responses and Stimuli.

world and nature and how small we are', 'peace and happiness', 'achievement', that 'the ocean is incredible', and 'that life can flourish on anything [the wreck];', 'gratitude for nature and the life experienced', feeling 'fortunate', that 'nature is amazing', 'marine life benefit from wreck sanctuaries', and 'we do little to preserve it [environment]'. For the whale tourism-related tourists who felt awe, they reported awe appraisals and resultant states such as being 'excited, amazed, and happy', 'amazed to be so close to such giants', 'just calm and relaxed', 'alive and grateful', 'like a very small creature in a big world', 'joy, pleasure, happiness', 'so grateful to be seeing them [whales] in their habitat', 'makes me want to treasure our environment more', 'engaged and connected to the experience', 'euphoric, calm, in awe', 'happy, in awe, respectful, majestic, fun, amazed', 'overwhelmed with awe and love for the whales', 'the size and grace of the whales was wonderful', 'delighted, at one with nature – love of the whales', feeling 'blessed to be in the presence of these beautiful mammals', 'breathless – was literally breath taking seeing such a magnificent site', and that 'I am really quite small in this world'.

Further, for those whale-related tourists who experienced awe, the overall impression they were left with when reflecting on their experience included: 'I love nature', 'whales are truly magnificent', that 'nature is beautiful', being 'amazed and so grateful', 'joy', 'wonder of the world', 'that nature is phenomenal', 'isn't nature the most amazing and inspiring thing!!', as well as feeling 'a greater love for marine wildlife' and experiencing 'a family connection'. Awe, as experienced by marine tourists in this study, is clearly a complex emotion that may be positive in nature ('awe', wonder, happiness, elation) or imbued with negative aspects (fear, disappointment). Key dimensions of awe were the perception of vastness (the size of the wreck and being a speck in a vast blue sea) and the need for cognitive accommodation, which relate to stimuli that overwhelms current ways of thinking and feeling (having the whale look me in the eye). Experiencing awe was described as transformative.

CONCLUSION

Awe appraisals and reported cognitions within the case study reflect the themes of vastness and the need for accommodation and demonstrate the potential of awe to contribute to individual and environmental wellbeing. For example, the appraisal of vastness and the need for accommodation were reflected in comments such as 'the majesty of such huge animals', 'amazed to be so close to such giants', 'breathless - was literally breath taking seeing such a magnificent sight', and being 'overwhelmed'. There were examples of cognitions of beauty and self-diminishment, such as 'beauty of the world and nature and how small we are', and several indicating a sense of being at 'one with the surroundings'. Awe appraisals and cognitions reflected the consolidation or modifications in individuals' views of themselves, connections to others and nature, and views of the natural environment evidenced with participants stating they felt 'really quite small in this world'. Additionally, awe's potency within the scuba and whale tourism experiences was related to a range of positive psychological outcomes, such as gratitude, peacefulness, humility, elation, joy, pleasure, happiness, relaxation and reduced stress, calmness, and rejuvenation. The value and influence of awe, as evidenced by these findings and the existing research, reiterate the need for those in the tourism industry to understand appraisals to cultivate awe.

Embracing awe within nature and providing genuine immersive experience design offers opportunities to grow the industry through social innovation. Immersive tourism experiences that facilitate strong emotional responses, such as awe, encourage connection with the environment, self and others, and reflection on current or new perspectives and appraisals. Thus, it is recommended that purposeful tourist experience design can increase the occurrence and intensity of awe, in turn, contributing to sustainable tourism offerings that benefit operators and communities, the wellness of tourists and the natural world.

While studies into awe are increasing, Coghlan et al. (2012) reiterate that there are limited empirical tourism studies of awe, particularly those exploring the generation of awe and how awe influences attitudinal and behavioural change. In addition, studies specifically focussing on the influences on cognitive change are few and there are limited studies on awe in tourism that have evaluated the appraisal of vastness within natural environments (Kimberley region, Western Australia, Pearce et al., 2017; Mount Emei, China, Lu et al. 2017). More research needs to focus on deepening the understanding of the appraisal of vastness and cognitive process of the need for accommodation within tourism contexts and how these core facets can be influenced to facilitate awe and its positive effects; particularly in respect to growing sustainable tourism for both individual and planetary wellbeing. Given the benefits of experiencing awe and nature, tourism and operators are in a unique and well-placed position to advance immersion in nature and awe.

It is assumed that being close to whales is essential to the watching experience. Seeing whales is essential, but their proximity to the boat may not be a critical factor. Increasing satisfaction was not simply about getting close to whales as other factors also influence guest experiences (Orams, 2000). In contrast to whale watching, swim with whale experiences, proximity to the whales correlated with passenger satisfaction (encounters with dwarf minke whales, Valentine et al., 2004). Encounters with whales have been described as spiritual, inspiring, awesome, and life changing. Previous studies reiterate the need to understanding the psychophysiological correlates of emotion, including awe, related to immersive experiences, is essential to tourism operators and destination managers who want to enhance economic growth through pathways such as differentiation, loyalty, and positive word of mouth. Knowing the emotion, the corresponding stimuli and appraisals, can inform the design of immersive experiences. Purposefully designed tourism and leisure offerings have been linked to improved health and wellbeing, both for people and places. In addition, inclusive, memorable, immersive experiences encourage tourists to respect, protect, and conserve what they have come to enjoy. Notably, awe and nature-based tourism have been shown to increase pro-environmental attitudes and perceptions (swim with humpback whale; Clark et al., 2019; Chirico & Yaden, 2018; Hoberg et al., 2021). Encounters with wildlife can contribute to conservation learning. Those having the most impact include observing animals in their 'natural' environment (includes zoos and aquariums), close encounters with and opportunities to observe animal behaviour, and those that engage tourists emotionally and connect them with their prior knowledge and experiences (Ballantyne et al., 2007). Further, learning can occur following such experiences of wildlife-inspired awe.

Tourism organisations call on operators to create a competitive advantage by delivering memorable, distinctive, unexpected, and exotic experiences. While aspirational, understanding peoples' cognitive and emotional responses within such activities or encounters are essential for effective experience design. Understanding of the appraisals within awe and nature experiences provides insight for the tourism industry and government to inform product development, enhance tourist satisfaction and wellness, encourage inclusiveness and connection, and advance environmental and destination management.

Recommendations

'Everyone is transformed by travel to some degree' (Lean, 2012, p. 169) and given awe's potential to be transformative, it is highly pertinent to tourist experience design (Gaggioli, 2016). Maximising the benefits of awe within tourist-nature interactions relies on understanding the experience. Importantly, from a tourism psychology perspective, strong emotions contribute to such experiences being consolidated in memory (they are memorable). The case study findings and extant literature highlight the recommendations to provoke further debate and future research regarding awe and tourism.

Awe-inspiring encounters can be designed by creating itineraries that focus on unique encounters in natural environments (Sheldon, 2020). Planning for serendipitous discovery to facilitate tourists in experiencing something new or unexpected can enable nature and social connection. Serendipitous discoveries are those moments when a tourist 'stumbles' comes across something 'special' or unexpected: a flower they have not seen before, a bird not seen since childhood, an ancient tree or colourful, brilliant sunrise.

Offering and encouraging opportunities for tourists to reflect on and talk about what they had seen, heard, experienced, accomplished, and to express the associated cognition, appraisals, and emotions, can offer more profound, meaningful, and memorable experiences.

Transformative change requires fostering emotional connections (Abson et al., 2017; Castree et al., 2014; Fazey et al., 2020), consideration for not only pre- and during but post-engagement with tourists, can establish a relationship and allow them to recall and reflect upon their experiences. It is essential to recognise that their experiences encompass pre-, during, and post-phases. For instance, tourists were motivated to participate in the immersive experiences to spend time with family/friends and engage in leisure. Others commented that the experience was something they have wanted to do for a very long time. When reflecting on their marine tourism experiences, participants indicated feelings of awe were recalled. Awe-related cognitions, emotions, and cognitive processing can occur during and after immersion in the leisure and tourist activity. Tapping into emotions felt, including awe, can increase participant loyalty and initiate opportunities to share (positive word of mouth) stories of their adventures and *awe*-some experiences. As information alone does not consistently result in transformative change (Blythe et al., 2021; Bradshaw et al., 2021), strategically presented content, with active participation is recommended in tourist experience design. These actions can be effective to facilitate loyalty, return visitation, and consumer recommendations/referrals.

A moderate emotional connection to nature was a prominent outcome in the case study. This finding substantiates tourism has increased the capacity to deliver tourist experiences that foster a connection with the natural environment. Human-nature connectedness enhances wellbeing (Pritchard et al., 2020) and subsequently, an opportunity for destinations to offer nature connectedness for a competitive advantage that can be further enhanced by the experienced of awe. By seeking ways to facilitate nature connectedness and awe will contribute to identifying avenues for operators to strengthen a destination's profile as an authentic hub for nature connection in tourism, as indicated by the presented marine tourism study.

Understanding and responding to tourism motivation and commitment can instil an affinity and attachment to a region or place (Giglio et al., 2018). Shaping tourist expectations and behaviour requires a deeper understanding of the conditions conducive to a satisfying experience that initiates intrinsic behavioural change (Gibson, 2010; Hill et al., 2014); the experience of awe within tourism activities may facilitate this occurring. Crucially, wildlife-inspired awe has been found to promote learning and positive memories (Hicks & Stewart, 2020). Further, active engagement can heighten the experience of awe and provide opportunities for inspiring future pro-environmental interests (Ardoin et al., 2015).

Social interactions between social groups or offering opportunities for tourists to make new social connections can aid experience consolidation. Social connectedness is a fundamental aspect of human life as belonging helps to gain a sense that our lives have meaning (Gössling et al., 2018). Encouraging tourists to reflect on their experience through open dialogue to cultivate within group discussions can increase positive emotional experiences, including awe, and processes of cognitive accommodation that aid in making sense of what has been experienced.

The Diversity Council of Australia states that 'inclusion occurs when a diversity of people (of different ages, cultural backgrounds, genders) feel valued and respected, have access to opportunities and resources, and can contribute their perspectives and talents' (https://www.dca.org.au/topics/inclusion). Thus, inclusive tourism should consider 'people with access requirements, including mobility, vision, hearing and cognitive dimensions of access, to function independently and with equity and dignity through the delivery of universally designed tourism products, services and environments' (Darcy et al., 2010, p. 34). While various terms are used, such as accessible and responsible tourism, designing for broader inclusion is a critical component of the sustainable agenda (Sica et al., 2020). The UNWTO (2020) identified accessible tourism as a 'game changer' for destinations emphasising the importance of placing inclusivity as the core of global pandemic recovery strategies for tourism. Redesigning products and services for diverse inclusion can generate business and regional advantage, improve customer service, and health and wellbeing of tourists, staff, and communities (UNWTO, 2020). Inclusive tourism 'can add value to tourism knowledge and understandings by seeking to explicitly overcome the exclusionary tendencies of tourism and to ensure that a wider range of people participate in and benefit from tourism endeavours' (Scheyvens & Biddulph, 2018, p. 590). Like awe experiences, inclusive tourism is considered transformative for 'addressing inequality, overcoming the separation of different groups living in different places, challenging stereotypes or generalised histories, and opening people up to understanding the situation of minorities' (Scheyvens & Biddulph, 2018, p. 592). Consequently, tourism experience design should be inclusive not only in terms of accessibility for all but in designing experiences that facilitate emotions such as awe.

Lastly, participants in the case study recalled the sounds of the whales singing, hearing the bubbles exhaled by scuba divers, seeing the joy on the faces and hearing the laughter of their children, and soaking up the silence of being underwater. As part of tourism experience design, operators should aim to incorporate multi-sensory elements to involve users in new ways to open avenues to elicit emotions. Multi-sensory elements (auditory, visual, tactile, smell, and taste) increase immersion by allowing users to be anchored to the present and focusses their attention on the activity being experienced. Incorporating multiple sensory elements can support tourists with visual or other sensory impediments to create meaningful and awe-inspiring experiences.

This chapter explored awe, a multifaceted and potentially powerful emotion with appraisal centred on vastness and the associated cognitive process. Awe, particularly within a nature tourism context, can contribute to individual wellbeing and adjustment of cognitive beliefs and schema, as well as intent to change behaviour (pro-environmental behaviour, product loyalty, word of mouth of tourism regions and activities). Tourism operations are in a unique position to facilitate and encourage awe within their offered immersive experiences, which may have a positive impact on their customer's wellbeing, their businesses, and connection with the natural environment.

Chapter 18

MINDFULNESS AND TOURISM

Lynn I-Ling Chen
National Pingtung University, Taiwan

Noel Scott
Edith Cowan University, Australia

Abstract: Mindfulness, a kind of mental perception, has a long history in Buddhist spiritual traditions, but has also a well-developed use in clinical psychology as a mental health intervention. Meditative mindfulness provides therapeutic benefits in the treatment of mental trauma. Recently, an interest in the outcomes derived from meditative mindful experience has emerged in tourism studies. Tourists experiencing a mindful-like event may get a sense of peacefulness and a harmonic relationship between the self and the environment. This chapter discusses how meditative mindfulness can be integrated into tourism experiences and create subtle therapeutic-like effects. Theoretically, this study bridges the relationship between a meditative mindful tourist experience and its perceived consequences. Practically, it explains how to generate mental health outcomes from tourism experiences. **Keywords**: Awareness; cognition; experience; meditative mindfulness; psychology; attention

INTRODUCTION

Living in contemporary society with rapid technological innovation, consumerism, and goal-oriented demands, the human mind is often occupied and distracted by external tasks. Sometimes, the mind will drift away from the present moment, and may generate spontaneous thoughts unrelated to the task in hand (in the here and now) that may result in poor performance of a task, and even cause psychological problems (O'Donnell, 2015). For

Cognitive Psychology and Tourism
Tourism Social Science Series, Volume 27, 227–236

Published under exclusive licence by Emerald Publishing Limited
ISSN: 1571-5043/doi:10.1108/S1571-504320240000027018

example, people often check emails every few minutes while working, leading to worry and procrastination. When travelling, if a tourist constantly thinks about office work during hiking, they may not notice the beauty of their surroundings. This implies that the distraction or proactive cognition unrelated to the current task can be a hindrance to an individual's awareness of and attention to an experience episode, and lead to negative affective responses (Smallwood & Schooler, 2006). Mindfulness is the capacity of noticing that one's mind has shifted away from processing current experiences, and bringing it back to the present (Remmers et al., 2016). This benefit requires understanding the mental processing associated with mindfulness (Siegel, 2009).

Mindfulness is a kind of mental perception (Wallace, 1999). It has become a trendy term used frequently in modern society as a means to promote mental clarity, greater wellbeing, and self-caring (Kabat-Zinn, 2002; Kang & Whittingham, 2010; Kuan, 2012). The concept of mindfulness originates in the ancient Buddhist teaching to describe simple bare awareness of moment to moment experience by sustaining attention on a familiar object (Kang & Whittingham, 2010). Sitting meditation is often seen as a formal mindfulness practice within Buddhism and as a means of optimal liberation from suffering, and to attain enlightenment (Kang & Whittingham, 2010; Nhât Hạnh & Laity, 2009). Later, Eastern meditative mindfulness practices have been applied as therapeutic treatments within Western clinical psychology contexts. Meditative mindfulness practice requires paying attention moment-by-moment in a nonreactive way and with nonjudgmental awareness (Kabat-Zinn, 2002). Mindfulness-based cognitive therapy and mindfulness-based stress reduction have been found beneficial for developing calmness and stability of mind (Kabat-Zinn, 2002). These formal mindfulness practices help people cultivating innate capacity through a deep, penetrative seeing/sensing of the awareness to both inner and outer experiences (Kabat-Zinn, 2003a).

There is also an emerging interest in the application of the concept of 'meditative mindfulness' derived from Eastern philosophy as a means to understand tourist experiences linked to mental ease and inner peace of mind (Chen et al., 2017; Pearce, 2016), in particular, for post-pandemic tourism (Chen, 2022; Ma et al., 2021; Stankov et al., 2020). Meditative mindful experience episodes have been identified empirically in tourism contexts (Chen et al., 2017). During such an episode, tourists were found to experience a sense of peacefulness and a harmonic relationship between the self and the environment (Chen et al., 2017; Kang, 2009; Macaulay et al., 2022). This chapter ponders a further question of interest to tourism: 'How does meditative mindfulness contribute to positive mental effects?' This chapter also discusses how meditative mindfulness may be integrated into tourism experiences to benefit tourists.

MINDFULNESS

This section introduces the theoretical manifestations of the Eastern notion of meditative mindfulness and its empirical applications in Buddhism, clinical psychology, and tourism. Meditative mindfulness has been used to improve personal mental health and greater wellbeing.

Mindfulness in Buddhism

The concept of mindfulness is rooted in ancient Eastern Buddhist traditions over 2,500 years old, primarily in the form of mindfulness meditation. Historically, the term 'mindfulness (Pali: sati or Sanskrit: smrti)' can be traced to 1881 when Rhys Davids translated the Buddhist term 'sati' and 'smrti' using the English word 'mindfulness'.

There are various rich discourses of mindfulness (sati) in Buddhist texts. Essentially, the concept of mindfulness can be defined as 'an awareness of being aware' (Hirst, 2003) or 'discernment' (Bodhi, 2011; Gethin, 2011). An operational definition of Buddhist mindfulness is proposed by Kang and Whittingham as 'non-reactive, non-elaborative, non-reified awareness that has meta-cognitive functions, monitoring ongoing awareness and discriminating wisely between aspects of awareness content, so that awareness and behaviour can be directed according to the goals of genuine happiness, virtue, and truth' (2020, p. 170).

Some interesting characteristics of awareness/attention associated with mindfulness in the Buddhist contexts have been noticed: simple bare awareness of moment-to-moment experience; gatekeeping awareness; remembering and sustaining attention on a familiar object, a process of systematically recollecting a sequence of ideas; a process of systematically recollecting a sequence of ideas; conjoined with introspective vigilance that monitors the stability and clarity of awareness; wisely directed attention that probes into the source of experiential content; and non-dual co-emergent awareness at the subtlest level of consciousness, free from all conceptual constructs and frames (Kang & Whittingham, 2010, pp. 163–164; Wallace, 2006, 2007).

Zen Master Thich Nhat Hanh, a Vietnamese Buddhist monk as well as a global spiritual pioneer of mindfulness, has brought the concept of the Buddhist mindfulness into daily practices for the general population as a way of living happily in the present moment. The main approach of mindfulness is about 'awakening'. This notion means that practitioners should keep their awareness shining, lighting each thought and feeling, recognising them, and being aware of them whether in forms or dissolution, without judging or evaluating them (Thích, 1988). During mindful meditation, full awareness of mindful breathing can take the practitioner directly into the living experience and its non-conceptual reality (Kabat-Zinn, 2002; Thích, 1976).

Traditionally, Buddhist meditation is a formal mindfulness practice and requires scanning of bodily sensations, observing aroused cognition and emotion, and paying attention to one's surroundings with a non-judgemental mind. Sitting meditation is a common way of practicing mindfulness. Based on the same discourse, meditation can also be embodied in different patterns, such as walking meditation, mindful breathing, total relation exercises, etc. (Chen et al., 2017). Zen Master Thich Nhat Hanh's mindfulness teachings indicate that meditation can be practiced through secular activities as an informal mindfulness practice, if one can be fully aware of all his/her actions nonjudgementally in the present moment. For example, while waking up, people know that they are waking up; while weeding the garden, they know they are weeding the garden and with thoughts remaining firmly rooted in awareness (Thích, 1976, 1988).

Eastern Buddhist meditative mindfulness practices have been used for centuries. With deeper mindful meditation, a practitioner's mind will be less reactive to external stimuli and become clear and more serene. The serene attentional state allows discernment about one's thoughts and feelings that can result in the ultimate goal of attaining enlightenment (Sayadaw, 1991). In Buddhist teaching, the perceived psychological benefits of mindfulness meditation include wellbeing with inner contentment, moments of inner peace, happiness, fulfilment, and freedom from suffering (Kang & Whittingham, 2010; Schmidt, 2011; Wallace, 1999).

Mindfulness in Clinical Psychology

Mindfulness in clinical psychology uses the practice of Buddhist mindfulness to provide therapeutic benefits derived from meditation and has developed various mindfulness-based interventions for clinical populations (1991, 2000, 2002a). Meditation as a core skill incorporated into mindfulness-based therapeutic programmes is a growing trend in the fields of mind-body medicine, clinical psychology, and neuroscience studies and applications.

Jon Kabat-Zinn, a clinical psychologist, brought meditative mindfulness into the mainstream of Western modern medicine. Kabat-Zinn proposed an operational working definition regarding the notion of therapeutic mindfulness as '... the awareness that emerges through paying attention on purpose, in the present moment, and nonjudgmentally to the unfolding of experience moment to moment'. The constructs of therapeutic mindfulness in psychotherapy applications link to three dimensions of awareness, including paying attention, present experience, and acceptance without judgement (Kabat-Zinn, 2003a, p. 145). Another definition of mindfulness in therapy is 'self-regulated, present-moment awareness that welcomes all experience without preconception or judgment, accepting "what is" with curiosity and compassion' (Gehart, 2012, p. 7). The core practice underpinning the therapeutic programmes is intentionally focussing and refocussing one's awareness on a particular object of present-moment experience.

Mindfulness-based psychotherapy treatments have been examined using neuroscience brain function studies (Siegel et al., 2009) and found to lead to relaxation (Wallace & Shapiro, 2006). Mindfulness-based interventions facilitate patients embodying 'affectionate awareness, mindful acceptance, openhearted presence, and mindful compassion' that can promote wellbeing in body and mind and help brain function (Siegel, 2010, p. 19). Even though these practices are adopted from Buddhist meditation, the clinical applications of therapeutic mindfulness go beyond the ideology of Buddhism. The skill of meditation and a focus on the direct experience of one's breath or bodily sensations can be used in clinical populations with limited or no previous meditation experience. Thus, mindfulness-based psychotherapies offer medical patients an achievable means of relieving sufferings and understanding of their mind/body connection (Kabat-Zinn, 2003a).

In clinical applications, mindfulness-based interventions have been adopted broadly in psychotherapy, behavioural, psychosomatic, and allied therapies (Bishop et al., 2004; Kabat-Zinn, 2003a). In general, the training involves an eight-week programme in which clients are advised to practice mindful meditation daily. The training is provided by a certified mindfulness-based stress reduction instructor who translates meditative concepts into a vernacular form which can be related to real-life issues. Mindfulness-based stress reduction and mindfulness-based cognitive therapy mindfulness-based cognitive therapy are well-used programmes and briefly exemplified below.

Mindfulness-based stress reduction is a secular mindfulness and non-religiously-rooted training programme that was launched in 1979 at the University of Massachusetts Medical Centre. It cannot cure diseases, but it can train patients to cultivate their self-awareness. It is based on intensive mindfulness training including mindful hatha yoga and is framed as a complement to patients' medical treatments with ancient but potentially transformative meditative practices (Kabat-Zinn, 2002, 2003a). The patients engage in the various mediation practices assigned as homework daily. The intention is to allow them to make more fully reasoned out and wise responses to challenges such as stress and pain in their lives (Do, 2011; Kabat-Zinn, 2003a,b).

Mindfulness-based cognitive therapy is a technique combining mindfulness-based stress reduction and cognitive behavioural therapy, developed to prevent recurrence or repeated bouts of depression and chronic unhappiness, defined as three or more episodes of chronic depression (Kabat-Zinn, 2003a; Kang & Whittingham, 2010; Segal et al., 2002). Therapeutic incorporates elements of cognitive therapy to help attentional control in noticing thoughts or feelings simply as mental objects, without elaborating on them, and allowing them to come and go in the mind, rather than in an attempt to resolve them (Baer, 2003; Kabat-Zinn, 2003a; Kang & Whittingham, 2010; Mason & Hargreaves, 2001). The intention is to stop the proliferation of negative cognitive patterns (Baer, 2003; Segal et al., 2002).

Through the professional clinical programmes, the mental processes developed in therapeutic mindfulness training can result in simultaneous relaxation and attentional stability, as well as reducing emotional distress and maladaptive behaviour (Baer, 2009, 2011; Bishop et al., 2004; Kabat-Zinn, 2002, 2003b; Wallace & Shapiro, 2006). Mindfulness-based clinical applications are effective in reducing patient's stress, anxiety, pain, and depression (Gehart, 2012). Therapeutic mindfulness can help people to face each experience, even negative challenges, with a more open and non-judgemental mindset. Recently, mindfulness therapies have also received attention as a means of improving psychological wellbeing during the COVID-19 pandemic for both clinical and non-clinical populations (Widha et al., 2021). Clearly there are more opportunities to use therapeutic mindfulness in people's everyday lives and hence address the challenges of contemporary society (Chen, 2022).

Mindfulness in Tourism

The term 'mindfulness' was developed in the mid-1990s and used in tourist behaviour studies on outcomes from tourism experience (Moscardo & Pearce, 1986). This approach to (socio-cognitive) mindfulness lies in social psychology (Langer, 1992) and was mainly adopted in the context of interpretation and learning (Chen et al., 2017). This concept refers to a flexible conscious awareness of external stimuli, where tourists tend to make thoughtful judgements about their actions and problem solving (Pearce, 2016). The nature of awareness embodied in this approach to mindfulness is essentially different from the Eastern notion of meditative mindfulness. This line of work differs from the core discourse of meditative mental processing underpinning either in Buddhist mindfulness or therapeutic mindfulness (Chen et al., 2017; Chen et al., 2014; Pearce, 2016). Hence, this notion is termed socio-cognitive mindfulness.

As already discussed, both Buddhist mindfulness and therapeutic mindfulness offer an approach to alleviate stress and restore inner peace for people. Similarly, tourists may experience a sense of peacefulness and build a harmonic relationship with the self and the environment in a tourism context. It is worth noting that the positive mental effects such as meditative-like state and therapeutic-like benefits can also be produced by tourist experiences associated with the Eastern notion of meditative mindfulness (Chen et al., 2017). Meditative mindfulness can be integrated into tourism experiences and create subtle therapeutic-like functions from the lens of cognitive psychology.

Tourists may engage more mindfully in tourism or travel experiences that evoke a sense of awareness, as well as relaxation or mental ease. The phenomenon of meditative mindfulness in tourism settings has been empirically identified in recent research. An operational definition of meditative mindfulness associated with an experience in a tourism context is proposed as 'paying attention to a present moment experience, without elaborating on it'. Specifically, the meditative mindful tourist experience can be characterised by three subtle psychological awareness-related constructs: paying attention to the experience, living in the present, and non-elaborative awareness (Chen et al., 2017, pp. 5–6).

The first psychological construct of tourist meditative mindfulness is paying attention to the experience, mainly because of awareness of the stimulation of the senses. During travel, tourists have more opportunities for relaxing, and being intuitively or intentionally engaged in recreational involvements, compared with their normal working life filled with pressure and goal-directed tasks (Li et al., 2013). That is, a tourist may be engaged in perception of their environment in an informal mindfulness-like manner (Carruthers & Hood, 2011; Chen et al., 2017) and an absence of goal-directed activity. At this moment, a mindful state can be activated by relaxed intentional awareness closely related to sight, hearing, smell, taste, and touch sensations (Chen et al., 2017). In other words, if a tourist's experience of present immediate sensations is aroused, then their attention may be captured in a 'gentle and stable' way. This means that they are less distracted by other thoughts, and instead their attention is on present moment occurrences (Chen et al., 2017; Weick & Sutcliffe, 2006). In this way, their mental processing mode can be directed into a more mindful state (Siegel, 2009).

Living in the present is the second psychological construct of a tourist meditative mindful experience. When one is paying attention to an experience that involves captivating sensory stimuli, one's awareness is not caught up with thinking about the past or the future problems of daily life. During the state of being mindful, a tourist is more immersed in their current engagements, and may a sense of emptiness sometimes described as having a 'blank brain' (Chen et al., 2017). In other words, this mental processing mode is in a state of 'an absence of conscious cognition'. The phenomenon may be experienced in a natural setting or nature-based destination. A number of studies have discussed that tourists often feel they are in harmony with nature, or the universe, when they are enjoying living in the moment (Agapito et al., 2014; Howell et al., 2011; Macaulay et al., 2022). The cognitive style of being immersed, like having a 'blank brain', influences how tourists think and results in a reduction of ruminative thoughts (Chen et al., 2017). It is interesting to note that such spontaneous 'mental ease' during a state of mindfulness means one is not preoccupied with past-oriented and future-oriented thoughts. Instead, one is simply and gently in the present moment without any elaborating on or judging it.

The third psychological construct of a tourist meditative mindful experience is non-elaborative awareness. At this stage, tourists are more aware of their mindful engagement, and often report having a vivid experience. The mindful state leads tourists to turn their attention inwards to the experience itself (Chen et al., 2017). Overall, the three distinct psychological constructs of tourist meditative mindful experiences interact with each other to produce a state of being mindful. This results in a state of being at ease (Chen et al., 2017). The operational definition, psychological constructs, and experiential benefits of meditative mindful tourist experience are summarised in Table 18.1 (after Chen et al. 2017, p. 10).

Results also indicate that the subtle therapeutic-like effect is akin to the functional focus of mindfulness-based therapies or interventions in clinical psychology (Kabat-Zinn, 2003a,b; Weick & Putnam, 2006) and brain studies (Siegel, 2009).

Tourism settings that encourage mindfulness contain contributing factors such as beauty and quietness, especially in contexts like wilderness, forests, seaside, national parks, adventure travel, or backpacking tourism (Chen et al., 2017; Cohen, 1979; Trauer, 2007). Aesthetic appreciation of beauty has found to create a sense of awe that captures a tourist's attention (Chen et al., 2017; Coghlan et al., 2012; Pearce et al., 2017). Being quiet without focussed cognitive information processing can help direct a tourist's attention to the present moment and facilitate a meditative mindfulness state (Chen et al., 2017). For instance, watching a beautiful starry sky in a quiet environment can promote a state of non-judgemental awareness, walking quietly in a

Table 18.1. Meditative Mindfulness in Tourism Contexts.

Meditative Mindfulness	Meditative Mindful Tourist Experience
Operational definition	• Paying attention to a present moment experience, without elaborating on it
Psychological constructs	• Paying attention to the experience • *Stimulating by sensory awareness*
	• Living in the present • *Attention not entangled with thinking* • *Being immersed, like 'blank brain'*
	• Non-elaborative awareness • *Internal dialogue merges awareness and cognition*
Experiential benefits	• Being at ease

Source: After (Chen et al. 2017, p. 10).

forest can evoke strong sensory awareness, and scuba diving in a deep ocean can lead to a sense of calmness. Chen et al. (2017) further illustrated, if a tourist pays attention without elaboration to the immediate sensations, the person experiences the 'now', by simply being aware that they are 'watching', 'walking', or 'diving'. The tourist pays attention to the experience and therefore they are not disturbed by past-oriented or/and future-oriented thoughts.

While respondents reported the meditative mindful episodes given above, this does not mean that all such tourists experience this mindful state. However, there does seem to be some commonality in the range of settings and activities that may settle a tourist's mind, and encourage them to be aware of the present moment (Chen et al., 2017). Several recent studies have examined meditative mindfulness and found it can be integrated into tourism experiences and create a mindful state. In the recent years, special interest operators and some individuals have sought to encourage mindfulness experiences in tourism contexts. Examples include mindfulness-based retreats in remote destinations (Chen, 2022), forest bathing (Farkic et al., 2021), engagement in nature (Macaulay et al., 2022), and wellness providers using technology-assisted e-mindfulness devices and mindfulness apps as hospitality services (Stankov et al., 2020). These innovations suggest that meditative mindfulness may be encouraged in various leisure, tourism, and hospitability contexts.

Several positive psychological outcomes derived from tourist meditative mindful experience have been noted. Chen et al. (2017) identified that the perceived psychological effects of meditative mindful experiences included a sense of emotional and cognitive balance, positive mental state, and re-evaluation of the importance of normal activities. The immediate positive mental states included a sense of serenity, relaxation, or stress release described as mental ease or psychological wellbeing. Other reported benefits include response flexibility, openness, and vivid memories of the event. Psychological benefits include mental resilience and meta-awareness (McGroarty et al., 2020), psychological restoration and connection with nature (Macaulay et al., 2022), and a sense of wellbeing through immersion in forest-bathing (Farkic et al., 2021).

CONCLUSION

Mindfulness has been practiced in Buddhism contexts for thousands of years. However, mindfulness-based applications have only become used in contemporary clinical, tourism, and leisure contexts over the past few decades. This may be due to in part to the recognition of the strong interaction between the mind and body in recent year and a trend towards wellness rather than health (Grénman & Räikkönen, 2015). The mental health and personal wellbeing benefits derived from meditative mindfulness practices have been recognised now, not only in religious contexts, and for clinical patients, but also in ordinary life and when travelling.

This chapter has interpreted the Eastern approach of meditative mindfulness from the lens of cognitive psychology. Its coverage provides evidence that tourism contexts can lead to mindful experiences with therapeutic-like benefits. Mindfulness is a means of preventing negative cognitive and affective responses. The value of mindfulness-based experiences has led to new products and services. One such niche market is travel to pursue meditative mindful experience and psychological resilience in the religious-social context of a Buddhist temple (Wang et al., 2021). In addition, meditative mindfulness has been advocated as a means to cope with the problems occurring in modern life during the COVID-19 pandemic (Chen, 2022; Ma et al., 2021; Stankov et al., 2020; Vione & Kotera, 2021; Widha et al., 2021). In turn, these developments should encourage cross-disciplinary studies to advance applications of the Eastern notion of meditative mindfulness in wellbeing and mental health. There is a need for further research on the cognitive processes supporting a meditative mindful experience in tourism episodes. This may support the design of effective wellbeing products employing technology and provide support during COVID-19 (or other pandemics), and also as a means to deal with a more complex, disturbing future.

Chapter 19

WHY STORIES WORK

Ana Cláudia Campos
University of Algarve, Portugal

Abstract: Stories are a part of being human. They are powerful tools to create social bonds, elicit emotions, and consolidate personal and shared memories. Their role, characteristics, and effects are studied in a variety of scientific fields. In tourism, they can enhance destination distinctiveness, as well as the tourist experience. However, the cognitive processes through which stories contribute to elicitation of emotions or memorability, the very reasons why they are considered important in tourism, are still little understood. This chapter discusses some of the cognitive processes involved in storytelling drawing insights from cognitive psychology and neuroscience, which may help in better understanding and managing the tourist experience. **Keywords**: Story; figuration; temporality; emotion; memory; attention; cognitive narratology; neuroscience

INTRODUCTION

Why do humans crave stories? Why are these combinations of fantasy and reality so compelling and effective in eliciting emotions? How are people mentally wired to interpret stories and what cognitive and neural mechanisms support this human ability to tell and listen to stories? Why are they important in tourism practice for both operators and tourists? This chapter discusses story effectiveness in relation to cognition and brain functioning by reviewing current evidence and theories of storytelling drawn from cognitive psychology and neuroscience. This chapter aims to provide insights for better understanding and managing the tourist experience.

The coverage introduces the theme by clarifying concepts (story, narrative, and storytelling), and subsequently presents the contribution of cognitive psychology and neuroscience to the understanding of stories and

Cognitive Psychology and Tourism
Tourism Social Science Series, Volume 27, 237–255

Published under exclusive licence by Emerald Publishing Limited
ISSN: 1571-5043/doi:10.1108/S1571-504320240000027019

storytelling that may be useful to tourism studies and practice (figuration, events and temporality, emotions, memory, and attention). It then reflects on why more research on stories based on a cognitive sciences approach is important to the study of the tourism experience.

Stories and Their Importance

Storytelling is a human activity found in every era and culture (Chaitin, 2003; McCabe & Foster, 2006). The telling of stories is deeply rooted in human oral communication (Crowley, 2018). Storytelling defines how humans survive and share knowledge and traditions. According to Popp-Baier, 'human beings live with stories or exist narratively and narratives are in one way or another constitutive of human reality' (2013, p. 1391). Stories are repositories of a shared past and the ingredients of daily life (McCabe & Stokoe, 2010). They are persuasive (Woodside, 2010) and instruments that evoke cognitive processes, such as attention, imagination, or memory, and can elicit profound emotions, including joy or sadness (Martinez-Conde et al., 2019; Zak, 2015). Most importantly, stories help to organise, store, and communicate information and knowledge (León, 2016; Moscardo, 2020; Woodside et al., 2008). Stories immerse and transport people to alternative worlds through a process of mental simulation (Richardson et al., 2020) and facilitate the interpretation of the present (Gerrig, 1993).

As shared elements of culture, stories are a means to entertain, educate, emotionally connect people, and convey moral values (Richardson et al., 2020; Sole & Wilson, 2002). Thus, they emerge from the social world, the collective experience of interacting with others, and reflect a 'social' brain which, in interaction with the world, enables the construction of stories (Zak, 2015). The idea that cognition develops from interactions among the brain, the body, and the social world, rather than in a disembodied mind, is called 'embodied cognition' (Armstrong, 2020), which combines the internal perspective of a brain that works with its own 'neural plumbing' with the external perspective that looks at how this 'plumbing' interrelates with the outer environment (Armstrong, 2019).

Disambiguating Story, Narrative, and Storytelling

It is not easy to disambiguate 'story' and 'narrative', in part because the narrative concept is itself elusive and diffuse, having different meanings in the different scientific fields (Schiff, 2012). Sometimes these two terms are themselves used interchangeably without any sort of disambiguation (Sarbin, 1986). However, for some, in relation to story, narrative seems to stand for a broader concept (Halverson, 2011; Zhong et al., 2017). In literary terms, a narrative is a type of discourse, a kind of speech with specific characteristics and structure that distinguishes it from other kinds. Thus, a novel as a work of fiction or a story told as an everyday conversation, both fall into the general category of narrative, whereas an instruction, a statistical analysis, an explanation, or a description would pertain to another category of discourse.

Despite debate in literary studies and other disciplines about what structure, speech elements and characteristics define narratives, in psychology and the other social sciences, narrative has been considered a metaphor to allude to the way humans make sense of life and the world (Popp-Baier, 2013). This idea has led scholars to advocate that, psychologically, humans have a narrative 'structure' (Crossley, 2000), corresponding to the way they experience the world, encompassing temporality (events' sequence) and connection (events' relationships).

According to a competing view within psychological research, narrative describes the function of the human process of meaning-making of life, rather than a structure of speech elements bearing some specific properties easily categorising narratives as opposed to other types of speech (Schiff, 2012). Arguably, it is not always straightforward to discern what is narrative from what is not, as subtle differences may be observed that refer to degree rather than type or form of speech (Schiff, 2012).

In this chapter, a story is considered a narrative that 'combines either real or imagined events that connect in such a way to provide a chain of events that are recounted to others' (Chaitin, 2003). Stories originated from oral communication (Richardson et al., 2020) and present specificities (Thorndyke, 1977) that some consider separating them from other types of discourse, such as argumentation, logical reasoning (Graesser et al., 2002), or even a plain report of a sequence of events (León, 2016). Drawing from Finlayson and Corman (2013), León (2016) defines story as a sequence of events, that are causally related, involving specific characters and times, and displays a certain level of organisation beyond basic coherence. In a similar way, a story is a narration of events and motivated characters in a situation, acting towards some goal and facing subsequent consequences (Thorndyke, 1977).

Stories may be fictional and non-fictional and presented using various narrative techniques and channels (Richardson et al., 2020). They are found not only in the form of myths and fables but also in historical records or autobiographical memoirs; they can be told through graphics, the written word, orally, or using mimicry; and they can follow a chronological order or else a reversed one, though sequentiality is one of their most acknowledged characteristics (Bruner, 1990; Crossley, 2000; Dancygier, 2011; Dominey, 2021).

They are said to be closer to everyday life playing a critical role in how one constitutes sense (Crossley, 2000; Popp-Baier, 2013), and this fact imposes on stories a natural character that facilitates memorability (Cormick, 2019; Woodside, 2010), and thus the building of identity and the transmission of culture (Graesser et al., 2002). Stories connect concepts to personal experience (Wilson & Desha, 2016), elicit emotions and linger in memory (Martinez-Conde et al., 2019; Zak, 2013), creating oral traditions that shape collective identity (Crowley, 2018). Moscardo highlights the entertaining and often hedonic character of stories, stating that they are 'a type of entertainment that involve a set of events, characters connected to these events, the decisions these characters make in response to the events, and the consequences of these responses' (2018, p. 94). Stories as narratives are considered cognitive structures that organise human action and experience (Popp-Baier, 2013).

Storytelling, a Compelling Narration

Stories are the subject matter of storytelling, and the latter is a practice best described as an act of verbalisation (Popp-Baier, 2013), embedded in social and cultural life (Lugmayr et al., 2016; Popp-Baier, 2013). Chaitin (2003) considers storytelling as an oral or written sharing of one's stories with others. For Delgadillo and Escalas (2004), it is an expression of interpersonal communication organised around a story. It makes use of language, written, oral, or a combination of both types of communication.

Storytelling usually involves application of techniques to grab attention or elicit emotion (like improvisation, modulation of voice) to appeal to an audience. As Mei, Hagensen and Kristiansen put it, 'the purpose of storytelling is not merely to convey facts and information, but to enrich, enhance and infuse facts with meanings to create a story around a concept' (2018, p. 2). Storytelling is a type of performance that puts into play a large variety of communicative resources. Telling a story or reading out loud a written story are known in many different contexts, and both practices provide evidence of the many resources used to make a story appealing, emotional, and meaningful. Storytelling is a means to conveying one's understanding of a particular aspect of the world, and answers to *how* it is done (Schachtner, 2020).

LINKING THE MIND AND THE WORLD WITH STORIES

The study of narrative (and thus also stories) crosses over different academic disciplines: linguistics and literary studies, philosophy, psychology and physiology, sociology, anthropology, and also more recently creativity, art, and neuroscience (Armstrong, 2020; Heinen & Sommer, 2009; Meuter, 2013). According to Meuter (2013) and Popp-Baier (2013), following the narrative turn, narratology has evolved as an interdisciplinary study which connects humanities with the other sciences, the social and the natural sciences.

Psychology is part of a wider group of sciences committed to the study of stories and storytelling as related to the human mind (Herman, 2009). Narrative psychology is an approach that acknowledges humans make sense of their world, their life and personal identity through stories (Richardson et al., 2020; Schiff, 2012), and that cognitive capacities come into play in story interpretation and mastering that surpass strict linguistic ones (Dancygier, 2011). Narrative psychology is a child of the 1950s' cognitive revolution (Popp-Baier, 2013), and evolved from the narratory principle that human beings think and act in consonance with narrative structures, and thus claims that narratives are a vehicle for understanding human cognitive and emotional processes, such as meaning construction, conceptualisation, comprehension, task performance, or emotion (Graesser et al., 2002; Meuter, 2013). It has been defined as 'the comprehensive designation for narrative approaches in a variety of research areas in psychology' (Popp-Baier, 2013, p. 1388), encompassing personality psychology, developmental psychology, social psychology, health psychology, and organisational psychology, under the general assumption that human psychology has a narrative structure (Crossley, 2000) or else *a narrative way of functioning* (Schiff, 2012). For example, developmental psychology has focussed on the study of narrative to explore children's processes of competence formation to progressively master it (Pinto et al., 2016).

Cognitive narratology delves further into the concept of 'structure', addressing 'the mental states, capacities and dispositions that provide grounds for ... narrative experiences' (Hermann, 2013, n.p.). It explores the nexus of language, mind, and world by combining the contributions of the cognitive sciences: philosophy; psychology; the neurosciences; computational intelligence; linguistics and language; and culture, cognition, and evolution. Narrative is a cognitive structure that works on specific principles that allow, for instance, linking abstraction and generalisation to the concrete and the situated by use of a story (Bruner, 1986; Popp-Baier, 2013). León states that 'part of human cognition is structured with narrative properties' (2016, p. 20), assuming that the cognitive processes of focus (= salience of story elements in the individual's mind) and inference (= knowledge and relationship about facts from incomplete information to create coherence) present that type of property.

Cognitive narratology developed from early attempts to account for the nature of narrative (Armstrong, 2020). Genetic approaches (= how narrative originated and changes through time and space) were superseded by structural ones which sought uncovering specific elements of language and the structural relations among them as found in grammar and syntax rules. The structuralist paradigm of narrative evolved from the key premise that mind, language, and narrative share innate, therefore universal and homogeneous, laws governing the logic of human life (Andrew, 1973). Posterior analyses have shown, however, that structuralist premises do not account adequately for some effective story features (Dancygier, 2011). The structuralist claim was replicated by the first wave of cognitive narratology scientists, who still endorse the view of the dependence of narrative from mental models described as frames, scripts, or schemata (Armstrong, 2020).

Two stages characterise the development of cognitive narratology, called 'first' and 'second generation' cognitive science, signalling changes in the way scientists understand the mind, departing from a computer-like mind programed with frames, scripts and schemata to an embodied and interactive mind (Armstrong, 2020; Kukkonen & Caracciolo, 2014). This first generation of cognitive narratology is described as the view of a mind 'independent from specific brains, bodies, and sensory modalities' (Kukkonen & Caracciolo, 2014, p. 261). In turn, the second generation is presented as 'a specific strand in contemporary cognitive science, one foregrounding the embodiment of mental processes and their extension into the world through material artefacts and socio-cultural practices' (Kukkonen & Caracciolo, 2014, p. 261). Dancygier (2011) argues that one can approach the study of narrative establishing a closer link between a cognitive perspective and linguistic analysis (cognitive linguistics), thus separating from the original structuralist paradigm through endorsement of the view that taxonomies and terminology do not apply to narrative whose meaning is contextually constructed (= dependent from situation) and involve more complex linguistic processes and choices which are absent from simple sentences (Thorndyke, 1977). In this line, linguistic analysis draws from cognitive constructs, such as 'mental space'. Neuroscience similarly considers that 'engaging with narrative involves a complex array of cognitive and affective processes' (Vaccaro et al., 2021). In neuroscience, stories play a role in the organisation and expression of thought, through existence of a story-based human brain that interprets the world through the lenses of stories (Willems et al., 2020).

Other fields, as in physiology, have studied how humans respond to stories. Richardson et al. (2020) have addressed the effect of a story at the level of the conscious experience (using self-reported engagement) and at the level of bodily responses (such as heart rate and skin conductance) using two different story channels (audio and video). They found a discrepancy between self-reported measures and physiological ones; more concretely, while video stories scored higher in self-reported engagement, audio stories reported higher in physiological measures, suggesting that listening to stories involves more active cognitive and emotional processing than the ones using video. A recent neuroscience study performed with children reported similar findings for picture-book reading and storytelling, concluding that oral storytelling engages the brain more intensely and actively, requiring more cognitive demands than reading a picture-book. Reading with visual aid stimuli alleviates the effort of interpretation (Yabe et al., 2018).

Bringing Closer Cognitive Psychology and Neuroscience

Neuroscience is the study of the neural basis of cognitive processes, and although cognitive processes are involved in stories (production, reception, transformation), the study of stories has been neglected in neuroscience research until recently (Martinez-Conde et al., 2019). Neuroscience is based in the parent disciplines of psychology and psychophysiology, and increasingly is combining with other academic fields to form the interdisciplinary area of cognitive sciences. Integrating narrative stimuli in the study of brain activity is necessary because stories are used to represent the ever-changing world, and organise and convey thoughts (Willems et al., 2020). Neuroscience research on narrative (Vaccaro et al., 2021) has recently empirically substantiated prior speculation and corroborated experimental studies on psychophysiological processes critical to embodied intersubjectivity, such as neural coupling in teller and listener communication or narrative engagement (Richardson et al., 2020; Song et al., 2021; Stephens et al., 2010).

For neuroscience, stories are narratives, defined as 'a discourse unit that includes events and one or more characters' (Willems et al., 2020, p. 271) that may be presented in different media format, written stories, plays, movies, travel narratives, or even journalistic writing. Neuroscientists acknowledge that stories improve information processing, increasing recall and interest (Martinez-Conde et al., 2019). This is assumedly because story comprehension prompts specific cognitive and affective mechanisms (Vaccaro et al., 2021). It appears that the human mind engages in multiple states, sometimes not easily grasped or described. However, from a narrative perspective, this complexity seems to dissipate. The narrative mode apparently mitigates the cognitive load necessary to disentangle one from another (Willems et al., 2020). The realistic content found in stories can shed light on processes in action when interacting with the real world.

Neuroscientists acknowledge that by simulating reality, stories are a fruitful context to study cognition and related brain processes (Willems et al., 2020), and findings using stories have been shaping new ways to understand language-related processes (Armstrong, 2019). For example, Martinez-Conde et al. (2019) explain how the study of brain activity through narrative has been expanding knowledge of the language network, traditionally viewed as narrower than what in fact seems to be. As these scholars emphasise, through stories greater understanding of the brain has been achieved on specific cognitive functions, such as language comprehension, social understanding, memory encoding, or event segmentation. Figuration is a key process occurring in the mind and brain when producing, listening, or reading stories (Armstrong, 2020), and as long as it relates to how stories mimic real-world experiences, it is possible to analyse the workings of perception or memory, and many other cognitive functions.

Neuroscience is advancing research on brain processes by means of examination of real-world experiences; in naturalistic contexts it is possible to access them through stories. The study of the neural mechanisms underlying narrative processing provided by neuroscience is argued to complement the contributions from other scientific fields (Armstrong, 2020; Willems et al., 2020).

Themes in Cognitive Psychology and Neuroscience

Figuration. It refers to the power of narratives to deliver an alternative perspective of the world (de Vries, 2016), thus contributing to the change and innovation of thought. This claim comes from Ricoeur's (1980) hermeneutic approach to narrative (Armstrong, 2020). Drawing from different theoretical backgrounds, such as philosophy (Heidegger's existentialism and Merleau Ponty's phenomenology), cognitive psychology and narratology, figuration has been identified with 'seeing X as Y', 'as-structure', 'as-if', or 'seeing-as' (Armstrong, 2019, 2020), encompassing dynamic, circular or recursive cognitive processes of categorisation (conceptualisation), configuration, and pattern building.

Armstrong (2020) argues that figuration, as a concept, joins cognitive sciences and narrative theory by pointing to a similar 'phenomenological process of configuring part-whole relations in a text or in life' (Armstrong, 2019, p. 412). As literary interpretation configures circularly part and whole of a story, so the brain categorises part and whole of sense perception and experiences with and in the world. Neuroscience points to fluidity and interactivity as characteristic of the relations between brain, body, and the world (Armstrong, 2020), and figuration in narrative captures this to-and-fro interaction. However, neuroscience does not support anatomically based mental entities that would perform the cognitive processes, such as schema (understood as 'structured mental representations embodied as organisations of related associations', Preston & Eichenbaum, 2013, p. 767), frames (standard situations), and scripts (standard action sequences) (Armstrong, 2019, 2020).

Nonetheless, figuration seems to be central to both stories and the embodied brain (Figure 19.1, adapted from Armstrong, 2020). Stories create a connection between different worlds, one prefigured, another configured (story), and the other refigured, by arranging events and synthesising heterogeneous elements, which seems consistent to the brain's activity of constructing (neuronal and cognitive) patterns as it evolves in its relationship with the body and the world (Armstrong, 2020).

Stories, Events, and Temporality. 'Human existence is embedded in a never-ending flow of time' (Dominey, 2021, p. 2) and storytelling is a specific

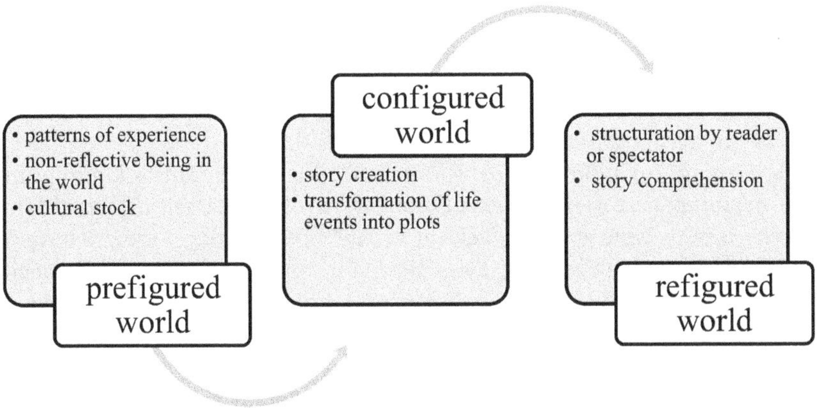

Figure 19.1. Figuration Stages.
Source: Adapted from Armstrong (2020).

form of human behaviour that coherently organises events over time (Dominey, 2021). Stories are set against a time reference, and use it in different ways (chronological order, reverse, or stream of consciousness), reflecting how subjective experiences of time relate to world events (Nelson & Spence, 2020). Time is a narrative element, as much as a phenomenological experience of being in the world, a self-experience (Crossley, 2000). Consciousness and time are closely interlinked phenomena (Crossley, 2000) and neuroscience research has uncovered 'the decentred, asynchronous temporality of the brain' (Armstrong, 2020, p. 54), making conscious awareness 'inherently out of balance' and in need of constantly catching up with itself. Consciousness is evidence of the to-and-fro movement of neurons that dynamically engage in synchronised and desynchronised activity; and according to findings, neuronal activity related to consciousness demands that it is spatially and temporally distributed (Armstrong, 2020; Baars & Gage, 2010).

Just as one's mind 'automatically assumes temporally extended forms in which future, present, and past mutually determine one another as parts of a whole' (Crossley, 2000, p. 534), the nervous system performs a segmentation and organisation of the spatiotemporal flow of perception and action so that a coherent structure emerges to allow effective behaviour (Dominey, 2021). Neuronal activity illuminates the complexity of the human brain as it simultaneously strives for stability (pattern) and change, synchrony and asynchrony. The riddle of time consciousness – 'how can we be conscious of that which is no longer or not yet?' – is practically resolved through storytelling (Armstrong, 2020, p. 27): stories use 'emplotment', the cognitive ability to configure actions adopting a narrative sequence (Armstrong, 2020), to structure a succession of events in a logical way, in which anticipation, representations of the unknown, and memory constantly interplay with one another (Nelson & Spence, 2020), leading to an understanding of the surrounding world that previously was not fully comprehended. This is called 'the event structure of the story' (Dominey, 2021, p. 2).

According to Kukkonen (2014), a plot facilitates the translation of events into meaningful stories. Event structuring involves segmenting continuous activity into discrete events (Zacks et al., 2007). In addition to matching the brain's time mechanisms and dynamics, stories organise events but concurrently leave room for narrative distortion, variation, surprise, and refiguration (Armstrong, 2019, 2020).

Stories are invoked by neuroscientists to study cognitive abilities, such as event structuring and segmentation, on the ground that stories work by editing events, sometimes editing them out, other times expanding and detailing them (Willems et al., 2020). Looking at how the brain works and represents the world, the story's event segmentation is seen to occur in different timescales due to the different demands for processing incoming stimuli, revealing the brain's temporal structure of neurophysiological signals (Dominey, 2021; Willems et al., 2020).

Stories and Emotions. The link between stories and emotions is widely recognised (Negrete, 2021; Vaccaro et al., 2021; Zak, 2014). Stories appeal to our social nature and the need to succeed in life (Völlm et al., 2006). They often feature compelling characters going through events that call into question the needs of survival and happiness, and lead to experiencing strong emotions. This is because the human brain evolved to respond emotionally to events (Brosch et al., 2013; Crowley, 2018), and stories are fabricated with them.

Emotions and cognition are interrelated dimensions of human experience (Brosch et al., 2013). The theory of mind posits that humans can attribute mental states to other humans (Armstrong, 2020; Frith & Frith, 1999; Völlm et al., 2006), a capacity that facilitates social cognition and successful interaction. Another version of this theory of mind, known as Simulation Theory, explains this mental attribution by assuming the innate ability 'to imaginatively "put ourselves in another person's shoes"' (Völlm et al., 2006, p. 90) using 'simulation routines' based on one's own mental experiences. Yet another closely related hypothesis states that humans are capable of inferring others' feelings and emotions, known as 'empathy' (Völlm et al., 2006). Psychophysiological studies indicate that storytelling plays a role in the production of oxytocin, thereby inducing positive emotion and empathy (Brockington et al., 2021; Zak, 2014). Neuroscience research has been in the forefront of the description of the neurobiological processes of emotions (Barrett et al., 2007). Emotions are content-rich events that may be psychologically described but that need to be understood in the light of neurobiological processes (Barrett et al., 2007). Recent neuroscientific findings indicate that emotional responses to events depend on the neural circuitry implicated in concrete experiences, and thus that emotions are instances of embodied cognition, meaning that they too reflect the brain's plasticity and figurative processes (Armstrong, 2019).

Additionally, individual success depends upon cooperation with others, consequently, emotional empathy is important to understand their feelings and states of mind and sharpen our own intersubjectivity skills (Kidd & Castano, 2017). Stories enact events around characters towards whom readers, viewers, or listeners may empathise. They help people understand emotions and emotional changes, which is important from the social point of view, as this ability is critical to develop emotional empathy and pro-social behaviour (Kidd & Castano, 2017).

Empirical studies have shown that emotions experienced with story characters and events are real, though people are aware of the fictional nature of stories and characters. Further, they have also revealed that empathy works similarly in the real world and the fictional one (Nomura & Akai, 2012). Neuroscience findings indicate that neural circuitries of cognition (memory) and emotions are closely interrelated (Armstrong, 2020; Barrett et al., 2007; Crowley, 2018). For example, the amygdala is involved in both emotional and memory processes (Crowley, 2018). Emotions are psychological and neurological experiences (Barrett et al., 2007), and physiological too. They are intentionally oriented towards objects, people, events, and stories mix effectively all these elements. Research has also examined the relationship between stories and emotions through the process of 'neural coupling' (Crowley, 2018; Madsen & Parra, 2021; Stephens et al., 2010). This process describes individuals' temporal alignment (coupling) of brain activity and response during natural communication interaction, as that occurring during telling and listening to a story, and ultimately may help explaining the mechanisms underlying the formation of human bonds and societies (Madsen & Parra, 2021; Stephens et al., 2010). Moreover, recent findings in neuroscience put into question the universality of neural structures of emotions (Armstrong, 2019; Lindquist et al., 2012), instead they hint at embodied cognition, by this opening way to the understanding of emotion as a combination product of learning processes and biology, and thus always present intersubjective elements that connect one to the world and other people (Armstrong, 2020).

The neural basis of emotions, according to neuroscience findings (Armstrong, 2020; Robinson, 2005), seems to match stories recursive characteristic of going to-and-fro among events, characters, times, and responses. Additionally, this neural basis of emotions indicates that they connect to cognition (through intentionality and judgement) in two ways, one by allowing bodily based reactions (non-cognitive affective appraisals) to occur prior to consciousness, and the other, by giving way to a subsequent cognitive monitoring, thus two non-simultaneous, though interrelated, cognitive processes occur recursively within the brain.

Stories and Memory. Memory is the glue that allows humans to follow stories (Armstrong, 2020). According to Schank (1999), human memory is based on stories, so that the telling of stories as an oral practice connecting elements of the community and building cultural identity precedes the advent of written language, and particularly, when established as a social practice among the majority of human societies. Once memory, individually and collectively, accounted for the transmission of myths, legends, and all sorts of stories that humans felt important to pass on the wisdom and legacy of the ancestors (Richardson et al., 2020). This feat was achieved by human long-term memory with the aid of mechanisms that facilitated encoding (mnemonics, repeated exposure, chunking). Today, lessons from studies on memory processes and effectiveness tell that in a world based on the written word, meaning comprehension prevails over wording, thus that stories are a better way to recall information, and accordingly that effective communication, even for science communication, is achieved more easily through stories (Negrete, 2021).

A large amount of information people hold comes from episodic memory (Schank, 1990). Memory has been described as an active process of reconstruction of the past in the shape of episodic and autobiographical stories (Roediger & Marsh, 2003). Either by recollecting personally lived events (a memorable Christmas dinner) or only by knowing personal factual information (the date of birth), individuals process them as narratives of changing detail and complexity. Research has pushed forward this view, concluding that the brain operates under a default mode, which is known by the 'narrative construction' mode (Armstrong, 2020). It has been argued though, to comprehend stories and connect the elements (events or characters), working memory needs to extend in time and with intervention of consciousness (Armstrong, 2020).

Script theory, as popularised by Schank and Abelson (1977, 1995), proposes a close link between memory and stories by postulating that the relationship between language, the world, and human knowledge is based on stories as originated from episodic memories. So, the link between humans and stories are much more about a mind and brain that work in the form of narrative rather than about a creative literary phenomenon (León, 2016). By the definition of Abelson, scripts are a type of schema, a higher-level knowledge structure, 'conceptual representations of stereotyped event sequences … activated when the understander can expect events in the sequence to occur in the text' (1981, p. 715). Scripts derive from stereotypic situations as found in direct life experiences dealing with routinised actions and activities.

In this line, these hypothesised conceptual representations, as fed by personal stories, facilitate long-term memory acquisition, being part of the human memory structure (Bower et al., 1978), but also the dynamic interaction between past experience and new knowledge (Crowley, 2018). Empirically, it has been demonstrated that accurate memory retrieval depends on prior successful encoding (Ferreri et al., 2013). Neurobiologically, again evidence points to an active brain, whose memory mechanisms constantly articulate with experience in search of fine-tuning, as if negotiating with the world what needs to be remembered and what needs to be forgotten. By all accounts, the ability of the brain to forget is also a functional quality that it shares with the storytelling experience (Armstrong, 2020). This particular is relevant from a neuroscience perspective, as well as from a narrative experience perspective, because what is believed to be a fallible memory is in fact the brain's flexibility and adaptability at work (Schacter, 2013). Ultimately, it is this flexibility that produces memory and stimulates imagination (remembering the past and envisioning the future), now believed to be different aspects of the same cognitive processes (Armstrong, 2020; Schacter, 2013).

Research findings relate stories to higher memorability as compared to raw factual information, regardless of the character of the story (history-based, autobiographical or purely fictional), since storytelling, either as a reading or listening experience, cognitively engages people more interactively (Crowley, 2018). Arguably, memory encoding works differently for stories and other types of narrative based on facts, due to the intervention of more brain areas in the process (Berns et al., 2013). The workings of the brain as to memory involve the hippocampus, which stimulates the connection between neurons, and the prefrontal cortex, in charge of deciding which information is relevant to store in the long term, processes that require to direct attention to it (Crowley, 2018). What, in particular, draws the attention in this selection process may be a combination of subjective factors (like personal interest) or non-subjective (a salient stimulus). Additionally, emotions are critical to long term memory effects, thus 'dramatic, eye-catching' narratives are more easily remembered. This has been ground for claiming 'the fiction superiority effect', according to which, to humans, the mere labelling of 'fictional' attached to a narrative makes it more memorable (Crowley, 2018).

Stories and Attention. Attention is a scarce cognitive resource that needs to be used parsimoniously. The term 'spotlight' is a metaphor for focussed attention on some part of the world (Zak, 2013). Stories provide a naturalistic approach to studying attentional phenomena, especially those involving longer periods of time. Focussed attention involves separating 'the continuous nature of thought and attention' into parts (Hinde et al., 2018, p. 2). Thus, a story may 'capture' attention. A study conducted by Hinde et al. (2018) revealed that stories using film as the *medium* impact attentional load during time, providing some ground to the claim that stories immerse and transport people by superiorly sustaining attention. This means stories create cognitive engagement and attention. Additionally, either through personal report or brain activity, their impact on people's attention and memory is related to their emotional content (Song et al., 2021).

Engagement with stories is defined as the experience of being immersed in one with heightened emotional arousal, attention, effort, and agency in task performance (Richardson et al., 2020; Song et al., 2021). A study by Song et al. (2021) shows that in particularly engaging stories, top-down sustained attention operates as in an everyday context, but also that the neural circuitry involved in top-down sustained attention and stimulus-related narrative engagement predict subjective engagement and posterior memory.

Tourism Research on Stories and Storytelling

The use of stories and storytelling in tourism is attracting increasing attention from scholars and recent research has called for the need to further elaborate on the theme by exploring scientific and theoretical foundations able to better inform future studies in tourism (Moscardo, 2021). In the past, the theme has passed mostly unnoticed (Moscardo, 2020; Servidio & Ruffolo, 2016). There is an obvious link between stories and the travel (Järv, 2010) and tourists share their travel experiences with relatives, friends, tourist communities, or transient strangers (Bosangit et al., 2015). Tussyadiah and Fesenmaier (2008) highlight the importance of storytelling as a process of communicating a tourism experience. Also, digital technologies are further facilitating this process by allowing to expand the sharing of stories, reaching wider audiences of readers (Hughes & Moscardo, 2019; Moscardo, 2020; Shen et al., 2020; Tussyadiah et al., 2011). Moreover, this practice is believed to impact others' perception of activities, companies, services, and destinations, thus leading to posterior behaviour, such as destination offering, selection, purchase, recommendation, or other (Bosangit et al., 2015).

Stories are a part of the destination experience (Tussyadiah & Fesenmaier, 2008), used to communicate destination representations (Moscardo, 2017a), draw attention to attractions (Kim, 2012), and help tourists to interpret and understand culture, symbols, and places, thus becoming critical to a meaningful destination experience (Mathisen, 2013; Moscardo, 2017a; Mossberg et al., 2010). These stories are the link between tourists and the destinations, as they effectively combine tools that allow interpretation and meaning

(Moscardo, 2017a). In this light, stories are a well-known strategy to promote destinations and their distinctive attributes (Moscardo, 2020).

In turn, Mossberg's work (2007, 2008) on storytelling highlighted how businesses can thrive and build differentiation and competitiveness for themselves and meaning for customers by investing in stories. In this context, a story is defined as the 'verbal and visual metaphor, which shows the total offering, the total package, which for tourists hopefully, is received as a positive experience' (Mossberg, 2007, p. 71). The power of stories to facilitate access to meaning and emotions has been understood by tourism companies and destinations, that are progressively seen to use storytelling as a communication tool and a way to emotionally engage customers (Moscardo, 2018).

The use of stories in tourism destinations and businesses has been in practice for some years (Mossberg, 2008). The recent developments in immersive technologies are further helping destinations in the design of experiences that involve tourists through stories of the sites combining elements of education and entertainment, by this providing access to a deeper onsite experience. These technology-based developments require the active involvement of participants through immersion in the experience environment, thus showing the growing relevance of a co-creation design approach to tourism development (Moscardo, 2020; Neuhofer & Buhalis, 2014).

Despite recognition of the critical importance of the cognitive sciences to the understanding of the tourism experience, particularly cognitive psychology, and that of stories in both perspectives of destinations/businesses and the tourist, the cognitive approach to storytelling has been generally ignored in research. The work of Moscardo (2017a, 2020) on storytelling is an exception to this, as most research of hers has been calling attention to the need of combining psychology science, the tourism experience, and interpretation to better propose and design memorable experiences. Specifically, the cognitive sciences (psychology but also neuroscience, Moscardo, 2020) contributed to establish the relation between stories and cognitive functions, such as knowledge construction, memory organisation, sense of self and identity building, but also the three basic principles that support the study of stories and storytelling practices in the context of tourism stories are the way people think about their experiences; they are the way people store information in their memory, and (3) they are the way people make sense of the world (Moscardo, 2018). Cognitive sciences have also studied the social role of stories, and how they shape the relations between individuals (Mar, 2018; Moscardo, 2020).

CONCLUSION

Overall stories matter in tourism, and concretely in tourist experiences, business companies, and destinations. Considering the narrative/story turn, Moscardo (2020) challenges researchers to become better informed on stories and the core sciences that have studied them and use this knowledge and extensive literature to effectively apply in tourism research. This is in line with the claim that by integration of the science behind stories, a better approach to the design of tourism experiences is possible (Moscardo, 2018). Cognitive psychology and more recently neuroscience are examples of those core disciplines that have contributed substantially to current understanding of human psychology and neurobiology and their relation to the theme, and accordingly this chapter focussed on them.

Psychological research and theories have found an interest in stories to understand psychological processes and functions. Language, comprehension, meaning, or figuration have been studied according to dominant paradigms. Despite divergencies, psychology and the cognitive sciences have stressed the link between language, the mind, and the world, which has facilitated stories as a means to study psychological functions and processes. Interestingly, only recently has neuroscience acknowledged this relationship long postulated by psychologists and literary scholars.

Neuroscience is a recent discipline in the study of stories but increasing awareness of contributions from this field are stimulating more collaborative work among the cognitive sciences. This is important for researchers interested in the study of stories and the tourism experience to invest in the core sciences that have a word to say on human psychology, behaviour, and the brain. Findings from neuroscience are changing the way one understands the brain and how it relates to human cognition, action, and the world. It is important to acknowledge that research results indicate that the brain is 'a bushy ensemble of anatomical features' (Armstrong, 2019, p. 413), a plastic organ that learns through experience and interaction with the world, so despite its need for constancy, it is constantly open to change.

This has implications for the way cognitive processes, such as language, interpretation, or emotions, are described and explained. Findings translate into a changing paradigm, both for neuroscience and the other cognitive sciences, that have to let go of formalist or structuralist premises that posit the brain as functioning with rigid modules and structures under fixed rules. What evidence tells about the brain is that neurons build networks and develop connectivity according to stimuli resulting from experience and repeated action and contexts, so whatever neural patterns are found in it they are emergent, depending more on that experience and repetition (along with its statistical regularities) than on anatomical and genetic determinants. In this sense, it is confrontation with new experiences that creates new avenues for brain structure and functionality (Armstrong, 2019).

Neurobiologically, figuration is not localised in a particular region of the brain, however it is a characteristic of stories that, by enabling people to access new worlds and perspectives, drives personal enrichment (Armstrong, 2019, 2020). Person characteristics, such as mental states, traits, or autobiographical memories, and cultural differences, are recognised factors that may affect how people react to stories or mentally simulate reality (Appel et al., 2015; Richardson et al., 2020; Vaccaro et al., 2021); however, this characteristic of the brain illuminates the need to constantly mix the dimensions of stability and novelty in stories.

Time is an element of a story and a literary device to tell it (León, 2016; Nelson & Spence, 2020). Stories feature different temporalities, some of them relating to time as measured by clocks or time as experienced by subjects, as a psychological phenomenon, which may distort clock time by contracting, expanding, or reversing it. As stories capture the subjective experience of time (and the consciousness of it), they are true repositories of the heterogeneity of human experience, an opportunity to learn about other subjects and cultures. The effectiveness of stories is manifold, and research is providing evidence of assumptions that for many years remained speculative and philosophical. Neuroscience research emphasises how the brain intertwines with time, that the brain itself needs time to process incoming stimuli, but most importantly that needs different times to do it depending on their characteristics (Dominey, 2021; Willems et al., 2020). As narrative devices are employed to convey how long or how short an event is perceived by a subject, time fluidity, and elasticity are characteristics to explore in stories.

That stories engage people in emotional experiences is widely accepted, and there is empirical evidence from human psychophysiology and neuroscience to this. Moreover, regardless of the fictional or historical character of the narrative, emotional responses are similar in both cases. Research results also indicate that human brains are capable of coupling. This suggests that analogous emotional experiences (of mental representation) are shared by different individuals engaged in the same activity (Richardson et al., 2020). Given the brain's dynamic circuitry connected to different cognitive functions, such as emotions, attention and memory, stories are a context for a rich and complex activity of engaging the whole human being.

The long tradition of oral communication worldwide is a testimony of the power of stories to persist in time. So, they are closely connected to human memory, individually and collectively. Stories help the formation of long-term memories, and oral tradition, long before the advent of written language, widely established stories as effective vehicles for communication. Empirically, research also shows that stories involve higher attentional engagement, which helps the brain to better encode, store, and retrieve information (Hinde et al., 2018; Song et al., 2021). By use of language, written or oral, stories facilitate this process by structuring discourse through mnemonics or chunking, which have been proven effective strategies for retention (Crowley, 2018). Neurobiologically, stories seem to engage the brain in complex processes and processing through intervention of different brain areas. But also here, evidence highlights the dispersion in the brain of the memory function, as different areas are implicated in action depending on the type of memory at work (short-term versus long-term), and of how time is differently required to perform specific memory work (Armstrong, 2020). How time is used to perform neural activity is, again, similar to the works of time during story comprehension. It has been demonstrated that the channels used to tell the story (visual, auditory as in watching a movie or listening to a story) play a part in the comprehension process, thus that mentally and neurobiologically different experiences actually take place. Some of these findings may help justify why destination promotion so heavily depends on visual stimuli (requiring less cognitive effort in attention capture), and otherwise provide guidance to the design of tourist experiences cognitively more engaging by relying preferably on oral communication.

What are tourism researchers to do with cognitive psychology and neuroscience findings about how mind and brain work with stories? How useful for tourism can theory be or develop from here, and, perhaps more importantly, what are practitioners to learn from scientific evidence? Psychology research matters for tourism experience design, as well as business and destination management (Moscardo, 2018, 2020, 2021). However, not all theories and assumptions from cognitive psychology are corroborated by neuroscience evidence. Neuroscience studies provide a dynamic picture of the brain arranged through neural networks that, though reinforced by experience, are not fixed. Cognitive functions occur in a complex manner, with one brain area involved in different cognitive functions, and different brain areas combining to perform one cognitive function (Crowley, 2018). Armstrong (2020) asserts that what is currently known about brain's activity compromises the vision of the first generation of cognitive narratology, developed from the representation of the human mind as an orderly product of an orderly brain. Cognitive psychology, and particularly in what concerns the role of stories as illuminating processes and functions of the mind, is challenged to propose an alternative view of the narrative experience integrating the new contributions from neuroscience. A more phenomenological approach to the study of the mind and the brain is needed to capture the fluidity of embodied cognition. These may lead to a change of paradigm from the mind as a detached entity working on its own rules and structures to a relational mind linked to a body and the surrounding world (embodied).

Cognitive engagement in general is good for the brain (Seider et al., 2016), and stories are particularly stimulating. These psychological rewards are complemented by the emotions aroused through reading or listening to a story. Stories are contexts for experiencing emotions (Negrete, 2021), and figuration as an 'as-if' narrative process enables people to look at the world differently, leading to a new understanding of events and others, and often doing this using the characters' own eyes, a process of doubling self ('me') and other ('not me'), called 'the paradox of the alter ego' (Armstrong, 2020, p. 156). This power of stories is widely recognised among scholars from different academic backgrounds and scientific fields, and here too cognitive psychologists and neuroscientists are bridging the gap between them. Cognitive hypotheses have been discussed to account for the power of understanding the human mind in others, as well as the positive social effects of storytelling. Despite the still undergoing debate on mirror neurons (Völlm et al., 2006), research has observed neural coupling evidence during cognitive processing, suggesting psychological attunement during storytelling (Stephens et al., 2010).

From this, a collaborative research agenda is postulated as necessary to integrate the tourism context into cognitive psychology and neuroscience research, given the amount of evidence in tourism studies that the practice of tourism, from the tourist perspective and from the businesses and destination perspectives, is increasingly making use of stories. In view of the above, some lines of inquiry on storytelling applied to tourism are suggested with the support of collaborative research teams. Researchers could explore and test cognitive psychology theories and neuroscience methods and tools applicable in the tourism field. Psychophysiological studies are already known (Scuttari, 2021), despite this more evidence is needed on the opportunities afforded by the combination of the cognitive sciences that may help improve knowledge and design of the tourism experience. Therefore, researchers could apply self-report, physiological measures, and brain activity using different channels for destination promotion, to probe the effects of different types of stimuli. Previous research has addressed the multisensory nature of tourist experiences (Agapito et al., 2013a). However, there is little research on how tourists cognitively and neurobiologically respond to sensory modalities used in promotional stories.

Research on the effect of stories (empathy) that tourists tell others in naturalistic settings and reports may help to infer behavioural intention towards the businesses or destinations depicted. These effects include stronger memories. The recommendations for future research link the tourist experience, stories, and the cognitive sciences. Such topics provide fertile ground for a better understanding of tourist experiences with implications for their design, marketing, and management.

SECTION 4
METHODS

Chapter 20

EYE-TRACKING IN TOURISM

Biqiang Liu
Griffith University, Australia

Brent Moyle
Griffith University, Australia

Anna Kralj
Griffith University, Australia

Yaoqi Li
Sun Yat-sen University, China

Abstract: Visual stimuli are integral for the destination selection process, as well as for the delivery of unique, novel and compelling tourist experiences. Emerging techniques, such as eye-tracking, are effective for mapping tourists' visual interests and paths, presenting an opportunity to identify patterns of visual attention, which provide insights into the underlying cognitive processes which underpin experiences. Building on a systematic review of the progress and development of eye-tracking in tourism field, this chapter summarises five main current research contexts for application and five future research directions. It also narrows the gap between eye-tracking and cognitive psychology by critically examining bottom-up and top-down attentional mechanisms. **Keywords**: Eye-tracking; attention; literature review; cognitive psychology; experimental design; heat map; eye movement

Cognitive Psychology and Tourism
Tourism Social Science Series, Volume 27, 259–272

Published under exclusive licence by Emerald Publishing Limited
ISSN: 1571-5043/doi:10.1108/S1571-504320240000027020

INTRODUCTION

Herbert Simon, the 1978 Nobel laureate in economics, was adamant that in the information era, the ability to attract consumer attention is the key to successful marketing (Simon, 1988). Vision is a primary channel for people to obtain external information, with an estimated 80% of the collective information retained by the human brain arriving via the visual system (Liu & Li, 2021). Further, an estimated 40% of brain functioning is used to process this visual information (Cavanagh, 2011). Consequently, how to attract and maintain consumer attention has generated substantial interest in academia, with growing momentum in the field of tourism (Scott et al., 2019).

A touristscape is a multi-dimensional landscape, which includes visual landscapes (Mordue et al., 2020), soundscapes (Liu et al., 2018) and smellscapes (Dann & Jacobsen, 2003). However, the visual landscape has been identified as exceptionally important for the delivery of memorable tourism experiences (Li et al., 2022; Scott et al., 2019; Wang & Sparks, 2016). While intangibility is a critical feature of tourist's experiences, a destination can become tangible through postcards, maps, photos, videos and websites which attract tourists' attention (Hillman, 2007). Compounding the increasingly complex visual landscape, tourists also create visual materials which are released on social media, not only sharing experience with others but also constructing and improving their self-image (Liu et al., 2022). Subsequently, it is critically important for both scholars and practitioners to develop an understanding of the visual displays and cues which capture tourists' attention.

Tourism stakeholders can utilise eye-tracking to improve their visual marketing material and learn how visual cues in marketing materials influence tourists' attention. Traditional self-report data collection methods (such as survey, interview, etc.) may lead to respondents' bias (social desirability bias, recall bias and impression management bias), which may mislead tourism business practices (Graham & Jeffery, 2011; Wästlund et al., 2015). The development of eye-tracking technology provides an objective method to track people's attention directly through non-invasive measurement of eye movements which measure normal attentional responses (Wedel & Pieters, 2017). In recent years, researchers in the tourism and hospitality fields have begun to use eye-tracking techniques as an important step forward, embracing the 'cognitive wave'.

EYE-TRACKING AND COGNITIVE PSYCHOLOGY

Tourism has been widely accepted as an interdisciplinary field of scholarly enquiry (Liu et al., 2022; Weiler et al., 2018). Among its various parent disciplines (psychology, economics, environmental studies and anthropology), psychology is important for the study of tourism marketing and product development (Skavronskaya et al., 2017). As discussed in Chapter 2, cognitive psychology is devoted to offer insights into the underlying mental process of cognition and behaviour.

Stimulus information from the external environment is received through a variety of sensory channels (Agapito et al., 2013; Krishna, 2012). Le et al. (2020, p. 20) argue that 'researchers using eye-tracking methods must have a sound understanding of perception and attention as cognitive processes in data analysis'. Concomitantly, Mayr et al. (2009, p. 198) write that 'interpretations of eye-tracking data ... are often based on assumptions and heuristics about underlying cognitive processes'. People's visual attention is a multilevel process where people actively or passively process a subset of relevant information from available visual cues (Kastner & Pinsk, 2004). In addition, individuals are limited in attention to stimuli due to their information processing capacity (Campos et al., 2017). As a result, visual attention is more likely to be directed towards those stimuli which are high in salience. Campos et al. (2020) note that rarity and novelty attract visual attention. A rare or extraordinary stimulus attracts attention and is considered to be exciting. Eye-tracking measures including fixation time, fixation count and the area of interest enable scholars to examine numerous cognitive functions, such as memory, attention and learning (Eckstein et al., 2017).

Bottom-Up and Top-Down Attentional Processing

Attentional processes are affected by beliefs or knowledge, interests or motives of the consumer (top-down factors) and characteristics of the visual marketing stimulus (bottom-up factors) (Skavronskaya et al., 2017; Wedel & Pieters, 2017). The difference of these two attentional processing mechanisms lies in the point of origin of the stimulus which is responsible for activation of the attentional response (Campos et al., 2020). The bottom-up attention mainly refers to the stimulus driven/extrinsic attention induced by the physical characteristics of the visual stimulus itself, which occurs subconsciously. Conversely, top-down attention is influenced by people's goals, such as personal preference, previous experience and personal characteristics (Wedel & Pieters, 2017). The eye movement process is driven by an internalised cognitive model, which runs in a top-down processing mode and remains stable in multiple exposures. Although a large body of literature has discussed the bottom-up and top-down attention as independent processes, there is still ongoing discourse which seeks to reconcile this debate. Campos et al. (2020) consider that bottom-up factors directly impact attention, while top-down factors are more likely to moderate such influence.

Background to Eye-Tracking

Eye-tracking is based on the movement of eyes when people obtain external visual information (Bai et al., 2008; Liu, & Li, 2021). Eye movement plays an essential role in processing a specific object clearly. Along the visual processing, the most sensitive visual region in human eyes is the fovea in the retina, which has the highest concentration of cells and is responsible for high resolution (Zhang et al., 2019). The closer to the fovea, the more cone cells which can sense colour and strong light in the retina there are, the higher the visual acuity is (Provis et al., 2013). Subsequently, to clearly see the object, it is necessary to focus the image of the object in the fovea using eye movements (Wedel & Pieters, 2017). These movements enable the attention mechanism to select only a small amount of relevant information for in-depth processing (Wästlund et al., 2015). Therefore, focusing on a specific object or position means improved processing of the focal stimuli with processing of the other stimuli not selected being inhibited. In this way, eye movement is an indicator of visual attention during information search.

There are three basic types of eye movements used in eye-tracking research: fixations, saccades and smooth pursuits (Yan et al., 2013). A fixation occurs when the fovea of the eye fixates on an object, for a period generally lasting 200–300 ms (Wedel & Pieters, 2017). Information derived from visual processing is mainly obtained during a fixation (Duchowski, 2017). A saccade refers to the rapid movement of the fovea of the eye between fixation points (Yan et al., 2013). It is a movement controlled by the human nervous system involving a rapid change in the fixation point. A saccade results in the fovea focusing on a new fixation point. A smooth pursuit is when the eye and the object are moving relative to each other, the eye will follow the object in order to keep fixated on the object (Larsson et al., 2015). The result of a smooth pursuit is the same as a saccade, to make the stimulus focus near the fovea (Le et al., 2020).

A large number of studies have shown that eye movement, as a physiological response, is an accurate reflection of the cognitive processing activities of an individual (Krugman, 1965; Rosbergen et al., 1997; Zhou et al., 2016). To accurately capture this physiological response, eye movement observation equipment has evolved to its present non-invasive state. Due to the limitations of technology, early eye-tracking research (such as direct observation method and mechanical recording method) were prone to bias, which made it difficult to obtain accurate and reliable results (Kim & Jeong, 2016). However, in recent years, eye-tracking technology has matured, allowing for the non-invasive, objective measurement of attention. Given its merits, eye-tracking technology has been widely used in psychology, clinical medicine, computer science, systems and human factors, engineering and marketing research, with studies emerging in the tourism field (Scott et al., 2019).

Eye movement indicators are classified according to the three basic eye movement types. They are selected depending on the specific questions of the research. Often a researcher will be interested in fixations on a particular area of interest. This can be any part of the visual material that researchers are interested in, and analysis software can be tailored to visualise the number or time of fixations occur in aa specific area of interest. These can be used to generate a heat map or gaze plot which intuitively demonstrate the subjects' attention to visual cues (examples in Table 20.1; and Figure 20.1).

Table 20.1. Eye-Tracking Metrics and Specification.

Eye-Tracking Metrics	Specification
Time-to-first fixation First fixation duration	This indicator refers to the time/duration of the subjects' first gaze at area of interest when a stimulus is presented to the subjects for the first time. This indicator is mainly used to measure the visual importance of stimulus information.
Fixation count Fixation duration	The number of fixation times/fixation duration is mainly used to measure the processing depth and cognitive effort. Some scholars believe that the more the fixation times, the more unreasonable the layout of area of interest; the longer the fixation time, the more difficult it is for the subjects to obtain information from area of interest.
Visit count Visit duration	The number of visits/total visit time to stimuli is mainly used to measure the degree of interest in stimuli.
Revisit count	The number of times to visit area of interest again after the first visit to area of interest. It shows the visual importance of area of interest and the reprocessing process.
Fixation frequency	The ratio of fixation count to fixation time. The higher fixation frequency shows, the more interested or important the subjects feel.
Percentage of fixation duration	The ratio of fixation time to total observation time. The larger the percentage of fixation duration, the longer the observation time of the area of interest compared to others.
Saccade frequency	The number of saccades per unit time. It can reflect the search behaviour of the subjects. In the process of information searching, the higher the saccade frequency is, the larger the amount of searching is. It also negatively affects the visual salience of stimulus information.

(Continued)

Table 20.1. *(Continued)*

Eye-Tracking Metrics	Specification
Percentage of saccade duration	The ratio of the total saccade time to the total observation time. It mainly reflects the time taken by the subjects to select the fixation point.
Average saccade amplitude	The index is used to reflect the scope of access to information. The larger the average saccade amplitude is, the more distinct the stimulus characteristics are, with showing subjects can visually reach area of interest directly.
Average saccade velocity	The index is to evaluate the scope of information obtained by the subjects. It reflects the distinctive characteristics of the area of interest information.
Pupil diameter	It reflects the subjects' interest in stimulation and cognitive load.

Source: Casado-Aranda et al. (2023).

Figure 20.1. Some Examples of Heat Map (Author's Research).

Applications of Eye-Tracking in Tourism

As a visual experience activity, the presentation of visual components plays an integral role in optimising tourism marketing activities and shaping a destination image (Feighey, 2003; Scott et al., 2019). A review of the relevant literature of tourism eye-tracking research found five main current research applications in tourism (Table 20.2).

Table 20.2. The Profile of Eye-Tracking Research in Tourism.

Authors (Year)	Stimuli	Sample/N	Apparatus	Metrics
Babakhani et al. (2020)	Menu	Students/54	Tobii TX-300	Fixation duration, time-to-first fixation
Babakhani et al. (2020)	Website	Students/31	Tobii-TX300	Fixation duration, gaze plot
Li et al. (2020)	Tourism pictures	10–13 years old children/ 45	SMI REDn Scientific	Fixation duration, fixation count, heat map
Kong (2019)	Website	Students/30	EyeLink 1000	Fixation duration, heat map
Liu and Upchurch (2020)	Tourism pictures	Students/55	Tobii Pro X3-120	Fixation duration, fixation count, heat map
Ngan and Yu (2019)	Service providers' photos	Students/ 112	Gazepoint GP3	Fixation duration, heat map
Scott et al. (2019)	Tourism pictures	Students/66	Tobii T60	Fixation duration, fixation count, heat map
Wang et al. (2019)	Hotel pictures	Real consumers/ 80		Fixation duration, fixation count
Kim et al. (2018)	Menu	Students/95	Eyetech VT2	Fixation duration, fixation count
Babakhani et al. (2017)	Tourism pictures	Students/20	Tobii-TX300	Fixation duration, visit duration
Li et al. (2016)	Tourism pictures	Students/37	EyeLink 1000	Fixation duration, fixation count, heat map

(Continued)

Table 20.2. *(Continued)*

Authors (Year)	Stimuli	Sample/N	Apparatus	Metrics
Scott et al. (2016)	Tourism advertisement	Students/25	Tobii T60	Fixation duration, fixation count, gaze plot, heat map
Wang and Sparks (2016)	Tourism pictures	Chinese and Australian students/30	Tobii T120	Fixation duration, fixation count, gaze plot, saccade
Pan et al. (2013)	Website	Faculty members/16	GazeTracker	Fixation count
Yang (2012)	Menu	Students/25	iScan EC501	Fixation duration, fixation frequency, gaze plot

Eye-tracking technology is designed to capture and record the entire process of the consumer being exposed to a marketing simulation in real time. However, it also enables researchers to grasp the sequence of a viewer's gazes while providing metrics such as fixation duration, saccades and pupil diameter changes at any time. This technology can be used to analyse the audience's attention and interest in the stimuli, thus assessing the effectiveness of tourism marketing strategies. The eye-tracking technique provides a way to understand if the marketing activities of tourism enterprises are effective (Feighey, 2003; Scott et al., 2019).

Tourism Website Layout Optimisation. Tourism websites have a substantial impact on tourists' information search and the decision-making process (Hernández-Méndez & Muñoz-Leiva, 2015). Eye-tracking has advanced the understanding of the individual's gaze on the tourism website interface, with strong implications for the effectiveness of website design (Scott et al., 2019). For example, Kong et al. (2019) explored the effect of different combinations of advertising pictures, introduction text and price information of tourism products on the effectiveness of tourism websites, and found that the advertising of websites containing introductory text and price information tended to attract the attention of tourists. Conversely, websites containing only advertising pictures and price information captured less attention. This study did not provide a theoretical reason why the combination of image, text and price was more effective than other combinations (colour, format and framing).

The field of hospitality is also embracing eye-tracking research. Pan et al. (2013) examined the influence of hotel reservation websites and the quantity of information, finding tourists' attention to online hotel information is related to not only the specific location displayed on the website but is influenced by the number of hotel options to select from and the format of presentation of pictures. Similarly, Babakhani et al. (2020) found that customers, when overloaded by visual stimuli, often ignore aspects such as pro-social information (social responsibility attributes) which are published on the hotel reservation website. To combat this, hotel websites could consider presenting social responsibility information prominently to guide customers' reservation decisions and reducing the total amount of information provided.

Tourism Advertisement Design. Tourism advertisements are fundamental for marketing, and whether technique can attract the attention of the audience is an important indicator of its effectiveness (Li et al., 2020). Eye-tracking in advertising effectiveness research can measure the interest and attention of users (Wang & Sparks, 2016). Xu et al. (2014) found that compared with the text-based pictures, illustration-based tourism pictures not only reduce the time spent on scanning but also increase fixation time. The relationship between pictures and text in advertisements has often been a subject of controversy in eye-tracking studies. Li et al. (2016) found that landscape pictures with words attract more fixation time and frequency, and that landscape pictures with indigenous people increase fixation time compared with pictures with foreign characters.

In addition, academic attention on stimulus material, scholarly inquiry is also shifting to explore a diversity of the samples. Chinese and Australian students are compared as a cross-cultural comparative study to explore the different visual pattern on low arousal or natural pictures (Wang & Sparks, 2016). Emerging groups, such as youth or child tourists (Li et al., 2020), are increasing in scholarly attention in eye-tracking research.

As research evolved, existing studies not only consider different types of tourism or hotel pictures but also have commenced focusing on broader aspects associated with the management of enterprises. The information demand differences of potential consumers in online hotel reservation decisions are a topical area of interest. The subjects are found to pay more attention to the tangible facilities and equipment information of budget hotels and the intangible service and atmosphere information of high-end properties (Yuan et al., 2019). Compounding this, the theories which are utilised to assess attention are also evolving, with attention restoration theory, applied to compare hotel image which highlights cultural and natural landscapes, noting that portraying the natural landscape can significantly improve the customer's fixation duration, cascading into perception of the restorative qualities of the hotel (Wang et al., 2019). In addition, the paradigm of top-down attentional processing has also been expanded to include task orientation, with studies demonstrating task orientation can significantly affect the gaze of tourists.

Menu and Food Label Design. Menu and food labels can provide useful information for customers in the hospitality sector to help with decision-making around food (Drichoutis et al., 2012). However, previous studies have contradictory conclusions on the effectiveness of menu labels. Studies have shown that calorie labelling can effectively reduce consumers' consumption of junk food (Burton et al., 2009; Dowray et al., 2013), and gluten-free labelling can significantly reduce consumers' purchase intention (Shin & Mattila, 2018). However, some scholars doubt the effectiveness of these practices (Swartz et al., 2011), as the so-called 'sweet spots', the areas that customers pay attention to first, last or often, have not been empirically confirmed when customers browse the menu (Yang, 2012). When focusing on carbon labels and those communicating local community benefits on restaurant menus, such area of interest's only attracted a small number of fixation time and fixation duration, neither the first area for customers to gaze when browsing menus (Babakhani et al., 2020). In addition, healthy food labels which display exercise and calorie labels (vs. numerical and colour calorie labels) are more attractive to consumers in terms of fixation time and fixation count, and thus can trigger their healthy diet choices (Kim et al., 2018).

Tourism Service Marketing. Tourists' evaluation of personnel during and after service encounters is fundamental for service quality, crucial to the success of service enterprises (Liu & Li, 2022). In the process of communication between tourists and service personnel, the external factors of service personnel (such as facial expression, clothing and appearance) are the primary information that consumers are exposed to, and consumers will pass on initial judgment according to the external information of service personnel (Li et al., 2021). Existing literature demonstrates this subjective judgment continues throughout the whole process of service interaction and generates an overall impression of service personnel (Li et al., 2022; Wan & Wyer, 2015). Eye-tracking technology can observe the real-time viewing behavior of tourists and avoid the influence of recall bias and impression management motivation. For example, the roles of different smile types on service recovery in the service failure scenario have been studied. Specifically, when the service staffs wear smiles showing the teeth and displaying crows' feet wrinkles, they could attract more attention and improve the possibility of getting customers' forgiveness (Ngan & Yu, 2019).

Evaluation of Tourism Resources. The evaluation of the attraction and aesthetic feeling of tourism resources is the basis for reasonable planning and development (Guo et al., 2017). An increasing number of scholars try to break through the relatively subjective limitations of traditional evaluation methods of resources, and then explore the relationship between more objective eye movement index and traditional methods. The research has used eye-tracking techniques in the context of wind farm landscapes (Liu & Upchurch, 2020), rural landscapes (Guo et al., 2017), port landscapes (Li & Dong, 2019) and coastal natural landscapes (Le et al., 2019). It is found that eye movement data have a certain indication effect on the visual quality of landscape, and then provides a new way of thinking for the evaluation of tourism resources aesthetic feeling.

CONCLUSION

Based on the review and analysis of the existing literature, this chapter identifies five areas for future research, which are guided by enhancing the understanding of the psychological mechanisms behind people's eye movement.

Exploring the Psychological Mechanism of Eye Movement

Analysing the psychological mechanisms which underpin tourists' visual change presents an opportunity to open the black-box of the psychological process of eye movement. In fact, assessing only the eye movement index does not adequately reflect the cognitive characteristics of the subjects' information processing process (Scott et al., 2019). Subsequently, research should focus on evaluating the mediating effect of visual perception process. Existing research of tourism eye-tracking mainly focuses on the direct impact of visual stimulation on people's perception, and analyses the audience's attitude, emotion and behaviour response to visual stimulation. However, there is limited focus on how to process the information of visual stimulation by the audience. Scholars are recommended to combine eye-tracking technology with self-report research methods, such as using survey to measure the psychological variables (emotion, attitude and memory) of the target population, so as to explain the psychological mechanism of eye-tracking behaviour, rather than only discussing group differences by regarding demographic variables as the classification criteria. Moreover, the use of the hybrid methods can verify and effectively triangulate research conclusions, thus improving robustness of approaches.

Eye-tracking research in combination with other neuroscience techniques, such as electroencephalogram, could also be quite beneficial. Although eye-tracking technology can effectively predict the internal cognitive process of individuals, it cannot directly reveal the physiological mechanisms utilised to process information. In addition, scholars are recommended to adopt eye-tracking technique to open the black-box of the tourism experience (emotional experience). Further engagement with neuroscience technologies, including eye-tracking, can provide valuable insights into uncovering potential elicitation of emotion through indication of activation of autonomic nervous system. Eye-tracking can be combined with neuroscience technologies, including electrophysiological/brain imaging, such as galvanic skin response, functional magnetic resonance imaging and event-related potential. Combining with other approaches would advance the understanding of the psychological significance of eye movement characteristics and deepen the understanding the eye-brain consistency, and further revealing the physiological mechanism of a psychological phenomenon.

Focusing on Visual Material Processing Mechanisms

With the scope of research on eye-tracking expanding, it is difficult to provide valuable insights into the psychological mechanisms of eye movement by only relying on the bottom-up physiological-driven processing mode of tourists (Campos et al., 2020). In order to analyse the psychological mechanism of eye movement in a deeper level, the focus on the top-down motivation-driven processing mode is required (Wedel & Pieters, 2017). The latter is mainly guided by human beings according to specific behavioural objectives. The visual objects often have high-level visual characteristics, and usually occur in the late stage of deep processing of visual materials (such as the process of individual work memory representation guidance attention capture) (Orquin & Loose, 2013). This provides a critical breakthrough to supplement the depth of the existing eye-tracking research. Therefore, the visual object design in the field of tourism eye-tracking should focus on the refinement of visual materials (the arrangement and sequence of internal elements) and grasp the dynamic changes of the whole process of tourists' gaze in different periods (early, medium, and late), and increase the consideration of the task status and time of tourists, so as to deepen the understanding of eye-tracking processing mode.

Considering the bottom-up attention processing mode, research which explores the cognitive mechanism of dynamic visual materials is limited. Unlike traditional static visual materials (text and pictures), dynamic visual materials (dynamic pictures and videos) are widely used in the tourism marketing in the new media era with its strong vivid and infectious advantages (Hernández-Méndez & Muñoz-Leiva, 2015). Due to the fierce market competition, how to design them creatively to better transfer consumers' attention to tourism products is a common concern and focus. However, dynamic visual materials are characterised by high level of diversity, dynamics and interaction, which not only needs more suitable research design but also puts forward higher requirements for the collection and analysis of eye-tracking data.

It has been found that dynamic visual materials play an important role in helping potential tourists improve destination cognition, form a positive evaluation of destination and induce intention to visit the destination. For example, virtual reality (VR) has been demonstrated to provide an alternative setting that combines the benefits of mobile eye-tracking with the flexibility and control provided by lab experiments. Compounding this, the usage of VR stimuli is rapidly emerging as a new approach for creating innovative products and services in businesses (Meißner et al., 2019). Considering the complexity of dynamic visual materials, it is necessary to consider the use of existing mathematical statistics, temporal and spatial indexes of eye movement, visual analysis, geographic information system time series data modelling and other relevant techniques, to qualitatively and quantitatively analyse the dynamic process of eye movement and visual cognition of tourists.

Explore the Eye Movement Index

Insufficient understanding of the psychological significance of eye-tracking indicators could result in important information overlooked, with difficulties surrounding generalisation of research findings from data analysis. The identification and critical examination of meaningful eye-tracking indicators (such as first fixation time and revisit count) will assist to explain more effectively and comprehensively assess the psychological mechanisms behind eye-tracking behaviour. Most of the existing tourism eye-tracking research only takes the temporal index such as fixation time, fixation count and heat map generated by the two indices, neglecting the importance of spatial eye-tracking index (such as the fixation point coordinates, saccade distance). Increased scholarly attention should be paid to the study of other eye-tracking indicators, especially spatial ones and eye-tracking path, to summarise the visual clues of tourists, and to combine multiple eye-tracking indicators to explain more complex tourism eye movement behaviours.

Improve the External Validity of Eye-Tracking Research

It is of upmost importance to focus on the external validity of the demonstrated eye-tracking psychological mechanisms from the parent discipline into the tourism field. External validity is related to sample diversity and research scenarios. The current research of tourism eye-tracking mainly uses desktop-based eye-tracker to carry out laboratory studies. Although the research method is helpful to clarify the causal relationship between variables and improve the internal validity of the results, it is difficult to simulate the real situation of tourism behaviour, which leads to the lack of external validity.

Compounding this issue, most of the existing research recruits students as the sample and have a small sample size. Lab experiments to some extent can reduce the interference factors caused by individual differences, but also greatly reduce the generalisation of the research conclusions (Li et al., 2020, 2019; Liu & Li, 2022). To address this, scholars have begun to explore the role of eye-tracking in the tourist wayfinding, museum tour and other related situations (Le et al., 2020). Research should not only be limited to students' samples and laboratory experiments but also the use of diversified and large number of samples and equipping with wearable ambulatory eye-tracking instruments for field study should become a new trend.

Standardise the Research Procedure of Eye Movement

A substantial gap between tourism field and the discipline of psychology pertaining to the use of experimental materials and experimental paradigm exists. Not controlling for interference in experimental eye-tracking research in tourism may lead to low reliability and validity in results. Moreover, if the understanding of the experimental paradigm is not thorough, the experimental procedure may not be standardised, with the conclusion not able to provide valid answers to research questions. Consequently, tourism eye-tracking experiments should draw from the experimental paradigm of psychology and the related research in consumer behaviour (such as using E-prime and other experimental software to standardise the stimulation organisation and time precision), to make up for the advantages and disadvantages, and improve the validity and quality of research results.

Chapter 21

NEUROSCIENCE FOR TOURISM RESEARCH

Marcel Bastiaansen
Breda University of Applied Sciences, Netherlands; School of Social and Behavioural Sciences, Tilburg University, Netherlands

Ondrej Mitas
Breda University of Applied Sciences, Netherlands

Wim Strijbosch
Breda University of Applied Sciences, Netherlands

Hans Revers
Tilburg University, Netherlands

Abstract: There is an emerging interest in understanding the cognitive, emotional and motivational processes that drive tourists' behaviour using neuroscientific research methods. This chapter briefly reviews the main methods of interest to tourism researchers, to then focuses on electroencephalography, which reflects electrical activity from the brain. Event-related potentials or electroencephalography oscillations reflect cognitive and affective processes. Components of the former can index emotional brain responses, and alpha oscillations are related to attention and approach/withdrawal. Existing tourism literature/using electroencephalography are reviewed. This is a promising tool for studying a range of phenomena that are of interest to tourism scholars, but require careful use of methods and interpretation. **Keywords**: Neuroscience; skin

Cognitive Psychology and Tourism
Tourism Social Science Series, Volume 27, 273–289

Published under exclusive licence by Emerald Publishing Limited
ISSN: 1571-5043/doi:10.1108/S1571-504320240000027021

conductance; functional magnetic resonance imaging; electroencephalography; event-related potentials; emotions

INTRODUCTION

Tourism is an academic domain that by its very nature embraces many scientific disciplines. It constitutes one of the largest economic sectors worldwide, so it should come as no surprise that there are many studies which address its economic aspects and impacts of tourism (Calero & Turner, 2020; Song et al., 2012). Similarly, as tourism involves travelling to other countries and experiencing other cultures, the cultural (Richards, 2018), sociological (Cohen & Cohen, 2019; Dann, 2005) and anthropological (Stronza, 2001) dimensions of tourism have extensively been studied. However, in the past two decades, there has been a steady increase in attention towards the psychological aspects of tourism (da Silva deMatos et al., 2021; Pearce & Packer, 2013; Ulker-Demirel & Ciftci, 2020), which broadly speaking addresses the 'lived and reported mental state of tourists' (Pearce & Packer, 2013, p. 387).

Psychology informs tourism research on such topics as consumer behaviour and decision-making, motivation, attitudes towards tourism, satisfaction and destination image (Pearce & Packer, 2013). Psychology itself is a broad discipline, encompassing different subdisciplines that bear relevance to tourism research, such as social psychology (how one perceives and interacts with others), environmental psychology (perceive and interact with the physical environment) or positive psychology (what drives well-being). Cognitive psychology as a subdiscipline of psychology has more recently gained the attention of tourism scholars (Scott, 2020; Skavronskaya et al., 2017). It was developed in the early 1970s, when psychologists started to realise the limitations of the then-dominant behaviourism tradition, which treats the human mind as a black-box, only measures observable behaviour, and exclusively focuses on stimulus–response relationships. Cognitive psychology puts emphasis on the black-box by studying the structure, function and organisation of individuals' mental processes. This, broadly speaking, come in two categories: cognitive processes (thoughts), which are considered predominantly rational in nature, and are relatively well accessible to consciousness, and affective processes (moods and emotions), which are more elusive to conscious access. Cognitive psychology informs tourism research on questions such as the role that novelty plays (Mitas & Bastiaansen, 2018) or what defines and characterises the experience (Bastiaansen et al., 2019; da Silva deMatos et al., 2021; Pearce & Zare, 2017).

Rapid developments in neuroimaging technology and its application to human cognition (the last decade of the 20th century was termed the "decade of the brain") have led to the emergence of a new branch of science: cognitive and affective neuroscience (Albright et al., 2000). This has given a tremendous impetus to cognitive psychology, which has been relying more and more on neuroscientific research methods to measure, study and understand cognitive and affective processes. This chapter briefly describes the major neuroscientific research tools and discusses the extent to which these tools can contribute to understanding human cognition and affect. It focuses on electroencephalography (EEG) and describes the extent to which it is useful in addressing topics that are of interest to tourism scholars. EEG is helpful in conjunction with existing tourism research methods in addressing issues that have remained elusive for tourism scholars. The present introduction into EEG will help, facilitate and stimulate scholars in successfully expanding their methodological toolbox to include EEG.

NEUROSCIENCE RESEARCH METHODS FOR TOURISM

What exactly can neuroscience bring to tourism studies? This question is relevant, as researchers are not usually acquainted with neuroscience methods, and will critically weigh the required investment (in equipment, software, expertise and the efforts involved in data collection) against the potential benefits. Neuroscience techniques allow one to measure emotions, more precisely emotional arousal, in a more or less unobtrusive way, with validity and with high (sub-second) time resolution. Emotion measurement is crucial to tourism research, as it plays a key role in understanding many different aspects of tourists' behaviour. Emotions are crucial in tourists' decision-making (Damasio, 2003; Kahneman, 2011). They play a crucial role in the social interactions important in tourism (Keltner & Kring, 1998) and help tourists respond effectively to challenges (Tooby & Cosmides, 1990). The different roles that emotions play in cognitive-affective processes are of key importance in such decision-making. They are crucial elements in delivering the core value of tourism as they provide memorable experiences (Bastiaansen et al., 2019), as shown in Figure 21.1.

From a cognitive psychological perspective, the notion of experiencing is strongly connected to conscious awareness, and to what is often called the stream of consciousness. This is the continuous flow of what 'fills the mind'

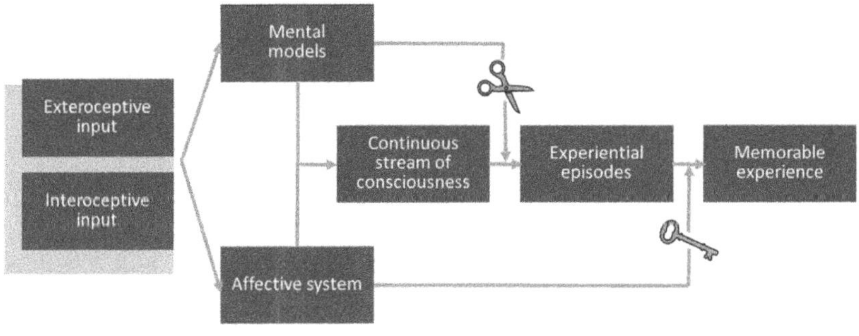

Figure 21.1. The Breda Model of Memorable Experiences.

from the moment one wakes up until the moment of falling asleep. There are basically two sources of input to this stream of consciousness. One is the input that is received from one's senses (exteroceptive input), which includes everything heard, seen, felt, smelled and tasted. The other source is interoceptive information, loosely defined as one's sense of the internal state of the body. This includes, among others, information about one's energy levels (how energetic, active, tired or sleepy one is); visceral information about whether one is hungry, thirsty or nauseated, and information about one's current affective state (whether one feels joyful or sad, elated or depressed).

Both sources of input activate stored information in our declarative memory system about what we know from the world based on previous experiences. Such stored information is often referred to as a mental model (Figure 21.1): a representation of external reality, hypothesised to play a major role in cognition, reasoning and decision-making (Johnson-Laird, 2010). In addition, interoceptive and exteroceptive input activate our affective system, subserving the emotional 'colouring' of the input (Barrett & Russell, 2014). Together, the activated mental models and affective system are integrated to constitute our stream of consciousness.

However, human beings do not typically experience a never-ending and temporally unstructured blur of experiencing. Rather, the mental models that we have of the world around us work to segment the stream of consciousness into identifiable episodes, such as 'having breakfast', or 'commuting to work'. In Figure 21.1, the segments that are thus identified in the stream of consciousness are referred to as experiential episodes (the scissors in Figure 21.1 indicate cutting of the stream of consciousness into episodes). Note that even when we engage in novel situations, our existing knowledge of the world will generate predictions (a mental model) of what to expect. For example, readers who have never spent a day on a tropical beach do have a more or less loosely defined idea of what the experience of 'being on a tropical beach' would be like. Note also that experiential episodes can be as short as a few seconds (e.g. when experiencing the aesthetic value of a painting), they can span minutes (e.g. a ride in a roller coaster), hours (e.g. visiting a museum or a theatre show), days (e.g. visiting a theme park) or even weeks (e.g. taking a summer holiday). One can conceive of the longer episodes as consisting of multiple shorter episodes, leading to the notion of multi-episodic experiences (e.g. Strijbosch et al., 2021).

The succession in time of experiential episodes comes close to how most of us experience life in general. However, not all experiential episodes are considered equally important. It is estimated that over 90% of all experiential episodes are so-called 'ordinary experiences', which are most of the time forgotten as soon as they have unfolded (Duerden et al., 2018). For example, it is unlikely that one remembers a commute from home to work three days ago. However, one might remember one trip, a few months ago, which stands out from the others, because of a strong sense of joy when a deer was spotted close to the road. This particular commute was memorable, because it triggered certain emotions (indicated by the key symbol in Figure 21.1). More generally, when a given experiential episode is emotionally engaging, it will be stored in episodic memory and become a memorable experience. In short, emotions make experiences memorable.

The experiencing of emotions while humans interact with their environment is associated with varying degrees of physiological arousal (Levenson, 2003). This is where neuroscience enters the equation. There are different neuroscience research methods that allow for a continuous monitoring of physiological arousal levels. When properly analysed and cautiously interpreted, different physiological signals provide different yet complementary, continuous measures of emotional engagement (for a review, see Kreibig, 2010). This allows for a temporally fine-grained view on how emotionally engaged tourists are at each moment in time during a tourism experience. Admittedly, there are challenges involved in interpreting physiological signals as indices of emotions. Although most measures provide a clear view on emotional arousal, none of them unequivocally translates into discrete emotion 'categories' such as joy, awe, fear or sadness (Barrett, 2017; Kreibig, 2010; see) for a related, comprehensive theoretical discussion on the nature of emotions. Hence, neuroscience methods often need to be complemented by self-report measures that address the qualitative nature of the emotional experiences that tourists have. However, neuroscience methods do overcome many of the validity issues that are associated with solely relying on self-report in emotion measurement (for discussion, see Bastiaansen et al., 2019; Mauss & Robinson, 2009). In conclusion, neuroscience methods provide a very useful addition to the toolbox of the tourism researcher, but do not replace the existing quantitative and qualitative research methodologies that are commonly used in tourism research.

The Toolbox of Cognitive and Affective Neuroscience

Human neuroscience involves measuring physiological activity of the human body that is a direct reflection of, or under the immediate control of, the human nervous system. The human nervous system can be divided into a central part (the brain, brain stem and spinal cord) and a peripheral part (the afferent nerve cells that pass on sensory information to the brain, and the efferent nerve cells that drive our movements and other bodily functions). As a result, one can distinguish between neuroscience methods that measure peripheral physiological processes such as heart rate and skin conductance, and neuroscience methods that measure central neurophysiological processes (i.e. brain activity). Here we briefly review the major neuroscientific research tools that are used for measuring emotions, and the processes that they measure.

Skin conductance responses: The skin has electric properties that change over time, on a time scale of seconds. Electrodermal activity is the general term that refers to these properties. Changes in the skin's electrical conductance stem from changes in sweat excretion from the sweat glands, which are under the direct influence of those parts of the brain that are primarily involved with the processing of emotions (the hypothalamus and the surrounding limbic system). Sudden, relatively rapid changes in electrical skin conductance, so-called skin conductance responses, therefore reflect an increase in emotional arousal (for extensive background and review, see Boucsein, 2012), at least when properly analysed. Bastiaansen, Oosterholt, et al. (2022) provide a specification of an analysis pipeline that is suitable for a tourism research context.

It should be noted though that skin conductance response is only a measure of emotional arousal, not of emotional valence. Therefore, it can index whether, and when emotions are being felt, but it cannot differentiate between positive and negative emotions. Because of its ease of recording, and the availability of unobtrusive wearable recording devices (e.g. Burns et al., 2010; Milstein & Gordon, 2020), skin conductance responses measures are increasingly being used in tourism research, resulting in over 20 published papers in the past few years. As a recent, comprehensive review is available from Li et al. (2022a, 2022b, 2022c, 2022d, 2022e), those studies are not addressed here.

Heart rate and heart rate variability: Just like electrodermal activity, heart rate (the frequency with which our heart beats) is under the influence of the autonomous nervous system (that part of the nervous system that is not under volitional control). However, skin conductance responses are exclusively driven by the sympathetic nervous system, which is primarily activating in nature, and serves to stimulate the body into fight-flight-or-freeze responses. For heart rate, activity in the sympathetic nervous system typically serves to increase heart rate, at a relatively slow pace. In addition, heart rate is also under the influence of the parasympathetic nervous system, which primarily subserves relaxation (rest-and-digest, or feed-and-breed). Typically, activity in the latter system rapidly decreases heart rate. In normally functioning individuals, the sympathetic and parasympathetic nervous systems interact antagonistically to produce varying degrees of physiological arousal.

In addition to being an index of physical activity, increases in heart rate have been consistently related to increased emotional arousal, at least in carefully controlled laboratory settings. For a review, see Kreibig (2010). Another index that has been of interest in affective neuroscience is the extent to which heart rate varies over time, that is, heart rate variability (HRV). HRV predominantly indexes parasympathetic activity, and has been robustly related to cognitive and affective phenomena (Laborde et al., 2017). HRV can also be used to signal episodes with increased emotional arousal. For example, HRV (or more precisely, the ratio between high-frequency variability and low-frequency variability (Berntson et al., 1997)) is higher during emotionally arousing compared to non-arousing episodes. Finally, and importantly, although there is no firm evidence, there are reasons to hypothesise that HRV can to some extent distinguish between negative and positive valence (that is, whether positive or negative emotions are experienced; Kreibig, 2010). Similar to skin conductance response measures, HRVs are easily recorded with wearable equipment. Despite the ease of recording and analysis, and despite the promise that HRV may differentiate between positive and negative valence, there is surprisingly little empirical work to date that uses HRV as a measure. To the best of our knowledge, only two studies have addressed HRV in tourism and hospitality research (Mitas et al., 2020, 2021).

Functional magnetic resonance imaging: Functional magnetic resonance imaging (fMRI) is a neuroimaging technique that has been developed in the 1990s, and which has had a significant impact on the field of cognitive and affective neuroscience. This technique produces three-dimensional images of the living human brain, and visualisation of the brain areas that are active at a given moment in time. Technically speaking, fMRI does not reflect neural activity, but it can visualise the concentration of oxygen in the blood that flows to specific brain regions. The idea behind this so-called blood oxygenation level-dependent (BOLD) response is that whenever neural tissue becomes active, its metabolic activity increases, and more oxygen needs to be transported to that tissue in order to support the metabolic processes (Gauthier & Fan, 2019).

fMRI BOLD responses can be overlayed with MRI-based structural images of participants' brains and gives a view of brain activity with a very high spatial resolution (up to one mm3, depending on the type of scanner and the specific scanning protocol used). Three decades of very active fMRI research (e.g. a Google Scholar search with the search term 'fMRI' yields greater than one million hits) have uncovered a wealth of scientific information. This research has documented which brain areas are involved in specific cognitive functions (Logothetis, 2008), and which brain areas are involved in processing emotions (Phan et al., 2002). A downside to fMRI is its poor temporal resolution. Because of the sluggishness of the BOLD response (which typically takes up to 20s to fully develop), the technique is blind to the timing of brain processes, which in the case of emotion processing is a crucial aspect to know. Other drawbacks to fMRI research are that it requires very expensive equipment (an MR scanner costs several million USD) and operating the scanner and analyzing fMRI data requires very specific expertise. Finally, fMRI research can only be done in a laboratory environment and imposes severe restrictions on the participant (who has to lay on his or her back in a narrow bore hole, and cannot move the head by more than a few millimeters) and on stimulus presentation (e.g. visual stimuli have to be presented through projection on a mirror just above the participants' head). Probably owing to these restrictions, to date only a handful of fMRI studies have been performed in a tourism context (for a review, see Lei et al., 2022). Altogether, although the fMRI methodology is very useful for a fundamental understanding of cognitive and affective processes, in our opinion, its use in tourism research will remain limited.

Electroencephalography (EEG), as the name implies, involves recording the electrical activity from the brain, and is predominantly sensitive to the activity (more specifically, the postsynaptic potentials) of the large, pyramidal neurons in the neocortex – the outer layers of the brain (Nunez, 2000). EEG as a neuroimaging technique dates back to the late 19th century, with the early pioneering work of Richard Caton and Adolf Beck, who first recorded electrical signals from animal brains in 1875 and 1890, respectively (Coenen et al., 2014). EEG was first applied to humans by the German physiologist Hans Berger to measure cognitive processes almost a century ago (Berger, 1929). Berger demonstrated that EEG was sensitive to cognitive processes by measuring EEG from his children and showing that the typical A (or alpha) waves that dominated the recordings when his children were resting disappeared and transformed into B (or beta) waves when asked to perform complex arithmetic tasks. After these initial and seminal publications, interest in EEG faded as a result of World War II, and of the emergence of behaviorism as a dominant paradigm in psychology. In the 1970s, when the scientific interest in cognitive and affective processes revived, EEG recording equipment became less expensive, more widely available, and computers became powerful enough to support careful analysis of EEG data; EEG became a prominent research tool for cognitive psychologists (Sutter & Kaplan, 2017). Nowadays, EEG equipment is affordable, data collection is relatively easy and widely available and easy-to-use data analysis software allows for analyzing EEG data without requiring extensive expertise and time investment.

Although there are currently only a few EEG studies that address topics relevant for tourism scholars (Lei et al., 2022; Li et al., 2022a, 2022b, 2022c, 2022d, 2022e), the combination of affordability and ease of use on the one hand, and the potential for objectively measuring the cognitive and affective processes that are important for a cognitive-psychological perspective on tourism research on the other hand, lead us to believe that EEG is a promising research tool for tourism scholars, which has the potential to make a relevant contribution to tourism research in the years to come. Therefore, in the remainder of this chapter, we more explicitly address the possibilities and the impossibilities of EEG research for tourism.

The Best of Both Worlds? EEG vs fMRI

EEG measures brain activity from the neocortex. This implies that with EEG one cannot reliably record activity from the subcortical brain structures (unlike fMRI, which allow for a full three-dimensional view on brain activity) that are often implied in the brain networks that subserve emotion and motivation (Pessoa, 2018). On the other hand, EEG does provide a measure of brain activity that has millisecond precision. This is a big advantage over fMRI. Furthermore, EEG is a direct reflection of neuronal activity, as opposed to fMRI, which relies on hemodynamics (changes in blood oxygenation) as an indirect reflection of which brain areas have been active. In sum, EEG and fMRI are complementary to each other, as fMRI measures with great precision where in the brain activity takes place, whereas EEG precisely indicates when the brain is active. It would seem then that the combination of these techniques ideally captures where and when brain activity takes place. Unfortunately, simultaneously measuring EEG and fMRI is technically extremely challenging, and requires extensive expertise in both conceptualising and running the appropriate data analyses (Philiastides et al., 2021). In sum, EEG and fMRI both have advantages and disadvantages, which should be combined across rather than within empirical studies.

Neuroscience for Tourism Research: The Case of Electroencephalography

As noted above, EEG reflects the electrical activity of neurons in the cortex, the outer layer of the brain. EEG signals consist of very tiny changes in electrical potential over time, in the order of microvolts (μV; one millionth of a volt), with a frequency content of ~0.1 Hz (so-called ultraslow potentials) to ~100 Hz (so-called high-frequency gamma activity). These signals are typically sampled 200–1,000 times per second in order to store them digitally for later processing and analysis. EEG signals are typically recorded from multiple (e.g. 32, or 64) locations on the scalp, in order to sample from different brain areas simultaneously. These electrodes are typically mounted in an elastic cap to ensure they cover the same standard locations on the scalp (Figure 21.2A) for each study participant. Thus, a typical EEG recording consists of tiny electrical potential changes over time recorded from multiple locations (Figure 21.2B; (a) typical electrode lay-out across the scalp (top view) for an EEG recording from standard locations at the scalp; (b) typical EEG record, consisting of electrical potential changes – ain μV – over time from multiple scalp locations, each line represents the EEG signal from one location at the scalp).

The potential changes over time that are recorded at the scalp represent the summed activity of large sets of neurons that fire more or less in synchrony. For an excellent and comprehensive introduction into EEG methodology, see Luck (2014). As these neurons participate in different brain networks at different times, the activity that is picked up at the scalp is a complex mixture of brain activity related to a wide multitude of brain functions, ranging from basic bodily processes such as regulating blood

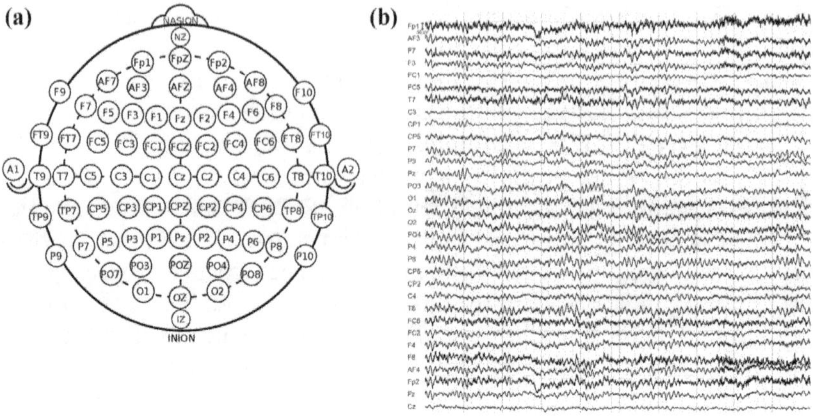

Figure 21.2. Electrical Potential Changes.

pressure, driving hormone secretion and maintaining body posture, to the cognitive and affective processes that are of interest to the cognitive psychologist. Isolating that part of the EEG that reflects a particular cognitive or affective process (the 'signal') from the EEG that is not related to the process of interest (the 'noise') requires careful signal analysis, as the noise is typically about 50–100 times larger than the signal (Luck, 2014). There are essentially two ways to improve the signal-to-noise ratio in EEG data analysis to levels that are sufficient for meaningful interpretation: by computing so-called event-related potentials (ERPs) (Luck, 2014), and by isolating specific EEG oscillations from the overall EEG (Bastiaansen et al., 2012). The next sections briefly explain the logic behind these two approaches, and discuss how these approaches may inform tourism researchers in studying phenomena that are of interest.

EEG, ERPs and ERP components: One way of extracting that part of the EEG that is related to the cognitive or affective process under study is by devising an experimental paradigm, which often consists of presenting stimuli to the study participant, or engaging the participant in a brief task, which is considered to engage the cognitive or affective process under study. EEG segments recorded while the participant is viewing the stimuli or is engaged in the task are then selected from the ongoing EEG and averaged across a number of repetitions (termed trials) of the stimulus or task. The logic behind this is that the part of the EEG that is related to the experimental event is the same ('stationary') across all trials, whereas that part of the EEG that is not related to the eliciting event is different across trials (non-stationary). Given a sufficiently large number of trials (repetitions), say 20 or 30, the non-stationary part of the EEG (the noise) cancels out, as positive and negative electric potentials will average to zero. Eventually, only the event-related, stationary part of the EEG (the signal) will survive the averaging procedure. The term event-related potential is used to refer to such averages across trials (Luck, 2014).

In short then, ERPs are averaged EEG segments which are precisely time-locked to their eliciting events. The events consist of carefully designed stimuli, responses or simple tasks that are thought to elicit, and ideally isolate specific cognitive or affective responses (e.g. attention, or the emotion 'disgust'). ERPs consist of (average) deflections from the baseline (the zero-potential point just before the eliciting event), that is, positive and negative 'peaks' (see Figure 21.3 for a typical ERP in an emotion-elicitation task). These peaks have a specific latency (the time from the eliciting event to the respective peak) and have either descriptive labels (e.g. P1, referring to the first positive peak in a series, N2, i.e. the second negative peak), labels that refer to polarity and latency (P300, a positive peak occurring 300 ms after the onset of the eliciting event, or N400, a negative peak occurring 400 ms after the eliciting event), or labels that refer to their cognitive function (e.g. error-related negativity, or ERN, which occurs after a participant has produced an erroneous response in a task). Typically, the functional significance of ERP components is determined by contrasting different situations, tasks or stimuli, say, an emotionally valanced versus an emotionally neutral stimulus. ERP components that are sensitive to what is being manipulated (that is, becoming larger, smaller or occurring later or earlier after a valanced vs. a neutral stimulus) are then considered to be an indication of brain activity related to the processing of emotional valence, and can serve as empirical markers for the detection of valance.

The initial report of the use and usefulness of ERPs (Walter, 1968) has triggered a very extensive field of research in cognitive psychology that is still very active and relevant today. Over half a century of active ERP research has yielded a wealth of insights and information on ERPs, and on which ERP components are sensitive to specific cognitive and affective processes, including, but not limited to memory, language, attention, emotion,

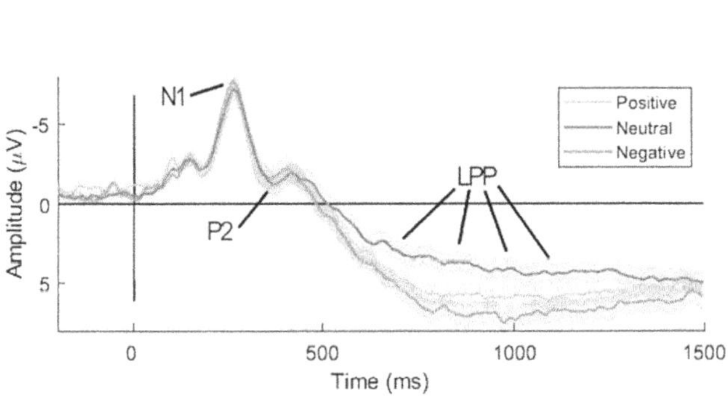

Figure 21.3. Typical ERP Response.

decision-making and many more. In this section, we selectively describe some of the ERP components that are unequivocally related to emotion, as we consider affective processes to be of prime interest to tourism scholars. For more information on the measurement, analysis and decomposition of cognitive and affective processes with ERPs, the reader is referred to the comprehensive review of the state of the art of this field of research in Luck and Kappenman (2011).

Emotion-related ERP components: Many empirical efforts have been devoted to identifying ERP components that are related to the processing of emotions – see Hajcak et al. (2012) and Olofsson et al. (2008) for reviews. A typical ERP response to an emotionally valanced vs emotionally neutral stimulus is given in Figure 21.3 (ERPs, author's unpublished data, elicited by a standardised picture set known to elicit emotional responses, Lang et al., 2005; traces show ERPs averaged over 97 participants, for three different types of pictures: emotionally neutral, positively valanced and negatively valanced). Shaded areas around the traces indicate standard errors around the mean. Well-established ERP components (N1, P2, LPP) are indicated in the figure. The LPP differentiates between emotionally neutral and emotionally valanced pictures, but not (convincingly) between positive and negative valence.

As Figure 21.3 shows, there are several well-established ERP components (positive and negative peaks), N1, P2 and LPP (for late positive potential), and that are known to be sensitive to emotions. Of those, the peaks labelled N1 and P2 show differences related to emotional arousal under certain circumstances (Hajcak et al., 2012), not in the data shown in Figure 21.3. Further, there is a relatively long-lasting difference labelled LPP that also typically differentiates between valanced and emotionally neutral stimuli. Also note that, although there are clear differences in the LPP between the neutral and valanced conditions, there is no observable difference between positively valanced stimuli (eliciting positive emotions) and negatively valanced stimuli (eliciting negative emotions). This is a well-established and robust finding in the emotion ERP literature: ERPs cannot discriminate between positive and negative emotions. Applications of ERP research in tourism are briefly presented later in this chapter.

EEG oscillations: A different way of approaching EEG recordings is to consider that they consist of a superposition of oscillations (sine waves) with different frequencies. The frequency-specific oscillations were already identified in the very first report on human EEG (Berger, 1929), which spoke of A-waves and B-waves. These phenomena later became known as alpha waves, or alpha oscillations, and beta oscillations. With the focus in the 1970s and 1980s being mainly in ERP research, analysing EEG oscillations grew in popularity in the late 1990s (Pfurtscheller & Da Silva, 1999), and rapidly became a very active field of study (e.g. Neuper & Klimesch, 2006). The basic idea is that EEG oscillations reflect activity in brain networks that subserve cognitive and affective processes (Varela et al., 2001). Different types of oscillations have been identified, and the waxing and waning of these oscillations can be indicative of different cognitive and affective processes. Globally, oscillations are defined by their frequency, and labelled accordingly. Typically one differentiates between delta (<4 Hz), theta (4–7 Hz), alpha (8–12 Hz), beta (12–30 Hz) and gamma oscillations (>30 Hz), but there is no one-on-one mapping between oscillation frequency and its functional significance (Bastiaansen et al., 2012). It is beyond the scope of the current chapter to give an extensive overview of the literature on EEG oscillations (Bastiaansen et al., 2012; Engel et al., 2001; Pavlov & Kotchoubey, 2022; Vecchio & De Pascalis, 2020). Instead, here we focus on an oscillatory EEG phenomenon that is of potential use to tourism scholars: frontal alpha asymmetry as a potential index of approach and withdrawal motivation.

Alpha oscillations were first identified by Hans Berger in 1929 (Berger, 1929), and consist of relatively large sine-wave-like oscillations with a frequency of 8–12 Hz. Alpha oscillations are readily identifiable in EEG recordings (Figure 21.4). They result from synchronous firing of large populations of neurons in the neocortex, and are commonly associated with a state of idling, or active suppression of the brain area that generates the activity (Jensen et al., 2012; Klimesch et al., 2007). As a result, the size (or amplitude) of alpha oscillations is inversely related to active processing, and has been robustly associated with the allocation of attention (Foster & Awh, 2019; Foxe & Snyder, 2011; Hanslmayr et al., 2011; Peylo et al., 2021). For practical purposes, this means that the amplitude of alpha oscillations can be used as a continuous index of increases and decreases in attention (Figure 21.4; typical EEG trace from one scalp location, containing clearly visible waxing and waning alpha oscillations. Episodes characterised by high-amplitude alpha oscillations alternate with episodes of low-amplitude alpha. These decreases in alpha amplitudes are associated with increased levels of attention). Such a continuous measure of attention can be useful for

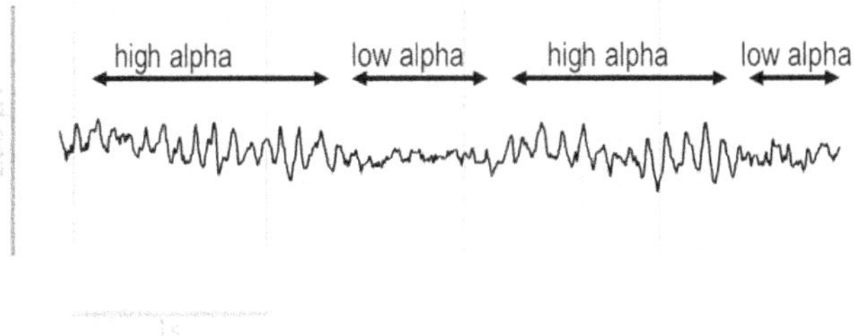

Figure 21.4. Alpha Oscillations Identifiable in EEG Recordings.

tourism scholars, and tourism practitioners in different ways, for instance, for establishing which destination promotion materials draw most attention.

Frontal alpha asymmetry: An index of emotional valence? Alpha oscillations are typically the largest over the back of the head, which overlays the part of the brain that deals with visual information. However, such oscillations can also be measured over frontal brain areas, and then it is typically referred to as frontal alpha. In itself, frontal alpha has been associated with creativity (Lustenberger et al., 2015) and mental simulation (De Lange et al., 2008). More interestingly, however, the extent to which frontal alpha differs between the left side of the brain and the right side of the brain (known as frontal alpha asymmetry, or FAA) has been related to affective processes. For example, FAA is an established biomarker for aggression (Peterson et al., 2008) and depression (Gold et al., 2013), and FAA-based neurofeedback is used to reduce negative emotions and anxiety (Mennella et al., 2017). Interestingly, FAA has been convincingly associated with approach-withdrawal motivation: larger alpha oscillations recorded over the left frontal parts of the brain are associated with approach motivation (wanting to move towards something or someone), whereas larger oscillations over the right frontal part of the brain are associated with withdrawal motivation (Harmon-Jones, 2003). This phenomenon has been used to study a wide variety of topics, ranging from product design to marketing, and from brain-computer communication to mental health studies (Briesemeister et al., 2013). The possibilities for applying FAA to tourism research are manifold, and some of them are quite obvious: it could be used to measure approach or withdrawal motivation in destination choice settings, destination promotional materials, from more or less crowded areas at a destination, or from different types of tourism activities. In sum, FAA as a tool opens up a myriad of possibilities to measure unobtrusively and objectively tourists' preferences, on which management decisions can be based.

A Review of the Literature: EEG Studies in Tourism

Despite the fact that EEG offers unprecedented possibilities to measure emotional responses by emotion-related ERP components, and approach-withdrawal motivation using FAA, to date there are only eight published studies that have actually used EEG in a tourism (or rather, a broader tourism, hospitality and leisure) context. Before we briefly review them here, let us note that most of these studies used a very small number of participants (less than 15), meaning that they can be seen at best as initial pilots, or feasibility studies. Typically, EEG studies should contain sample sizes of 30–40 participants at the very least, in order to yield robust, reliable and validly interpretable results.

Out of these eight studies, four used ERPs, and specific ERP components as dependent measures. A pilot study with only three participants (Ma et al., 2014) found smaller amplitudes of the P2 component in response to beautiful landscapes compared to ordinary landscapes. Note that the opposite effect would have been expected based on the functional interpretation of the P2 component (which usually shows larger amplitudes in response to positively or negatively valanced stimuli compared to neutral ones). However, the extremely small sample size employed in this study does not warrant a clear interpretation of the findings. Two other studies used ERPs to study the effect of destination marketing on affective destination image, using either an organic destination movie (Bastiaansen et al., 2018) or a TV commercial (Bastiaansen, Straatman, et al., 2022). In both studies, ERPs were elicited by pictures of a tourism destination before and after viewing destination marketing materials, and the amplitude of components N1, P2 and LPP were compared. In both studies, the P2 and LPP were larger after viewing marketing materials compared to before, and this effect was not observed in control groups watching comparable destination marketing materials that were unrelated to the destination under study. In the latter study (Bastiaansen, Straatman, et al., 2022), self-reported ratings of valence of the destination pictures did not show such effects, suggesting that ERPs may be more sensitive to detect the influence of destination marketing than traditional self-reports.

The four other studies focused on changes in EEG oscillations as dependent measures. Three of them were hospitality-oriented and aimed at understanding decision-making processes in choosing hotels. One brief report (Tosun et al., 2016) describes an EEG study in which participants decided on hotel preferences while their EEG was measured. The report is not sufficiently elaborate to understand the details of the study and EEG analysis, but the authors identify theta oscillations and beta oscillations as being meaningfully related to different aspects of the decision-making process. A second study (Hsu & Chen, 2020) tested the effect of the subliminal presentation of a smiling emoji in hotel advertising on consumers' choices and their EEG. The subliminal stimuli did affect consumers' hotel choice, and lead to larger theta oscillations and smaller beta oscillations in the EEG compared to advertisements without subliminally presented smileys. Unfortunately, the authors' speculative interpretation of the EEG findings were not based on existing literature about the functional significance of beta and theta oscillations. A third study (Fronda et al., 2021) looked at the EEG correlates of exploring different areas of a hotel, and report increased theta and beta oscillations during the exploration phase compared to a two-minute baseline recording preceding the exploration phase. These findings were interpreted to be related to cognitive elaboration processes and attentional response (for the beta oscillations), and to 'motivational and emotional response and affective arousal' (Fronda et al., 2021, p. 15523), or to visiospatial working memory processes (for the theta oscillations). Both interpretations do not appear to be based on extensive knowledge of the literature on beta and theta oscillations.

Finally, in a cultural tourism context, a recent study set out to establish the effects of the aesthetic experience of watching paintings and other visual artworks on ERPs and oscillations (Strijbosch et al., 2022). Participants judged 148 artworks on the extent to which they were moved by them, and EEG responses were divided into three categories: highly moving, moderately moving and lowly moving. Contrary to expectation, the authors did not find any effects of being moved on ERP components (including P2 and LPP). They did however find stronger reduction in alpha oscillations (reflecting stronger attentional capture) for artworks that were rated as highly and lowly moving compared to those that were moderately moving. In addition, increased gamma oscillations were observed in response to viewing highly and moderately moving artworks compared to lowly moving ones, which the authors interpreted as 'sustained sense-making' (Strijbosch et al., 2022).

CONCLUSION

In conclusion, there is an emerging interest in using neuroscientific research methods in understanding cognitive and affective processes that are relevant for tourism researchers. While there is a growing number of studies that use skin conductance to study emotional arousal in tourism contexts, we see many possibilities to use measures of heart rate, and HRV as tools in tourism studies. fMRI seems less suitable for tourism scholars, owing to the large investments (in money, time and expertise) this technique requires, and its limited ecological validity. The interest in EEG as a research tool is rapidly rising. For one part, this is due to the increasing availability of affordable and easy-to-use equipment for data collection, and ready-to-use software for data analysis. For another part, this is due to the fact that EEG holds promise as a tool that can measure such processes unobtrusively and objectively, with millisecond precision. The fact that four out of the eight existing EEG studies in a tourism, hospitality and leisure context were published in the past two years underlines the rapid pace with which tourism scholars have become interested in using EEG as a tool.

Initial studies underline the potential of EEG to contribute to theoretical knowledge development and to practical applicability of tourism research. At the same time, tourism scholars need to be careful in designing studies that are suitable for EEG measurement and should be especially careful in interpreting the findings they obtain in ways that are in agreement with existing EEG literature. Data collection and data analysis may have become easy, interpreting findings still requires a substantial time investment in reading and understanding the relevant parts of the large body of literature on the neuroscientific correlates of cognitive and affective processes. Too much loose interpretation, or overinterpretation of findings in these early stages of exploring the usefulness of neuroscience methods in tourism research could unnecessarily compromise the credibility of neuroscience research methods in tourism research.

Chapter 22

DIRECT EFFECTS OF TOURISM COMMERCIALS
Skin Conductance Methods

Shanshi Li
Xiamen University, China

Billy Sung
Curtin University, Australia

Xinyan Wei
Xiamen University, Malaysia

Noel Scott
Edith Cowan University, Australia

Abstract: This chapter documents initial results of using skin conductance measures to compare the relative effectiveness of three commercials for destinations and one beer advertisement. Data were collected from 12 respondents in a laboratory-based study. Analysis of skin conductance data showed that the beer commercial evoked a greater intensity of emotional response than the three tourism commercials. Moreover, the time trace of the tourism commercials indicated a downward trend, and this is interpreted as a lack of overall emotional arousal. The results suggest that skin conductance measures may be used for testing tourism advertising effectiveness and also a need for further research to examine the emotional content of tourism advertising. **Keywords**: Advertising effectiveness; skin conductance; tourism commercial; attention; electrodermal activity; arousal

Cognitive Psychology and Tourism
Tourism Social Science Series, Volume 27, 291–301

Published under exclusive licence by Emerald Publishing Limited
ISSN: 1571-5043/doi:10.1108/S1571-504320240000027022

INTRODUCTION

Tourism advertising is used to attract the attention and interest of potential tourists to a destination (Park et al., 2013; Pratt et al., 2010). Many authorities make considerable investments in destination advertising on the basis that it is effective in attracting tourists (Ruhanen et al., 2012). The magnitude and importance of investment in marketing indicate that evaluating its effectiveness is an important issue for both academics and practitioners (Pratt et al., 2010; Scott et al., 2016; Stienmetz et al., 2015).

Evaluation of tourism advertising effectiveness is a complex process and existing studies have demonstrated several commonly used approaches, such as conversion studies (Messmer & Johnson, 1993; Pratt et al., 2010; Woodside & Reid, 1974), use of an advertising tracking model (Siegel & Ziff-Levine, 1990), true experiments (Woodside, 1990) and quasi-experiment approaches (Messmer & Johnson, 1993; Mok, 1990). Whichever of these methods is used, they all measure the final effects of an advertisement, such as inquiry of travel information, actual visits and spending, after promotional materials have been used. Thus, the tourism marketer is able to evaluate the effectiveness of an advertisement only after the costs of advertising have been incurred.

In the wider literature of advertising effectiveness, there is a significant discussion of the direct effects of an advertisement on the viewer. The literature examines the effect of an advertising stimulus on a viewer when watching it and includes cognitive and emotional responses to the advertisement. Such research allows advertisers to improve their performance based on an evaluation of direct effects before the cost of mass advertising is incurred. This research is supported by theories of the cognitive and emotive mechanisms of the advertising process (Vakratsas & Ambler, 1999). In particular, much interest has surrounded the role of emotion in consumers' responses to advertising (Baird & Wahlers, 2007; Grappi & Romani, 2012; Teixeira & Wedel, 2012). Several studies have demonstrated that emotional responses dominate cognitive evaluations in their ability to predict advertising effectiveness (Morris et al., 2002), and this is particularly true for hedonic products and service (Sirgy, 1991).

A few studies, however, have examined the direct effects of the tourism advertising, especially the viewers' emotional responses evoked by the advertising materials. Instead, most of the evaluation literature has focused on the assessment of the knowledge or cognitive outcomes of tourism ads. For example, Kim et al. (2005) examined the cognitive effect of advertisement, such as viewers' top-of-mind awareness and advertising awareness. Other studies have examined the effects of advertisement on the formation of destination image (Govers et al., 2007; Pan, 2011) with results indicating that advertising does not have a major impact upon tourist perceptions (Govers et al., 2007). There is however a curious lack of research on the emotional outcomes from viewing tourism advertising despite the fact that promotional campaigns are often emotional in nature and are designed to evoke consumers' feelings (Kandampully et al., 2001). This lack of research on the emotional outcomes of tourism advertising is a significant gap in the evaluation of its effectiveness.

SKIN CONDUCTANCE MEASURES

Skin conductance (SC), also known as electrodermal activity, is a commonly used physiological technique. It is usually measured in micro-siemens and can be used to reflect the activation of the autonomic nervous system that controls most of the human organs and muscles (Dawson et al., 2007; Jacobs et al., 2012). The autonomic nervous system is composed of two parts: the sympathetic nervous system and the parasympathetic nervous system. The former originates from the spinal cord and represents muscle activation, which regulates the body's emotional arousal and behavioural response to dangers or threats. On the other hand, the parasympathetic nervous system represents muscle relaxation and adjusts the body's self-balance state after rest and eating.

SC is used to measure an individual' arousal level. Such an arousal level is related to the activation of sympathetic nervous system (Mauss & Robinson, 2009). In particular, when individuals are stimulated by an external event or experience, the system will be activated, and the sweat gland activity will increase, resulting in higher skin conductivity (Grabe et al., 2000). The degree of arousal can be measured by either the conductance, resistance, impedance or admittance of the skin, and the adoption of a particular method depends on the nature of the recording apparatus (Boucsein, 2012). The capability of SC to measure arousal has been reliably confirmed in the past study (Klebba, 1985). Bradley and Lang (2000) found that stimuli with different arousal levels, whether positive or negative, would lead to changes in skin conductance levels (SCLs).

Skin conductance measures produce two types of data, phasic and tonic data. The former refers to the skin conductance responses (SCRs) of the respondent, which is a short-term change in skin conductance caused by a specific external stimulus, usually lasting only a few seconds. Tonic data, also known as SCL, reflect the basic level of the individual's skin conductance. Tonic data provide a relatively long-term response to external stimuli, usually at least 30 seconds (Algie, 2005). The measurement of electrodermal activity is achieved by placing two electrodes on the distal phalanx of the index finger and middle finger of the respondent's non-writing hand. To control the hydration level of participants, participants are often asked to wipe the palm surface and fingers before connecting the electrodes (Bolls et al., 2001). Skin conductance measures have strict equipment, environmental, data acquisition, recording, processing, conversion and analysis requirements. Caruelle et al. (2019) conducted a detailed review of the recording and analysis methods for electrodermal activity data. They emphasise a need for transparency when reporting the research results from using electrodermal methods so as to allow evaluation of the reliability and validity of findings.

Most of the advertising research uses self-report questionnaires to collect consumers' emotional responses to advertising stimuli. Self-report measure is a retrospective method that attaches importance to the interpretation and expression of introspective reflection on the subjective feelings component of ones' emotions (Poels & Dewitte, 2006). Respondents are usually asked to fill out a questionnaire to report their cognitive and emotional responses to the advertisement, using either an open-ended question or scoring on a series of emotional items (Walters & Sparks, 2012). The advantages of self-report method are convenient, simple, cheap and allows researchers to conduct research with a large sample size. However, self-report questionnaire has been criticised because of its inherent limitations, and recent tourism studies begin to use SC in measuring emotional responses (Hadinejad et al., 2019; Li et al., 2014).

Skin conductance can overcome limitations of the self-report method in measuring emotional responses. First, self-report is easily influenced by cognitive biases and memory biases. Specifically, self-report is used to measure an individual's emotions after the advertisement is displayed. Relying on recalled emotions may result in problems as individuals may not remember them (Poels & Dewitte, 2006) or provide processed responses that are not based on felt emotions (Plassmann et al., 2015). There may also be cognitive and memory biases that distort the recall accuracy (Ravaja, 2004). Skin conductance is an independent physiological process controlled only by autonomic nervous system (Sequeira et al., 2009) and does not rely on individual memory and cognitive processing. In other words, SC can capture emotional responses and eliminate the interference of language, memory or information processing (Ravaja, 2004). As a result, skin conductance can minimise cognitive and memory biases, and circumvents prejudice caused by psychological processing or social desirability constraints.

Second, self-report method cannot capture emotions which do not reach the threshold for consciousness. However, such reactions are worth measuring because they can affect individual behaviour (Winkielman & Berridge, 2004). Since emotions are generated unconsciously, individuals may not realise that they have experienced an emotional reaction. It is also difficult for individuals to explain their emotional reactions with words or emotional items scores through self-reporting, so important information may be missed. By contrast, skin conductance can collect data from processes that one is not conscious of.

Most importantly, emotions change dynamically, and self-reporting methods have difficulty in measuring continuous emotional reactions

(Winkielman & Berridge, 2004). Emotions are dynamic processes that change over time and during exposure to a video advertisement (Scherer, 2009). Therefore, it is useful to keep track of the subtle changes in the process of emotional stimulus exposure. However, a self-report questionnaire is a post-event report, so it cannot record the individual's immediate emotional reaction process in real time. In some studies, respondents are asked to report the intensity of their perceived emotions in real time using an instant rating (Poels & Dewitte, 2006); however, this method may seriously interfere with information processing as rating behaviour itself happens at the conscious level of the respondents (Ravaja, 2004).

Moreover, in situations where respondents fill out a questionnaire with large number of emotional scales, the scoring process may tyre the respondents, so the reliability and credibility of the collected data is difficult to guarantee. Alternatively, skin conductance allows continuous measurement and can continuously capture emotional responses (Dawson et al., 2007), and can provide a visualisation of the temporal emotional pattern including emotional peaks and troughs. Its recording equipment can perform fine-grained and continuous sampling of the electrical characteristics of the skin over a period of time, providing researchers with rich and valuable data.

While SC has several advantages, it also has several limitations. One, it only shows one's arousal level, but cannot indicate emotional valence (Groeppe-Klein, 2005; Groeppel-Klein & Baun, 2001). If researchers would like to further measure emotional valence, they need to combine skin conductance with other measures, such as self-report questionnaire (Groeppe-Klein, 2005; Shoval et al., 2018), post-research interviews (Hurley et al., 2015; Kim & Fesenmaier, 2015) or physiological measurements, such as startle reflex modulation (Walla et al., 2017), facial electromyography (Somervuori & Ravaja, 2013) and electroencephalogram (Boshoff, 2017).

Two, skin conductance not only indicates emotional arousal but also reflects the individual's attention level (Kroeber-Riel, 1979) and cognitive processing (Dawson et al., 2007). When it is used to measure emotional arousal, it may be confused with other mental processes, resulting in some errors in the measurement. This requires rigorous experimental design and strict control, avoiding other psychological processes to ensure a single variable is being measured each time. Last but not least, when an individual is exposed to a stimulus and generates arousal to cause a change in SCR data, there will be a one to four second latency (Caruelle et al., 2019). Therefore, it is difficult to accurately identify and verify the stimulus that generated individual arousal especially when there are multiple stimuli in a short time. This requires researchers to strictly control each stimulus, or to leave sufficient interval time before each stimulus, so as to reduce the residual effect by the previous stimulus and improve the reliability and credibility of the experimental data.

Through the above analysis, although skin conductance also has certain limitations in measuring emotions, these limitations can be compensated and controlled by combining with other emotional measurement methods, improving experimental design, and strictly controlling the experimental process. Based on the criticisms of self-report measures and the increasing use of SC measures in tourism studies (Li et al., 2016), this chapter therefore uses skin conductance to measure and compare tourism consumers' emotional responses evoked by a mixture of tourism and non-tourism commercials.

Study Methodology

Three tourism commercials and one beer commercial using skin conductance measure are evaluated in this chapter. Participants were a convenience sample of university students (male and female 18–25 years) who consented in writing to participate in the study and given a AUS$20 gift voucher. The data were collected in March and April 2015. The commercials were presented on the centre of a 17-inch monitor, and participants sat about 90 cm in front of the screen. Respondents' skin conductance data were collected through placing two standard nine millimetre diameter silver/silver chloride (Ag/AgCI) electrodes on the participant's fingertips of their non-writing hand. Before attaching electrodes to the participants' fingers, participants were asked to wipe their palmar surfaces and fingers to control hydration levels across participants (Bolls et al., 2001). These electrodes were connected to an amplifier manufactured by Becker Meditec™ and the sampling interval was set up as 96 ms during their exposure to the video clips. The respondents' continuous skin conductance data were recorded and saved using the Biometric Software Suite™ software for later data analysis offline.

Television (video) commercials were used in the study. Video is one of the most used stimuli to induce emotions under laboratory conditions (Kreibig, 2010). The commercials consist of three tourism destination commercials (Fiji, Australia and Queensland) The original commercials can be found in following links: Fiji, where happiness finds you: https://www.youtube.com/watch?v=AK1iV53Sbu4; Tourism Australia: https://www.youtube.com/watch?v=WMGuw1d2ojU; Queensland. Where Australia shines: https://www.youtube.com/watch?v=HWaMt3CLRt8 and one beer commercial (Guinness, basketball: https://www.youtube.com/watch?v=Au8Y98Rgxbk). These four commercials were selected after discussions with advertising and tourism marketing managers in Queensland to identify advertisements considered to produce an emotional effect in respondents. All the videos were less than 90 seconds in order to avoid causing participant fatigue.

Respondents were tested one at a time. Participants were asked to relax for around three minutes, while their skin conductance baseline level equilibrated. Participants then viewed four commercials in random order. As suggested by

Fredrickson and Levenson (1998), the interval between two video clips was set up about at 3 min, which allowed a return to baseline levels and hence minimise any 'carryover' effect. Once the advertisements had been viewed, participants were debriefed, and the data collection concluded.

The variables analysed from the skin conductance data collected included SCR and SCL. SCR refers to short-term fluctuations in the electricity activity of skin, which usually lasts for a few seconds (shown as blue symbols in Figure 22.1). SCL (Algie, 2005) measures long-term reactions to the stimuli, which last for over 30 seconds (demonstrated by the overall waveform in Figure 22.1). According to Algie, SCL is 'more relevant for reporting patterns of arousal within the commercial' (2005, p 4). In the study, these four commercials were compared based on both respondents' SCL and SCR.

The software this study used to analyse the SCL and SCR data was Acqknowledge™, manufactured by Biopac™. The software is designed to view, measure, transform and analyse psychophysiological data. The skin conductance data collected by Biometric Software Suite™ was imported into Acqknowledge™ for the further analysis and data filter. Also, analysis of variance was conducted using SPSS 22 to test if there were statistically significant differences in the SCL and SCR data recorded for these four commercials.

Study Results

The four commercials were compared in terms of the number of SCR and their average SCL value per commercial. In terms of the number of SCR, the

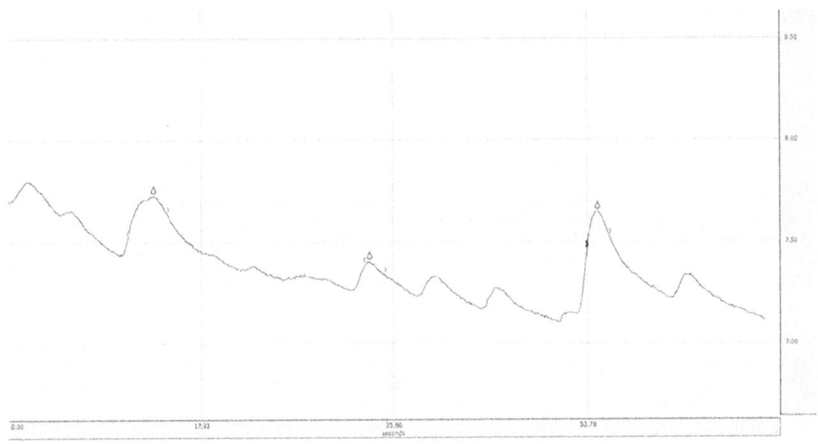

Figure 22.1. SCR and SCL of Respondent No.10 Towards Australian Commercial.

analysis of variance reveals that there are no significant differences among four commercials (Table 22.1). Since SCR is related with specific stimulus or event in the commercial, this finding shows that there are no differences in the number of stimuli embedded in the commercials.

In terms of SCL value, the beer commercial had an average count across the 12 respondents of 5.50 micro-siemens while the tourism advertisements received 4.48, 4.55 and 4.99 micro-siemens for Fiji, Australia and Queensland, respectively (Table 22.2). There is a significant difference in SCL, whereby this in the beer advertisement is significantly higher than Australia advertisement (p = 0.015). SCL is positively related to the intensity of emotional response (Boucsein, 2012). Therefore, this finding suggests that the beer advertisement could evoke significantly stronger emotional responses than Australian tourism advertisements. Furthermore, the differences in the mean scores shown in Table 22.2 indicate that beer advertisement seems to be superior to the other three tourism advertisements in terms of its ability to evoke emotional responses.

The SCL waveform reflects respondents‘ emotional flow patterns for the commercial. Here the Queensland commercial is taken as an example to

Table 22.1. SCR Number for Each Advertisement.

Advertisement	Mean N = 12	Standard Deviation
Australia commercial (tourism Australia)	1.1667	1.58592
Beer commercial (wheelchair basketball)	1.25	1.60255
Fiji commercial (where happiness finds you)	1.25	1.658312
Queensland commercial (where Australia shines)	1	1.2792

Table 22.2. SCL Value for Each Advertisement.

Group	Mean N = 12	Standard Deviation
Australia commercial (tourism Australia)	4.5531	2.4092
Beer commercial (wheelchair basketball)	5.4943	2.53923
Fiji commercial (where happiness finds you)	4.4859	2.65765
Queensland commercial ‘where Australia shines’	4.9977	2.53547

illustrate the real-time skin conductance data. SCL change score was computed by subtracting it of each commercial to participants' baseline. Referring to Figure 22.2, the overall SCL trend of Queensland commercial is declining in spite of several fluctuations in the middle. In marketing research, arousal level is also related with attention and interest (Bolls et al., 2003), and this might indicate that respondents were gradually losing their interests while watching the commercial.

Specifically, it can be seen that the SCL change was 0.9 micro-siemens at the beginning and over time this level kept on decreasing until around the 30th–35th second when it reached its baseline level. Then it experienced a slight increase and started to drop again and stayed at a relatively low level for the following 15 seconds. During the last five seconds of the commercial, the SCL saw a slight increase again. The commercial was replayed, in order to identify the link between the event in the commercial and the SCL data. This explorative analysis considered that there is a 0.5–2 second delay between the presentation of a stimulus and the SCR. The scenes during the first 34 seconds are repetitive and showed a series of water activities such as surfing, rafting, diving, snorkeling, which may result in viewers gradually losing their interest. Then the content of the commercial changes from water-based activities to other types such as bushwalking, hot-air balloon and dancing, which may have re-attracted respondents' attention. This is consistent with the slight increase in SCL in the middle of the commercial.

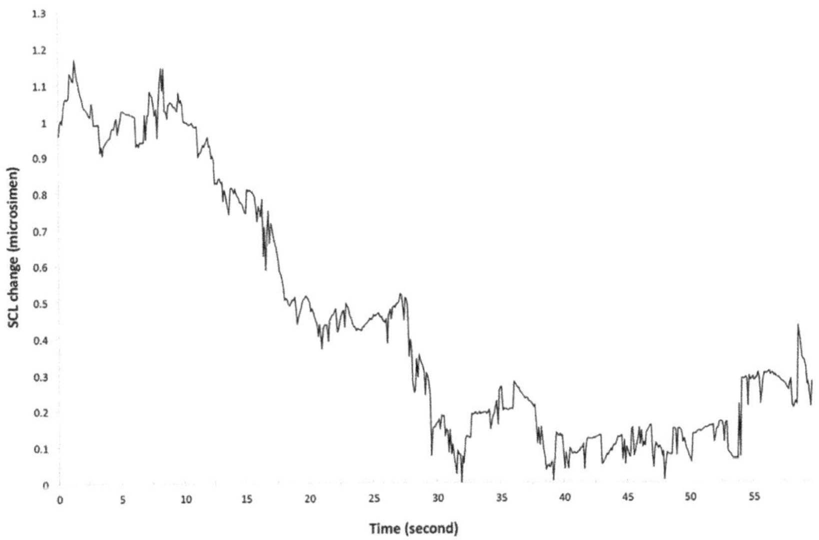

Figure 22.2. Average SCL Changes in Queensland Commercial Across Respondents.

Then the scenes return back to a series of water-based activities until the last five seconds, when the brand 'Queensland-Where Australia shines' slogan was presented in the commercial, which may explain the rise in SCL at the end of the waveform.

CONCLUSION

Overall, the study results demonstrate the usefulness of skin conductance method for the evaluation of tourism advertising effectiveness and its ability to compare different advertisements. Several analysis techniques such as SCR, SCL and changes in the latter from the baseline have been used to evaluate the four advertisements in the study. The advantage of skin conductance method over traditional self-report questionnaire lies in the fact that it cannot only measure respondent's average emotional response to a particular advertisement, but it is also capable of drawing the viewers' real-time emotional flow pattern and response during a period of time. The results indicate that the beer commercial provoked a higher int'nsity of emotional responses compared with the tourism advertisements used. A further analysis of the real-time SCL data for the Queensland advertisement illustrated a downward trend in spite of several fluctuations in the middle.

The relatively low SCL measured arousal t'wards tourism advertisements is surprising. The results suggest a need for further investigation to examine why viewers exhibit relatively lower scores in tourism advertisements. It appears that beer advertisement is telling a story about 'friendship' and describes a group of young people playing basketball. At the end of the story, all the people except one stood up. This revealed that only one of them needs the wheelchair to play the game. After the game, the group went to a pub for a few post-basketball drinks. Comparatively, the beer advertisement may attract more attention as the story unfolded gradually and the viewers may have become intrigued and interested in the end of the story. In this commercial, the people and their intense movements are also more likely to evoke stronger arousal responses. On the other hand, the tourism commercials lacked a storyline, and feature activities and the scenery in destinations. All the scenes of activities, attractions or landscape together in a tourism commercial may not be an ideal way to attract consumers' attention and evoke emotional responses.

The study is, however, not without its limitations and these provide avenue for future research. First, the respondents may feel unnatural or being manipulated in a laboratory setting, and their feelings about the commercials may have differed from what they would be in real life (Scott et al., 2016). Second, the sample size is small compared with traditional advertising research. However, the procedure behind this method is complicated and it takes time to collect data for each individual (Peacock et al., 2011). More importantly, this study aims to provide a potential method for evaluating tourism advertising instead of reflecting the general characteristics of a specific market. Third, the sample consisted of university students, and the reactions of students may be different from individuals in other age groups. More recent technology allows researchers to accurately measure skin conductance in field studies (Ming-Zher et al., 2010). Thus, future studies may extend the current study by measuring viewers' skin conductance to assess the effectiveness of tourism advertisements or commercials with a more representative sample in a naturalistic environment.

Despite these limitations, this chapter also has its strengths. Most tourism advertising studies still employ self-report questionnaires to measure and compare advertising effectiveness. Purely relying on the viewers recall may lead to inaccurate and unreliable results (Li et al., 2014). Given the criticisms of self-report method, a skin conductance technique was employed, the objective and visualised features of which may provide destination advertisers more valuable information for further refinement of their commercials.

This chapter documents initial results of using skin conductance measure to compare the relative effectiveness of four commercials. The approach used can also be used for evaluating other advertising stimuli, such as tourism websites or printed advertising (brochure). It may also be interesting to compare respondents' self-reported data with their physiological responses, and this may reflect nuances between their conscious and subconscious emotional responses. Furthermore, the psychophysiological method (skin conductance) can only reflect respondents' emotional intensity. Future studies can incorporate other measures such as heart rate and facial electromyography to determine the valence of emotional responses. Future studies may also record respondents' eye-tracking data while they are watching the commercials on the monitor. The combination of skin conductance measure and eye tracking method would enable one to analyse the correlations between the areas respondents are staring at and their corresponding emotional responses. This approach would be diagnostic and provide more detailed guidelines as to how to improve the commercials.

SECTION 5
CONCLUSION

Chapter 23

A CAUTIONARY TALE

Brent Moyle
Griffith University, Australia

Biqiang Liu
Griffith University, Australia

Siqi Lu
Griffith University, Australia

Abstract: This volume has sought to identify the benefits of using cognitive psychology in understanding the tourist experience. It discusses areas where tourism researchers should be cautious in applying cognitive psychology. Certainly, *t* scholars must understand the assumptions, criticisms and limitations of cognitive theories and associated data collection techniques. Cognitive psychology is a dynamic field with alternative theoretical perspectives and emerging debates. This volume discusses particular mainstream ideas, such as cognitive appraisal theory and seeks to identify their relevance and identifies difficulties and perceived limitations scholars discover in exploring this new field. **Keywords**: Cognitive psychology; cognitive appraisal theory; criticisms; limitations; future research; emotions

INTRODUCTION

The growing interest among researchers in cognitive psychology has precipitated discussion of the psychological mechanisms which underpin the tourist experience. This discussion about how people engage with and remember experiences is a worthwhile pursuit, but caution is required. While cognitive psychology is advancing rapidly, there is still much to learn about

Cognitive Psychology and Tourism
Tourism Social Science Series, Volume 27, 305–313

ISSN: 1571-5043/doi:10.1108/S1571-504320240000027023

how brain processes work. Further, the application of existing knowledge to study complex real word experiences is in its infancy. Use of the paradigm, adoption of appropriate theories and concepts can provide insight into theoretical discussions and the methods provide novel data for comparison with that from existing methods. However, the boundaries for the application of these theories, and the most useful topics for study require further work. As already noted, further developments require critical engagement with advances in the parent discipline. This chapter presents a cautionary tale on the application of cognitive psychological theories and approaches in tourism.

APPLICATION OF COGNITIVE PSYCHOLOGY

Cognitive psychology is developing rapidly. In mainstream cognitive psychology research, ideas from two decades ago may have been discarded but may still be current in folk discussions and applied fields. For example, 'early' cognitivist researchers used the metaphor of the 'mind as a computer' (Still & Costall, 1991). Some cognitive scholars familiar with the functioning of a computer proposed that the mind functions like a digital computer, with mental states similar to computational states, and mental processes functioning as computational processes (Searle, 1991).

The basic assumption underpinning the computer metaphor is that explanation should be determined at the programme or functional level, rather than the neuronal implementation level (Garnham, 2019). As such, the mind – much like a computer – collected information and followed a programme to produce an output. While this metaphor has some usefulness, it has now been discarded in mainstream cognitive psychology discussions. The metaphor is criticised as it is now evident that the brain has a generalised and flexible capacity that cannot be modelled than any current computer (Bongard & Levin, 2021). Reduction of the brain to a machine does not adequately account for the influence of human emotion and motivation on the cognitive system and how this may affect the processing of information (Plass & Kalyuga, 2019). Further, external information is not 'processed' in one common data format. It is not 'processed' in one 'system' (Cohen et al., 1993), but by multiple systems interacting recursively in a holistic way (Núñez et al., 2019).

However, the 'mind as a computer' is used in everyday science communication (Searle, 1990). If tourism researchers are unaware of the recent cognitive psychology literature, they may view it as reductionist, and potentially ignoring the complexity of human functioning. For instance, cognitive approaches have been criticised for being cold and mechanistic, assuming that all behaviour is the result of rational thought processing, with research focused on understanding the logical aspects of cognitive processing and with less emphasis on the creative, emotional and social aspects which affects thinking (Argyle, 2020). This criticism is therefore related to past approaches and not relevant today.

A second criticism is that brain processes are not directly observable, with multiple explanations for behaviour possible, and studies of the brain draw predominantly on inference (Gigerenzer, 2020). Behavioural psychologists reject the idea inferences based on assumptions about internal mental processes. This is based on a lack of objective measurement of the brain's operation (Bargh & Ferguson, 2000). Skinner (1985) criticised the cognitive approach, claiming that only external stimulus-response behaviour can be scientifically measured and subsequently mediation processes do not exist as they cannot be observed and measured. On the other hand, it was dissatisfaction with the prevailing behaviourist approach, and the simplistic emphasis on external behaviour rather than internal processes which shifted mainstream psychology towards trying to understand the mediation of the brain in processing external stimuli (Barsalou, 2014).

Despite this, cognitive psychologists currently find it difficult to interpret internal processes, such as attention and perception directly. Early studies are criticised as assuming external stimuli are related to internal processes without explicit observable evidence for this connection (Williams, 1987). This is essentially an empirical problem and advances in experimental methods. However, with the advent of neurophysiological instruments (electro-encephalography; event-related potentials), studies are mitigating this limitation. For example, the fundamental role of bottom-up and top-down processing mechanism has not only been supported by cognitive appraisal theory (CAT) but also endorsed by neuroscience evidence: top-down mechanism is modulated by the superior parietal lobe regions along the intraparietal sulcus, while bottom-up mechanism is mediated by inferior parietal lobe regions at the temporo-parietal junction (Ciaramelli et al., 2008). Within visual processing research, Ko et al. (2017) study sorted participants' electro-encephalography spectra with respect to response times, which indicated the level of visual alertness to randomly introduced visual stimuli, revealed significant changes in the brain oscillation patterns.

Likewise, this limitation is receiving growing attention in the marketing and tourism field (Li, Liu, et al., 2021; Li, Jiang, et al., 2021). Neurophysiological methods are emerging as an approach unveiling the physiological mechanism of how consumers and tourists cognitively process a servicescape or touristscape. Event-related potential waveforms representing electrical potential (voltage) changes at the scalp are characterised by a succession of peaks and troughs (considered as its components) that reflect the brain activity elicited by the stimuli (Bastiaansen et al., 2018). In global field power analysis, emotion is indicated when alpha (frontal electrodes 7.5–12.5 Hz) and theta (frontal electrodes 3.5–7.5 Hz) brain wave activity are simultaneously and significantly high compared to baseline resting state (Gordon et al., 2018). Imagination activity can be measured in electro-encephalography by low resolution brain electromagnetic tomography through right parietal beta (frontal electrodes 3.5–7.5 Hz) brain wave activity (Gordon et al., 2018). Ohme et al. (2010) identified frontal cortex activation in reaction to TV advertisements. They consider that the left frontal cortex is involved in a system facilitating approach behaviour, while the right cortex is involved in a system facilitating withdrawal behaviour from aversive stimuli.

A third criticism that has been discussed in other chapters is the scope of cognitive psychology in studying internal information processing. Cognitive psychology studies perception, attention, learning, emotion, consciousness and memory and how this subsequently affects behaviour (Hunt, 1989). In the tourism field, researchers have focused on cognitive concepts such emotion and experience, primarily using a behavioural approach but with a completely different set of definitions, terminology and associated assumptions (Weiler et al., 2018). For instance, behaviourists correlate external stimuli in the touristscape with emotional reactions, or intention to perform specific behaviours, such as revisitation. They do not consider the mediation effects of mental processes. When applying a cognitive approach, tourism researchers should exercise caution in correlating the effect of antecedent stimuli and consequences of hedonic consumption, and instead focus on advancing the conceptual understanding of the mental processes which connect the two together (Scott, 2020). For example, tourism researchers using CAT (Chapter 8) might seek to explain why a particular emotion was elicited, or behaviour performed based on an individual's goal and prior experience (Liu et al., 2022; Ma et al., 2013; Ma, Gao, et al., 2017; Ma, Scott, et al., 2017). This means different emotions will be expected if a group of consumers vary in their goals and prior experience. Thus, modelling stimulus-response links proposed by behaviourism requires understanding the mediation effects of goals and prior experience (novelty).

Most if not all tourism studies do not model behaviour in this way. Their assumption is that some input stimulus is associated with 'emotion', as a continuous variable. Instead, CAT indicates that emotions, their valance and intensity are determined by the mediation of goals. The implications for structural models are that they should investigate the mediating mental process of different dimensions of cognitive appraisals in the correlation between stimuli in the touristscape and emotions. As such, cognitive psychologists consider it essential to examine the underlying mental processes and their influence on behaviour. Without developing this understanding, researchers will not have a complete picture of the tourist experience nor consequent behaviour.

The tendency for cognitive psychologists to apply experimental design in highly controlled laboratory-based settings designed to enhance replicability has also been criticised. These studies do not represent the 'real world' (Gleaves et al., 2004). For instance, during tourism experiences, many cognitive processes operate simultaneously, and as such the validity of cognitive approaches applied in artificial settings like laboratories is potentially undermined. In laboratory-based environments, prospective 'tourists' may be influenced by pressure to perform well, and subsequently, memories may actually be weaker than if it was recalled after a 'real tourism experience' (Muschalla & Schonborn, 2021). This lack of ecological validity makes results of studies in laboratory-based settings less representative to everyday life, and subsequently if not replicated in the real-world findings may not be generalisable outside of artificial environments. Such a criticism is accepted, but what to do as a result is less clear. This argument does not detract from the importance of mental processes or the cognitive paradigm. Researchers should not throw the 'baby out with the bathwater'. In some cases, these criticisms are made to justify an approach that is equally able to be disproved.

Clearly, tourism scholars need to understand the theories and methods used in cognitive psychology and their associated debates and traditions. Adoption without understanding this background is fraught with danger and may be criticised as 'mere grabbing'. This position is where a researcher reads one paper and seeks to apply its concepts without understanding the relevant theory and its associated paradigm.

Criticisms of the Particular Cognitive Theories

There are also criticisms of particular concepts and theories from the cognitive approach. Leschziner and Brett (2021) view that the way information is processed based on past experiences from schemas is unclear. Other scholars consider schemata to be ambiguous and vague, citing inherent difficulties associated with measuring schema, while acknowledging its role as providing a plausible descriptive framework for understanding knowledge processing (Dahlin, 2001). When evaluating such criticisms, it is important to consider the disciplinary perspective of the author and the recency of the paper. For example, Leschziner and Brett (2021) are sociologists and may not necessarily be aware of current cognitive psychology research on schemata. Dahlin wrote in 2001 and thus must have based their study on schema theory from the 1990s. Therefore, when evaluating criticism of how schemata affect cognitive processing, it is necessary to read and understand the original literature and place new research in context (Zhang, Scott, et al., 2018; Zhang, Jahromi, et al., 2018). Despite this, research is needed into how people acquire a schema, select between schemata or process information using existing ones (Hunzaker & Valentino, 2019). While this has contributed to advancing the conceptual understanding of mental processes, tourism researchers should exercise caution and engage with broader debates in cognitive psychology on the efficacy of schemata for advancing knowledge on the mental processes which underpin tourists' experiences.

Another criticism from early opponents of the cognitive approach is that similar to other psychological approaches, the social reasons for behaviour are often overlooked (Eiser & Eiser, 1986). In addition, biological and cultural factors may not be considered from a cognitive perspective (Norman, 1980). These criticisms are correct, but their studies fail to understand the scope of cognitive psychology or ascribe to an alternative disciplinary paradigm. Taking the example of schema, cognitive psychology does not seek to explain (except in a general sense) why a person holds a particular one. Instead, it only states that people have schemata and that these are derived from generalised previous experiences. Thus, schemata organise memory, but it is necessary to determine what the schema of a particular person about a particular destination is. One consequence is that it is necessary to determine if schemata are commonly held. It would be expected that individuals may hold quite different schema as has been found in means-end analysis studies (Kang et al., 2012).

As to the CAT, in its basic form (Chapter 8), it is underpinned by the assertation that emotions are elicited by evaluations (appraisals) of events and situations (Omdahl, 2014). The theory explains why people experiencing the same event report different emotions (Garcia-Prieto & Scherer, 2016). However, CAT has been criticised due to differences in the number of appraisal dimensions identified. Scherer (1997b) identified eight appraisal dimensions, while others have noted further dimensions (Tong, 2015). Dimensions for appraisals include outcome desirability, fairness, certainty, coping potential, novelty, goal relevance, goal congruence and so on (Roseman & Smith, 2001). This limitation is acknowledged here. However, it is possible to identify three appraisal dimensions that are common. Goal relevance and goal congruence are critical determinants of emotional responses and degree of arousal.

Another criticism of CAT is that it focuses too much on higher level processing of external stimuli and does not include fast reactions to dangerous situations (Ziegler, 2001). It has also been criticised for the inability to explain quick non-conscious appraisals which are experiences which are triggered through an activation of the parasympathetic nervous systems (Schooler, 2002). This criticism may be seen as indicating that CAT denies the importance of stimuli in eliciting any emotional or behavioural response (Dimberg et al., 2002). It indicates a lack of wider cognitive psychology theory. There are two key pathways for processing stimuli. When a stimulus is perceived as dangerous, then appraisal is rapid and leads to a fight or flight response. Other stimuli that are not processed in this way are appraised against goals and novelty dimensions (Skavronskaya et al., 2017).

A second criticism is the difficulty of measuring brain functions using self-report methods (Jobe, 2003). A number of contemporary studies in the tourism field have sought to use self-report measures, such as questionnaires (Hosany et al., 2021). Mostly, these studies have categorised emotions using scales such as Plutchik's (1980) eight primary emotions, Izard's (1977) differential emotion scale, Mehrabian and Russell's (1974) pleasure, arousal and dominance scale and Watson et al. (1988) positive affect and negative affect schedule. While such self-reported measurements can have distinct advantages, they assume an individual is conscious of their emotions (Bargh & Morsella, 2008). This leads to the use of self-report measures of emotion. Data collection techniques, such as eye tracking, galvanic skin response and FMRi, are required to overcome the limitations associated with self-reported measurement, especially the application of scale instruments to assess cognitive appraisals (Hadinejad et al., 2019; Li et al., 2022a, 2022b, 2022c, 2022d, 2022e).

Reported emotions are based on feelings, which are the result of a coping processes. Additionally, feelings about past events are based on reconstruction of those from memories. In some situations, recalled memories are reinterpreted in the light of subsequent experiences and thus the elicited emotion will differ from the original. Thus, collection of data about emotions must be based on an understanding of these processes. Consequently, the impact of an emotional experience cannot be understood without deeper engagement with cognitive science (Novick et al., 2019).

Current Debates in Cognitive Psychology

There are a number of emerging theoretical perspectives in cognitive psychology. Even though not all can be explored, some areas are worth mentioning. First, while various aspects of cognition, such as perception, attention, consciousness, memory, learning, thinking and problem solving, have been studied in the laboratory, there has been growing interest in the application of cognitive psychology to situations in the 'real world'. Termed applied cognitive psychology, this approach is explicitly concerned with investigating how cognitive processes elicit emotion and affect behaviour outside of laboratory-based settings (Groome & Eysenck, 2015). By studying complex behaviour in complex outside of artificial environments, applied cognitive psychologists are not only concerned with core areas of cognitive psychology, such as perception and cognition, but also are required to be aware of and consider the social and organisational context within which specific behaviours take place (Alice et al., 2011). Given the uniqueness of tourism experiences, engagement with the applied cognitive psychology could be beneficial. For example, during complex tourism experiences, researchers could study the implications of inattention and multi-tasking (Liu, Moyle, et al., 2022; Liu, Li, et al., 2022).

Feature integration theory is another area of potential use. It is considered a landmark in cognitive psychology, particularly in visual perception, attention and cognition (Kristjánsson & Egeth, 2020). The theory postulates that low-level identification and recognition occurs without conscious attention. Attentional resources are needed only to process unusual or difficult stimuli (Sternberg & Sternberg, 2011). Feature integration theory may be a useful to assess attention to tourism marketing stimuli pre-visit, but also during complex 'real world' tourism experiences.

Cognitive load theory has been extensively applied in educational psychology but is yet to be integrated into the tourism field. This theory examines how much working memory is required to perform a specific task (Plass et al., 2010). Tourism experiences often involve novelty, but the effect of this on working memory for use on future tourism experiences has not been assessed. Cognitive load theory has implications for the design of tourism experiences with an educational component, as when cognitive load is increased, learning and transfer of knowledge is hampered (Sweller et al., 2019). The theory also considers the relationship between emotion and cognitive load, which has implications for the tourism field, given the emotional arousal of tourism experiences. Emotional experiences during learning have not been sufficiently covered in the context of cognitive load theory (Plass & Kalyuga, 2019). Attention and working memory are also linked (Angelopoulou & Drigas, 2021). Working memory is crucial for attention management because processes in working memory impact sustained attention to an event (Adam & DeBettencourt, 2019).

Finally, while there is a long history of research on working memory, the topic of nostalgia is currently fascinating cognitive psychologists. Working in the related area of neuroscience conducted under fMRI has shed light on the neural mechanisms of nostalgia, identifying parts of the brain which are active in self-reflection, autobiographical memory, emotion regulation and reward processing (Yang et al., 2023). Early work notes that nostalgia is rewarding and promotes positive affect, termed as pleasantness, activating the pre-frontal cortex to seek further rewards seeking congruent with the pursuit of goals (Azima et al., 2021). Subsequently, activating aspects, such as nostalgia in the recollection stage of tourism, may affect tourist decision-making (Chi & Chi, 2022). Future research on how the regions of the brain activate in response to recollecting tourism experiences, and subsequent emotional responses would be useful.

CONCLUSION

Given criticisms of the cognitive approaches used in early studies, it is not surprising the tourism academy may be slow to embrace the forthcoming 'cognitive wave'. However, failing to innovate and anticipate the future limits the tourism field. To successfully ride the 'cognitive wave', scholars may wish to revisit social science findings based upon new data from neuroscience and cognitive psychology. Theory-based prediction of ideas and behaviour should be emphasised, rather than the focus on descriptive studies or stimulus-response correlations. Tourism researchers must engage with emergent theoretical perspectives and cutting-edge methods in cognitive psychology to advance the tourism field. However, when doing so, tourism scholars must be exceptionally cautious to recognise and engage with criticisms of the cognitive approach and seek not only to acknowledge but also to mitigate associated limitations.

This chapter has examined criticisms of cognitive psychology, as well as suggesting some areas for further research based on emerging debates in cognitive psychology. Caution must be utilised by tourism researchers to understand the assumptions, criticisms and limitations of cognitive approaches and associated data collection techniques prior to integration into the tourism field.

Chapter 24

CONCLUSION

Noel Scott
Edith Cowan University, Australia

Brent Moyle
Griffith University, Australia

Ana Cláudia Campos
University of Algarve, Portugal

Liubov Skavronskaya
Griffith College, Australia

Biqiang Liu
Griffith University, Australia

Abstract: This chapter provides concluding comments on a cognitive paradigm and related concepts applied to tourism, as well as in studies of social and environmental sustainability. Adopting a new paradigm is difficult but worthwhile as it allows researchers to move on from a behaviourist worldview and open the black-box, that is, the brain. The main points of each of the previous chapters are provided. A number of areas for future research are discussed, including schema change, wellness and the mind-body problem, new methodologies, motivation and goals for travel and the importance of further research on emotion in tourism experiences. The use of cognitive psychology is vital for further progress into a theory of experience design. **Keywords**: Cognitive

Cognitive Psychology and Tourism
Tourism Social Science Series, Volume 27, 315–324

ISSN: 1571-5043/doi:10.1108/S1571-504320240000027024

paradigm; schema change; mind-body problem; wellness: experience design; tourism experience

INTRODUCTION

This volume demonstrates the applicability and usefulness of a cognitive paradigm and related concepts in research in applied fields such as tourism. It has argued and provided examples (see Chapter 4 on the Stimulus-Organism-Response Model) that much of the current literature is based in behavioural psychology with its assumptions about the brain as a black-box. Given the nature of the brain in this space, social science scholars have identified many concepts that purport to explain the way one's thinking operates in choosing a destination. They include perception, destination image, values, beliefs, motivation, rational conscious choice and planned behaviour and so on. This book argues that cognitive psychology provides an opportunity to contrast the existing understanding of these concepts with the experimental insights from an external perspective, not bound up in the complexity of a century of competing folk and social science theories.

This book provides offers much, but the discourse requires some schema change by the tourism academy. Many of the problems of importance to the world and to tourism as an economic sector are psychological in nature. Of these issues, the primary one must be social and environmental sustainability, as it is an existential threat to the existence of species. Understanding why individuals make decisions requires an understanding of subconscious biases and their current goals and related schema. Researchers should recognise the difficulty in schema change to explain why provision of information may not be effective in changing behaviour. Concurrently, schema change is embedded with 'mental models' in system thinking. The system needs to identify leverage points to make changes to existing mental models (schemas) – cognitive psychology/cognitive paradigm can be seen as a key/starting point for such changes. Cognitive psychology in a long term can help to make changes in mental models/schemas and improve tourist's behaviour towards being more sustainable. Cognitive psychology is worthwhile considering by academic interested in system thinking and sustainability as not only tourism research can benefit from it but also it can add value to find solutions for different tourism systems problems. This is an insight for behavioural researchers reading this book. A transformation in their schema about the basic assumptions of a field of study involves restructuring it to include extremely incongruent knowledge, a difficult task often associated with strongly emotional episodes.

The question is whether the pain of adopting a new paradigm is worthwhile. This question ultimately is answered by understanding an individual's goals in undertaking research. There is a group of academics for whom their goal is to receive a promotion, which in turn is achieved by publishing academic journal articles. For such, the adoption of a new paradigm is not likely to be goal congruent, at least in the short term. This is because the dominant behaviourist paradigm of a journal's reviewers is likely to result in rejection. Unfortunately, this is the fate of many innovative ideas and requires adoption of the long view. A second group of academics may have an idealistic belief in the importance of the pursuit of truth but lack the time or resources to spend to absorb new ideas. This book some insights on this subject. Finally, there is a nascent 'invisible college' who have begun to explore the fertile ground of cognitive psychology. For these, this book may provide an opportunity to expand the scope of their research and to strengthen links between tourism psychology and sustainability, as this book is the first to seek to bring together the literature and its application in tourism.

Evidence can provide the difficulty of schema change. One starting point is the common assumption that an individual's surroundings are indicative of their goals (stimulus-organism-response theory and theory of planned behaviour). This assumption is found in debates such as the definition of ecotourists (Beall et al., 2020; Sinay et al., 2020; Weaver, 2005), food tourists (Ellis et al., 2018; McKercher et al., 2008) or wellness tourists (He et al., 2022; Smith & Puczkó, 2014). Are ecotourists defined by their presence in a national park, or by their goal in visiting the park? Are food tourists defined by their presence in a gastronomical event, or by their goal in visiting that event? Are wellness tourists defined by their presence in a wellness destination, or by their goal in visiting the site? A behaviourist may say that attendance is sufficient to identify group membership. A cognitivist would argue that the goals of such attendees would vary and may be unrelated to the commonly assumed reasons for visiting a place. These are not trivial distinctions but underpin policy arguments, such as the value of national parks and justify government spending.

Adoption of a cognitive paradigm also prompts the examination of the relative influence of social and individual factors on behaviour. Individual schema derived from previous personal experiences along with tourists' current goals determine interaction with their surroundings in a particular situation and the valence of subsequent tourism experiences or emotions. There is no other mechanism available to a cognitivist. In a cognitivist paradigm, one's schema may be weakly or strongly congruent with social norms and culture. However, these operate through an individual's own experiences of those norms and culture. The association between typical social norms and an individual's behaviour must be tested and not assumed. Thus, theories such as planned behaviour, the stimulus-organism-response model and so on, should be questioned. A cognitivist approach attributes the causes of behaviour to tourists rather than to their environment or to their culture and its social norms. Tourists may share common behaviour because the voluntary choices determine a self-selecting sample that may share similar goals, but this cannot be simply assumed to be true. Research of this type cannot be generalised to the wider population.

One useful benefit of adopting the cognitive paradigm in tourism research is that it allows researchers to study mental processes that underlie tourist behaviour, such as perception, learning, problem-solving and decision-making, which utilise the same cognitive processes but differ in their content. While an individual's experience of a place visited may differ from that of others, the reasons why they differ are due to their goals at that time, and their previous experience with certain characteristics of the situation such as its novelty. This may allow seemingly distinct experiences but associated with similar goals to be compared. A necessary complication in using a cognitive approach is that it recognises the cognitive biases and heuristics that affect decision-making, and how these potentially may be overcome. These biases are in part due to the evolutionary history of the brain. This means that the brain is not a unified designed organ but has layers of development, one building on the others and is designed for survival in more primitive environments, leading to problems such as overeating in a context of food surplus. Change in the brain is quantitative and qualitative, because new layers being created eventually lead to new circumvolutions.

Finally, applying cognitive psychology requires the translation of previous research concepts into its language, known as disambiguation. One of the necessary projects involved in applying this knowledge to the tourism field is to 'translate' the concepts used in previous research into the concepts used in cognitive psychology. Translation is necessary because of the legacy of several psychological paradigms used in the past 100 years. This book assists the 'cognitive wave' to approach and permeate tourism field – discussing concepts and theories from cognitive psychology and to demonstrate its usefulness for understanding mental process underlying behaviours human exhibit, especially for the transfer of psychological paradigms to tourism research.

The Chapters

The chapters of this volume introduce the reader to complex concepts and theories and also provide some links to existing areas of research interest. They may contain misinterpretations and should be considered introductions only. They are a type of proof of concept, hoping to induce the reader to consider their contents. For this reason, there is some repetition of key ideas and arguments. This may be somewhat distracting but also may highlight that the brain and its processes are a complex and interacting system. Thus, different processes are affected by the same concepts, notably the current goal.

Chapter 2 introduced the cognitive paradigm and provided arguments for the adoption of a cognitive-based social sciences and in the field of tourism. Application to tourism is useful because of the hedonic and eudemonic, and hence emotional nature of tourism experiences, the novelty of travel outside one's usual environment, the importance of imagining, daydreams, emotions and desire for travel decisions and the importance of autobiographical memories because of travel.

Chapter 3 discussed the main psychological paradigms used in the past 100 years, psychodynamism, behaviourism, cognitivism based on an information processing paradigm, and later cognitivism based on complex interactive mental processes. It briefly introduced the main concepts of later cognitive psychology: consciousness, sensation, perception, attention, emotion and memory. Each of these concepts was then discussed in detail in later chapters along with their application to tourism.

Chapter 4 presented a case for updating the use and development of psychological theory in tourism through an examination of the psychological stimulus-organism-response model. A revised version of the model is provided based on importing new theoretical knowledge from the original discipline. It argues that tourism theory may benefit from adopting theory from the parental discipline rather than inventing its own.

Chapter 5 pointed to the need for definitional clarity when using psychological concepts in tourism research. The same term used in 'folk psychology', or behavioural or cognitive psychology can have quite different definitions and associated paradigmatic assumptions. This chapter argues that an understanding of mainstream cognitive psychology is needed when using concepts that are within its explanatory scope. Emotion is one such concept that has been extensively studied in cognitive psychology. The chapter discussed emotion and related terms such as feelings, novelty, unexpectedness and surprise to provide conceptual clarity.

The next chapters discussed each of the main concepts of cognitive psychology. It examined sensory experiences and subjective perception which are sources of information that influence people's cognition, emotions and behaviour. This can be useful, for instance, to improve accessible tourism, as understanding the mechanisms of sensory processing can help in the development of assistive technologies for individuals with sensory impairments.

Chapter 7 examined the definition and recent theories of consciousness, such as global workplace theory. Aspects of conscious experiences are useful in studying tourism decision-making, memory, emotions, tourism behaviour and other topics related to tourism psychology. For instance, studies on conscious experiences of decision-making are useful to better understand destination choice and tourists' emotions elicitation, unconscious biases help to understand tourism attitude formation, while explicit and implicit cognition research provides insights for understanding destination image. Future studies may consider using objective and subjective methods of studying consciousness in the tourism context.

Chapter 8 provides a discussion of the process of elicitation of emotions. Emotions play a vital role in tourism experiences and future orientated hedonic decision-making. Importantly, the valance and level of arousal of elicited emotions are mediated by the tourist's goals and prior experience. A majority of tourism research assumes a consciously deliberate decision-making process and ignores prospective emotional processes such as fantasy and dreaming.

Chapter 9 elaborated on goal-orientated behaviour. These play a critical role in everyday life and shape a person's life and lifestyle. Understanding tourists' goals is critical to optimise tourism experiences at different stages, as well as critical in studies on tourists' satisfaction.

Chapter 10 focused on novelty. Its detection in the environment is an important mental ability for evolutionary survival. It is also critical for understanding emotional arousal and its effect on memories in tourism. Tourism takes place, by definition, in an unusual environment and therefore intimately involves this concept.

Chapter 11 covered memory and how it is reconstructed, resulting in biases. It is important to study memory biases because such biases can influence how tourists remember and recall their travel experiences. Furthermore, memory biases can affect how tourists evaluate their experiences and make decisions about future travel. Better understanding memories is also crucial for tourism practitioners, as memory bias may also affect how information is recalled about destinations and the services received as it might influence destination reputation and loyalty.

Chapter 12 discussed schema, a concept used to describe the organisation of memory about an object. Schema are perceptual prototypes abstracted from numerous episodic events based on their similarities or commonalities. In tourism, a schema may be termed a network, mental model, social schema, stereotype, social role, script, worldview, heuristics or archetype. A schema is flexible. New information may be either assimilated or accommodated based on its degree of congruity. Schema incongruent information requires more mental effort to process. The concept can be used in research on destination brand image, storytelling, social relationships, attitude change and transformation during travel. The concept is useful in discussion of the change to a sustainable worldview.

Chapter 13 focuses on prospection. This concept refers to human ability of 'pre-experience' the future and hence how that future 'feels', which is why it is sometimes called affective forecasting. One key issue here is that prospection, and retrospection, use the same processes as thinking about the present. For tourism, prospection is needed for planning a trip and retrospection underpins nostalgia.

Chapter 14 examined the process of emotional coping and added how strategies can be used to cope with difficult situations. This concept can provide an understanding of customer/tourist emotional responses and the effects of dealing with challenging situations. Application areas include service failure, such as flight delays and interpersonal conflict.

Chapters 15–19 applied ideas covered in the previous chapters. It discussed their application to experience design, while Chapter 16 discussed co-creation of tourism experiences. Chapter 17 discussed awe, as a multidimensional emotional experience that has the potential to adjust one's beliefs and schema. Both an appraisal of vastness and the need for cognitive accommodation are essential to experiencing awe. Chapter 18 stated that mindfulness has been practiced in Buddhism contexts for thousands of years. Mindfulness emphasises the strong interaction between the mind and body and contributes to wellness The discussion provides evidence that tourism contexts can lead to mindful experiences with therapeutic-like benefits. Chapter 19 discusses stories, as the long tradition of oral communication worldwide is a testimony of their power. Stories help the formation of long-term memories. They involve higher attentional engagement, which helps the brain to better encode, store and retrieve information.

Chapters 20–22 discussed eye-tracking, electroencephalography and skin conductance as techniques for data collection. A field advances with the availability of data and these methods provide new data with which to test theories.

Chapters 23 provided some caution by discussing criticisms of cognitive psychology. Caution must be utilised by tourism researchers to understand the assumptions, criticisms and limitations of cognitive approaches and associated data collection techniques prior to integration into the tourism field.

NEW PERSPECTIVES

How should cognitive psychology be applied in tourism? At the paradigm level, cognitivism provides an alternative to behaviourism. In tourism research, there is a persuasive application of behavioural or social psychology without recognition of their paradigmatic assumptions. This book has highlighted the implications of the approach, and the advantages of the alternative cognitive view. It would be useful for future researchers to explicitly recognise their paradigmatic assumptions in their writings. Further, it would be useful to examine if, and how, alternative theories produce different and useful results to existing theories.

Wellness and the Mind-Body Problem

The mind-body problem and how the brain affects the body and vice versa is not usually acknowledged in the business or tourism literature. If one accepts that thoughts and actions have a neuronal basis, and are related to feelings and other body effects, then surely there is a basis for explaining how thoughts affect actions. These neuronal signals may be organised into processes, such as emotion, attention and so on, provide another layer of complexity, but this does not detract from the profound interaction between mind and body. For example, there is some evidence that travel reduces stress for some period after the return home but for how long is less clear. Similarly, the wellness sector began with a focus on the body using exercise, massage and spa treatments but embracing mental treatments, such as spiritual retreats or forest bathing. These more recent wellness treatments integrate mind and body interventions.

This book calls for future research endeavour focusing on the tourism–body relationship with a cognitive flavour. Compounding this, the relationship is supposed to be observed as a bidirectional functional system, rather than only treating it as a one-way direction that explains sensation channels affect psychological experience or vice versa. Therefore, the bidirectional body–mind effect requires further clarification in tourism.

Motivation

Motivation is of paramount importance in tourism studies, because travel and decision-making are inseparable from motivational forces and processes. The study of motivation has been linked to goal-oriented behaviour. Goals are the tools with which people engage in volitional behaviour. A study of how tourism relates to goal-orientated activity is central to emotion and experience design. There is a need to clarify the concept of motivation and related terms, such as interest, attention, desires, needs, wants, etc. Accordingly, it is worthwhile to notice the interconnection between motivations and other endogenous mental process, such as attention. It is suggested that these various terms were developed to discuss similar mental phenomena. However, there is a continuing effort to look at how these are different rather than similar. Concurrently, future tourism scholarship is expected to empirically explore the effect of goal-related appraisal dimensions on emotions elicited and experiential outcomes.

Experience Design

The chapters in this book identify several techniques that are important for the design of memorable tourism experience. The importance of goals means that a design should specify the cognitive psychological outcomes (emotion) expected rather than only behavioural outcomes. The emotional state of a tourist plays a central role in determining the summative evaluation of an experience reported such as satisfied or delighted. Emotions are subconscious but rational mental processes. Rational here means goal directed and can be translated into statements, propositions, judgements. Cocreation is a technique that stimulates attention in experience design. Increased attention to an activity caused by tourism experience design generates higher potential for memorability. Symbols and cues are techniques to activate relevant goals and schema. Stories are techniques to increase attention, activate relevant goals and schema. Design of a memorable tourism experience requires surprise, uncertainty and novelty. A designed experience perceived as novel and salient will increase attention, emotional arousal and memory recall. Therefore, the comprehensive framework of tourism experience co-creation from a cognitive perspective (see Chapter 15), conceptually linking sensation, perception, cognition and memory, is expected to be further extended and advanced in the future.

Taken together, future scholarship should focus on the meaning translating process in which experiences are initiated through sensing using basic human physical capabilities and then processing through three subsystems: perception, emotion and cognition. The chapters of this book provide the basis for the development of a theoretically informed science of experience design. While further descriptive research on experience design would be useful, an integrated theoretical framework must acknowledge the mental processes that explain an experience and its behavioural effects.

Cognitive psychology is useful for tourism to better understand the impact of technologies. For instance, it is also useful for tourism to better understand how to design immersive experiences and engage tourists into an immersive environment (virtual reality) in a more profound level (through senses, emotions, feelings and attention). Furthermore, metaverse is defined as an embodied internet; therefore, cognitive embodiment concept might be useful to better understand how to design immersive experiences for metaverse, because cognitive psychology helps to understand how people perceive, process and interact with digital information.

Metaverse researchers discuss creating digital twins of destinations (Parisi, 2022). Cognitive psychology concepts and principles may be useful for creating engaging tourism content for the metaverse and other digital environments. This could include how people navigate virtual environments, and how they form mental representations of digital spaces. By understanding these cognitive processes, experience designers can create more intuitive, user-friendly and accessible virtual environments that better align with tourists' expectations and goals, leading to a more immersive and enjoyable experience.

Methodology

The understanding of the mind-body problem ultimately will be improved through new sources of data. The emergence of new technology makes it possible to capture meaningful moments in real time that provide new essential information for tourism design. Given the physiological foundations of an experience, the techniques discussed in Chapters 20–22 provide data on which new explanations may be based. The existing dominance of self-reported measurement may be useful, but it is based on an individual's conscious thoughts. In contrast, cognitive psychology studies the unconsciousness/subconsciousness processes that underpin these conscious thoughts.

Further Research on Emotion

There remain ongoing debates in cognitive psychology. The cognitive appraisal theory, developed by theorists such as Richard Lazarus (Lazarus, 1991), suggests that emotions are generated by a cognitive process in which an individual assesses the personal significance of a particular event and how it relates to their goals and needs. The constructivist theory of emotions, developed by Lisa Feldman Barrett (Barrett, 2017; Gendron & Barrett, 2009), suggests that emotions are not fixed and universal, but rather are constructed by the brain based on an individual's past experiences and cultural context. This theory emphasises the role of cognitive processes in the construction of emotions but differs from the cognitive appraisal theory in that it suggests that emotions are not necessarily tied to specific events or stimuli.

One main difference between these two theories is that the cognitive appraisal theory focuses on the role of certain appraisal dimensions, while constructivist theory emphasises the role of past experiences and cultural context. Overall, both cognitive appraisal and constructivist theories offer important insights into the nature of emotions and how they are experienced. Further research is needed to fully understand the complex processes involved in the experience of emotions and to determine the relative importance of cognitive appraisal, past experiences and cultural context.

REFERENCES

Aarts, H., & Custers, R. (2012). Unconscious goal pursuit: Nonconscious goal regulation and motivation. In R. M. Ryan (Ed.), *The Oxford handbook of human motivation* (pp. 232–247). Oxford University Press.

Abelson, R. P. (1981). Psychological status of the script concept. *American Psychologist*, *36*, 715–729. https://doi.org/10.1037/0003-066X.36.7.715

Abrahams, R. D. (1986). Ordinary and extraordinary experience. In V. Turner & E. M. Bruner (Eds.), *The anthropology of experience* (pp. 45–73). University of Illinois Press.

Abson, D. J., Fischer, J., Leventon, J., Newig, J., Schomerus, T., Vilsmaier, U., von Wehrden, H., Abernethy, P., Ives, C. D., Jager, N. W., & Lang, D. J. (2017). Leverage points for sustainability transformation. *Ambio*, *46*(1), 30–39. https://doi.org/10.1007/s13280-016-0800-y

Adam, K. C., & de Bettencourt, M. T. (2019). Fluctuations of attention and working memory. *Journal of Cognition*, *2*(1), 1–4. https://doi.org/10.5334/joc.70

Agapito, D. (2020). The senses in tourism design: A bibliometric review. *Annals of Tourism Research*, *83*, 102934. https://doi.org/10.1016/j.annals.2020.102934

Agapito, D., Mendes, J., & Valle, P. (2013). Exploring the conceptualization of the sensory dimension of tourist experiences. *Journal of Destination Marketing & Management*, *2*(2), 62–73.

Agapito, D., Oom do Valle, P., & da Costa Mendes, J. (2013). The cognitive-affective-conative model of destination image: A confirmatory analysis. *Journal of Travel & Tourism Marketing*, *30*(5), 471–481.

Agapito, D., Pinto, P., & Mendes, J. (2017). Tourists' memories, sensory impressions, and loyalty: In loco and post-visit study in Southwest Portugal. *Tourism Management*, *58*, 108–118. http://doi.org/10.1016/j.tourman.2016.10.015

Agapito, D., Valle, P., & Mendes, J. (2014). The sensory dimension of tourist experiences: Capturing meaningful sensory-informed themes in Southwest Portugal. *Tourism Management*, *42*, 224–237. https://doi.org/10.1016/j.tourman.2013.11.011

Aggarwal, P., & McGill, A. L. (2007). Is that car smiling at me? Schema congruity as a basis for evaluating anthropomorphized products. *Journal of Consumer Research*, *34*(4), 468–479.

Agyeiwaah, E., Adam, I., Dayour, F., & Badu Baiden, F. (2021). Perceived impacts of COVID-19 on risk perceptions, emotions, and travel intentions: Evidence from Macau higher educational institutions. *Tourism Recreation Research*, *46*(2), 195–211.

Ahn, S. J. G., Bailenson, J. N., & Park, D. (2014). Short-and long-term effects of embodied experiences in immersive virtual environments on environmental locus of control and behavior. *Computers in Human Behavior*, *39*, 235–245. https://doi.org/10.1016/j.chb.2014.07.025

Ahn, Y.-J., Hyun, S. S., & Kim, I. (2017). Vivid-memory formation through experiential value in the context of the international industrial exhibition. *Journal of Travel & Tourism Marketing*, *34*(3), 291–311. https://doi.org/10.1080/10548408.2016.1156617

Aho, S. K. (2001). Towards a general theory of touristic experiences: Modelling experience process in tourism. *Tourism Review*, *56*(3/4), 33–37.

Ajzen, I. (2001). Nature and operation of attitudes. *Annual Review of Psychology*, *52*(1), 27–58.

Ajzen, I. (2005). *Attitudes, personality, and behavior*. McGraw-Hill Education.

Ajzen, I. (2015). The theory of planned behaviour is alive and well, and not ready to retire: A commentary on Sniehotta, Presseau, and Araujo-Soares. *Health Psychology Review*, *9*(2), 131–137. https://doi.org/10.1080/17437199.2014.883474

Ajzen, I., & Kruglanski, A. W. (2019). Reasoned action in the service of goal pursuit. *Psychological Review*, *126*(5), 774–786. https://doi.org/10.1037/rev0000155

Alba, J. W., & Williams, E. F. (2013). Pleasure principles: A review of research on hedonic consumption. *Journal of Consumer Psychology*, *23*(1), 2–18.

Albright, T. D., Kandel, E. R., & Posner, M. I. (2000). Cognitive neuroscience. *Current Opinion in Neurobiology*, *10*(5), 612–624.

Aldwin, C. M. (2007). *Stress, coping, and development: An integrative perspective*. Guilford Press.

Alessandroni, N. (2017). Varieties of embodiment in cognitive science. *Theory & Psychology*, *28*(2), 227–248. https://doi.org/10.1177/0959354317745589

Alexander, C., Ishikawa, S., Silverstein, M., Jacobson, M., Fiksdahl-King, I., & Angel, S. (1977). *A pattern language: Towns, buildings, construction*. Oxford University Press.

Algie, J. A. (2005). A comparison of viewers' cognitive and psychophysiological responses to threat-appeal TV advertisements. In K. Ghallachir (Ed.), *Academy of marketing: Marketing: Building business, shaping society conference*. Academy of Marketing.

Alice, F., Healy, L. E., & Bourne, J. (2011). Applied cognitive psychology. In P. R. Martin, F. M. Cheung, M. C. Knowles, M. Kyrios, L. Littlefield, J. B. Overmier, & J. M. Prieto (Eds.), *IAAP handbook of applied psychology* (pp. 559–572). John Wiley & Sons. Incorporated.

Amodio, D. M., & Berg, J. J. (2018). Toward a multiple memory systems model of attitudes and social cognition. *Psychological Inquiry*, *29*(1), 14–19.

Anderson, M. L. (2010). Neural reuse: A fundamental organizational principle of the brain. *Behavioral and Brain Sciences*, *33*(4), 245–266.

Andersson, T. D. (2007). The tourist in the experience economy. *Scandinavian Journal of Hospitality and Tourism*, *7*(1), 46–58.

Andrades, L., & Dimanche, F. (2014). Co-creation of experience value: A tourist behaviour approach. In N. K. Prebensen, J. S. Chen, & M. Uysal (Eds.), *Creating experience value in tourism* (pp. 95–112). CABI.

Andrades, L., & Dimanche, F. (2018). Co-creation of experience value: A tourist behavior approach. In *Creating experience value in tourism* (pp. 83–97). CAB International.

Andrew, J. D. (1973). The structuralist study of narrative: Its history, use, and limits. *Bulletin of the Midwest Modern Language Association*, *6*(1), 45–61. https://doi.org/10.2307/1314848

Angelopoulou, E., & Drigas, A. (2021). Working memory, attention, and their relationship: A theoretical overview. *Research, Society and Development*, *10*(5), e46410515288–e46410515288. https://doi.org/10.33448/rsd-v10i5.15288

Antón, C., Camarero, C., & Garrido, M. J. (2018a). A journey through the museum: Visit factors that prevent or further visitor satiation. *Annals of Tourism Research*, *73*, 48–61. https://doi.org/10.1016/j.annals.2018.08.002

Antón, C., Camarero, C., & Garrido, M. J. (2018b). Exploring the experience value of museum visitors as a co-creation process. *Current Issues in Tourism*, *21*(12), 1406–1425.

Appel, M., Gnambs, T., Richter, T., & Green, M. C. (2015). The transportation scale–Short form (TS–SF). *Media Psychology*, *18*(2), 243–266. https://doi.org/10.1080/15213269.2014.987400

Appel, M., & Richter, T. (2010). Transportation and need for affect in narrative persuasion: A mediated moderation model. *Media Psychology*, *13*(2), 101–135.

Arcodia, C., Novais, M. A., Cavlek, N., & Humpe, A. (2020). Educational tourism and experiential learning: Students' perceptions of field trips. *Tourism Review*, *76*(1), 241–254.

Ardila, A. (2008). On the evolutionary origins of executive functions. *Brain and Cognition*, *68*(1), 92–99.

Ardoin, N. M., Wheaton, M., Bowers, A. W., Hunt, C. A., & Durham, W. H. (2015). Nature-based tourism's impact on environmental knowledge, attitudes, and behavior: A review and analysis of the literature and potential future research. *Journal of Sustainable Tourism*, *23*(6), 838–858. https://doi.org/10.1007/s13280-016-0800-y

Arentze, T. A., Dellaert, B. G., & Timmermans, H. J. (2008). Modeling and measuring individuals' mental representations of complex spatio-temporal decision problems. *Environment and Behavior*, *40*(6), 843–869.

Argyle, M. (2020). A critique of cognitive approaches to social judgments and social behavior. In *Emotion and social judgments* (pp. 161–178). Garland Science.

Ariely, D., & Berns, G. S. (2010). Neuromarketing: The hope and hype of neuroimaging in business. *Nature Reviews Neuroscience*, *11*(4), 284–292.

Arimitsu, K., & Hofmann, S. G. (2017). Effects of compassionate thinking on negative emotions. *Cognition & Emotion*, *31*(1), 160–167.

Aristotle. (2001). In C. H. Gomes (Trans.), *Da Alma [De Anima]*. Textos Filosóficos.

Armitage, C. J., & Conner, M. (2001). Efficacy of the theory of planned behaviour: A meta-analytic review. *British Journal of Social Psychology*, *40*(4), 471–499.

Armstrong, P. B. (2019). Neuroscience, narrative, and narratology. *Poetics Today*, *40*(3), 395–428. https://doi.org/10.1215/03335372-7558052

Armstrong, P. B. (2020). *Stories and the brain. The neuroscience of narrative*. John Hopkins University Press.

Arnould, E., & Price, L. L. (1993). River magic: Extraordinary experience and the extended service encounter. *Journal of Consumer Research*, *20*, 24–45.

Åstrøm, J. K. (2017). Theme factors that drive the tourist customer experience. *International Journal of Culture, Tourism and Hospitality Research*, *11*(2), 125–141.

Åstrøm, J. K. (2019). Why theming? Identifying the purposes of theming in tourism. *Journal of Quality Assurance in Hospitality & Tourism*, *21*(3), 245–266.

Atadil, H. A., Sirakaya-Turk, E., Baloglu, S., & Kirillova, K. (2017). Destination neurogenetics: Creation of destination meme maps of tourists. *Journal of Business Research*, *74*, 154–161. http://doi.org/10.1016/j.jbusres.2016.10.028

Atalay, A. S., Bodur, H. O., & Rasolofoarison, D. (2012). Shining in the centre: Central gaze cascade effect on product choice. *Journal of Consumer Research*, *39*(4), 848–866.

Atance, C. M., & O'Neill, D. K. (2005). The emergence of episodic future thinking in humans. *Learning and Motivation*, *36*(2), 126–144.

Atkinson, R. C., & Shiffrin, R. M. (1968). Human memory: A proposed system and its control processes. In *Psychology of learning and motivation* (pp. 89–195). Academic Press.

Avnet, T., Pham, M. T., & Stephen, A. T. (2012). Consumers' trust in feelings as information. *Journal of Consumer Research*, *39*(4), 720–735.

Azevedo, A. (2010). Designing unique and memorable experiences: Co-creation and the surprise factor. *International Journal of Hospitality and Tourism Systems*, *3*(1), 42–54.

Azima, S., Jafari, P., Motie Nasrabadi, A., & Sammaknejad, N. (2021). Evaluating consumers' emotional, cognitive, and behavioral responses to nostalgic stimuli in advertising: A mix method study. *Consumer Behavior Studies Journal*, *8*(3), 140–171.

Baars, B. J. (1993). *A cognitive theory of consciousness*. Cambridge University Press.

Baars, B. J. (2002). The conscious access hypothesis: Origins and recent evidence. *Trends in Cognitive Sciences*, *6*(1), 47–52.

Baars, B. J. (2005). Global workspace theory of consciousness: Toward a cognitive neuroscience of human experience. *Progress in Brain Research*, *150*, 45–53.

Baars, B. J., & Franklin, S. (2003). How conscious experience and working memory interact. *Trends in Cognitive Sciences*, *7*(4), 166–172.

Baars, B. J., & Gage, N. M. (2010). Cognition, brain and consciousness. *Introduction to cognitive neuroscience* (2nd ed.). Elsevier Ltd.

Babakhani, N., Randle, M., & Dolnicar, S. (2020). Do tourists notice social responsibility information? *Current Issues in Tourism*, *23*(5), 559–571.

Babin, B., Darden, W., & Griffin, M. (1994). Work and/or fun: Measuring hedonic and utilitarian shopping value. *Journal of Consumer Research*, *20*(4), 644–656.

Baer, R. A. (2003). Mindfulness training as a clinical intervention: A conceptual and empirical review. *Clinical Psychology: Science and Practice*, *10*(2), 125–143. https://doi.org/10.1093/clipsy.bpg015

Baer, R. A. (2009). *Assessment of mindfulness* (pp. 153–168). Springer New York.

Baer, R. A. (2011). Measuring mindfulness. *Contemporary Buddhism*, *12*(1), 241–261.

Baggio, R., Scott, N., & Cooper, C. (2010). Network science–A review with a focus on tourism. *Annals of Tourism Research*, *37*(3), 802–827.

Bagozzi, R. P., Baumgartner, H., & Pieters, R. (1998). Goal-directed emotions. *Cognition & Emotion*, *12*(1), 1–26.

Bagozzi, R. P., Gopinath, M., & Nyer, P. U. (1999). The role of emotions in marketing. *Journal of the Academy of Marketing Science*, *27*(2), 184–206.

Bagozzi, R. P., & Utpal, D. (1999). Goal setting and goal striving in consumer behavior. *Journal of Marketing*, *63*(4), 19–32.

Bai, X., Yan, G., Liversedge, S. P., Zang, C., & Rayner, K. (2008). Reading spaced and unspaced Chinese text: Evidence from eye movements. *Journal of Experimental Psychology: Human Perception and Performance*, *34*(5), 1277–1287.

Bainbridge, W. A., Dilks, D. D., & Oliva, A. (2017). Memorability: A stimulus-driven perceptual neural signature distinctive from memory. *NeuroImage*, *149*, 141–152. https://doi.org/10.1016/j.neuroimage.2017.01.063

Baird, T. R., & Wahlers, R. G. (2007). Non-recognition of print advertising: Emotion arousal and gender effects. *Journal of Marketing Communications*, *13*(1), 39–57. https://doi.org/10.1080/13527260600942616

Baker, J. (1986). The role of the environment in marketing services: The consumer perspective. In J. A. Czepiel, A. Carole, C. A. Congram, & J. Shanahan (Eds.), *The services challenge: Integrating for competitive advantage* (pp. 79–84). American Marketing Association.

Baker, J. P., & Berenbaum, H. (2007). Emotional approach and problem-focused coping: A comparison of potentially adaptive strategies. *Cognition & Emotion*, *21*(1), 95–118.

Bakker, I., van der Voordt, T., Vink, P., & de Boon, J. (2014). Pleasure, arousal, dominance: Mehrabian and Russell revisited. *Current Psychology*, *33*(3), 405–421.

Balaji, M. S., Roy, S. K., & Quazi, A. (2017). Customers' emotion regulation strategies in service failure encounters. *European Journal of Marketing*, *51*(5/6), 960–982.

Ballantyne, R., Packer, J., Hughes, K., & Dierking, L. (2007). Conservation learning in wildlife tourism settings: Lessons from research in zoos and aquariums. *Environmental Education Research*, *13*(3), 367–383. https://doi.org/10.1080/13504620701430604

Ballantyne, R., Packer, J., & Sutherland, L. A. (2011). Visitors' memories of wildlife tourism: Implications for the design of powerful interpretive experiences. *Tourism Management*, *32*(4), 770–779.

Baloglu, S., & McCleary, K. (1999). US international pleasure travelers' images of four mediterranean: A comparisons of visitors and non-visitors. *Journal of Travel Research*, *38*(2), 144–152.

Baloglu, S., & McCleary, K. W. (1999). A model of destination image formation. *Annals of Tourism Research*, *26*(4), 868–897.

Balota, D. A. (2000). Cognition. In A. Kazdin (Ed.), *Encyclopedia of psychology* (Vol. 2, pp. 131–132). American Psychological Association.

Bandura, A. (1986). *Social foundations of thought and action: A social cognitive theory*. Prentice-Hall, Inc.

Bandura, A. (1994). Self-efficacy. In V. S. Ramachaudran (Ed.), *Encyclopedia of human behavior* (Vol. 4, pp. 71–81). Academic Press.

Bandura, A. (1997). *Self-efficacy: The exercise of control*. Freeman.

Bandura, A., & Schunk, D. H. (1981). Cultivating competence, self-efficacy, and intrinsic interest through proximal self-motivation. *Journal of Personality and Social Psychology*, *41*(3), 586–598.

Banich, M. T. (2009). Executive function: The search for an integrated account. *Current Directions in Psychological Science*, *18*(2), 89–94.

Bargh, J. A. (2002). Losing consciousness: Automatic influences on consumer judgment, behavior, and motivation. *Journal of Consumer Research*, *29*(2), 280–285.

Bargh, J. A. (2006). What have we been priming all these years? On the development, mechanisms, and ecology of nonconscious social behavior. *European Journal of Social Psychology*, *36*(2), 147–168.

Bargh, J. A., & Ferguson, M. J. (2000). Beyond behaviorism: On the automaticity of higher mental processes. *Psychological Bulletin*, *126*(6), 925–945. https://doi.org/10.1037/0033-2909.126.6.925

Bargh, J. A., & Gollwitzer, P. M. (1994). Environmental control of goal-directed action: Automatic and strategic contingencies between situations and behavior. In W. D. Spaulding (Ed.), *Integrative views of motivation, cognition, and emotion* (pp. 71–124). University of Nebraska Press.

Bargh, J. A., & Huang, J. Y. (2009). The selfish goal. In G. B. Moskowitz & H. Grant (Eds.), *The psychology of goals* (pp. 127–150). Guildford Press.

Bargh, J. A., & Morsella, E. (2008). The unconscious mind. *Perspectives on Psychological Science*, *3*(1), 73–79.

Barnett, L. A. (2013). What people want from their leisure: The contributions of personality facets in differentially predicting desired leisure outcomes. *Journal of Leisure Research*, *45*(2), 150–191.

Barrett, L. F. (2017). The theory of constructed emotion: An active inference account of interoception and categorization. *Social Cognitive and Affective Neuroscience*, *12*(1), 1–23.

Barrett, L. F., Mesquita, B., Ochsner, K. N., & Gross, J. J. (2007). The experience of emotion. *Annual Review of Psychology*, *58*, 373–403. https://doi.org/10.1146/annurev.psych.58.110405.085709

Barrett, L. F., & Russell, J. A. (2014). *The psychological construction of emotion*. Guilford Publications.

Barsalou, L. W. (2008). Grounded cognition. *Annual Review of Psychology*, *59*, 617–645.

Barsalou, L. W. (2014). *Cognitive psychology: An overview for cognitive scientists*. Psychology Press.

Barsalou, L. W. (2020). Challenges and opportunities for grounding cognition. *Journal of Cognition*, *3*(1), 1–24. https://doi.org/10.5334/joc.116

Bartels, K. (2013). Public encounters: The history and future of face-to-face contact between public professionals and citizens. *Public Administration*, *91*, 469–483.

Bartlett, F. C. (1932). *Remembering*. Cambridge University Press.

Barto, A., Mirolli, M., & Baldassarre, G. (2013). Novelty or surprise? *Frontiers in Psychology*, *4*(907), 1–15.

Bastiaansen, M., Lub, X. D., Mitas, O., Jung, T. H., Ascenção, M. P., Han, D.-I., Moilanen, T., Smit, B., & Strijbosch, W. (2019). Emotions as core building blocks of an experience. *International Journal of Contemporary Hospitality Management*, *31*(2), 651–668.

Bastiaansen, M., Mazaheri, A., & Jensen, O. (2012). Beyond ERPs: Oscillatory neuronal dynamics. In S. Luck & E. Kappenman (Eds.), *Oxford handbook of event-related potential components* (pp. 31–49). Oxford University Press.

Bastiaansen, M., Oosterholt, M., Mitas, O., Han, D., & Lub, X. (2022). An emotional roller coaster: Electrophysiological evidence of emotional engagement during a roller-coaster ride with virtual reality add-on. *Journal of Hospitality & Tourism Research*, *46*(1), 29–54.

Bastiaansen, M., Straatman, S., Driessen, E., Mitas, O., Stekelenburg, J., & Wang, L. (2018). My destination in your brain: A novel neuro marketing approach for evaluating the effectiveness of destination marketing. *Journal of Destination Marketing & Management*, *7*, 76–88.

Bastiaansen, M., Straatman, S., Mitas, O., Stekelenburg, J., & Jansen, S. (2022). Emotion measurement in tourism destination marketing: A comparative electroencephalographic and behavioral study. *Journal of Travel Research*, *61*(2), 252–264.

Batra, R., & Ahtola, O. T. (1991). Measuring the hedonic and utilitarian sources of consumer attitudes. *Marketing Letters*, *2*(2), 159–170.

Baumeister, R. F., Vohs, K. D., & Oettingen, G. (2016). Pragmatic prospection: How and why people think about the future. *Review of General Psychology*, *20*(1), 3–16.

Beall, J. M., Boley, B. B., Landon, A. C., & Woosnam, K. M. (2020). What drives ecotourism: Environmental values or symbolic conspicuous consumption? *Journal of Sustainable Tourism*, *29*(8), 1215–1234. https://doi.org/10.1080/09669582.2020.1825458

Beard, J., & Ragheb, M. (1983). Measuring leisure motivation. *Journal of Leisure Research*, *15*(3), 219–228.

Bec, A., Moyle, B., Timms, K., Schaffer, V., Skavronskaya, L., & Little, C. (2019). Management of immersive heritage tourism experiences: A conceptual model. *Tourism Management*, *72*, 117–120.

Becken, S., & Kaur, J. (2021). Anchoring "tourism value" within a regenerative tourism paradigm–A government perspective. *Journal of Sustainable Tourism*, *30*(1), 52–68. https://doi.org/10.1080/09669582.2021.1990305

Belk, R. W. (1974). An exploratory assessment of situational effects in buyer behavior. *Journal of Marketing Research*, *11*(2), 156–163.

Belk, R. W. (1975). The objective situation as a determinant of consumer behavior. *ACR North American Advances*, *2*, 427–438.

Bellman, S., & Rossiter, J. R. (2004). The website schema. *Journal of Interactive Advertising*, *4*(2), 38–48.

Ben-Zur, H. (2009). Coping styles and affect. *International Journal of Stress Management*, *16*(2), 87–101.

Benn, Y., Webb, T. L., Chang, B. P., Sun, Y. H., Wilkinson, I. D., & Farrow, T. F. (2014). The neural basis of monitoring goal progress. *Frontiers in Human Neuroscience*, *8*, 688. https://doi.org/10.3389/fnhum.2014.00688

Berger, H. (1929). Über das elektroenkephalogramm des Menschen. *Archives de Psychologie*, *87*(1), 527–570.

Berk, R. A. (2015). The greatest veneration: Humour as a coping strategy for the challenges of aging. *Social Work in Mental Health*, *13*(1), 30–47.

Berkman, E. T. (2018). The neuroscience of goals and behavior change. *Journal of Consulting Psychology*, *70*(1), 28–44. https://doi.org/10.1037/cpb0000094

Berkman, E. T., & Lieberman, M. D. (2010). Approaching the bad and avoiding the good: Lateral prefrontal cortical asymmetry distinguishes between action and valence. *Journal of Cognitive Neuroscience*, *22*(9), 1970–1979.

Berkman, E. T., Livingston, J. L., & Kahn, L. E. (2017). Finding the "self" in self-regulation: The identity-value model. *Psychological Inquiry*, *28*(2–3), 77–98. https://doi.org/10.1080/1047840X.2017.1323463

Bernard, H. R., & Killworth, P. D. (1977). Informant accuracy in social network data II. *Human Communication Research*, *4*(1), 3–18.

Bernard, H. R., Killworth, P. D., & Sailer, L. (1979). Informant accuracy in social network data IV: A comparison of clique-level structure in behavioral and cognitive network data. *Social Networks*, *2*(3), 191–218.

Berns, G. S., Blaine, K., Prietula, M. J., & Pye, B. E. (2013). Short- and long-term effects of a novel on connectivity in the brain. *Brain Connectivity*, *3*(6), 590–600. https://doi.org/10.1089/brain.2013.0166

Berntson, G. G., Thomas Bigger, J., Jr, Eckberg, D. L., Grossman, P., Kaufmann, P. G., Malik, M., Nagaraja, H. N., Porges, S. W., Saul, J. P., & Stone, P. H. (1997). Heart rate variability: Origins, methods, and interpretive caveats. *Psychophysiology*, *34*(6), 623–648.

Berridge, K. C. (2004). Motivation concepts in behavioral neuroscience. *Physiology and Behavior*, *81*(2), 179–209. https://doi.org/10.1016/j.physbeh.2004.02.004

Berridge, K. C., & Robinson, T. E. (2003). Parsing reward. *Trends in Neurosciences*, *26*(9), 507–513.

Bertella, G. (2014). The co-creation of animal-based tourism experience. *Tourism Recreation Research*, *39*(1), 115–125.

Bertella, G., Fumagalli, M., & Williams-Grey, V. (2019). Wildlife tourism through the co-creation lens. *Tourism Recreation Research*, *44*(3), 300–310.

Bettiga, D., Lamberti, L., & Noci, G. (2017). Do mind and body agree? Unconscious versus conscious arousal in product attitude formation. *Journal of Business Research*, *75*, 108–117.

Bidwell, L. C., Knopik, V. S., Audrain-McGovern, J., Glynn, T. R., Spillane, N. S., Ray, L. A., & Leventhal, A. M. (2015). Novelty seeking as a phenotypic marker of adolescent substance use. *Substance Abuse: Research and Treatment*, *9*, 1–10. https://doi.org/10.4137/SART.S22440

Bieger, T., & Laesser, C. (2002). Market segmentation by motivation: The case of Switzerland. *Journal of Travel Research*, *41*(1), 68–76. https://doi.org/10.1177/004728750204100108

Bigne, E., Chatzipanagiotou, K., & Ruiz, C. (2020). Pictorial content, sequence of conflicting online reviews and consumer decision-making: The stimulus-organism-response model revisited. *Journal of Business Research*, *115*, 403–416.

Bigne, J. E., Andreu, L., & Gnoth, J. (2005). The theme park experience: An analysis of pleasure, arousal, and satisfaction. *Tourism Management*, *26*(6), 833–844.

Binkhorst, E., & Den Dekker, T. (2009). Agenda for co-creation tourism experience research. *Journal of Hospitality Marketing & Management*, *18*(2–3), 311–327.

Bishop, S. R., Lau, M., Shapiro, S., Carlson, L., Anderson, N. D., Carmody, J., Segal, Z. V., Abbey, S., Speca, M., Velting, D., & Devins, G. (2004). Mindfulness: A proposed operational definition. *Clinical Psychology: Science and Practice*, *11*(3), 230–241. https://doi.org/10.1093/clipsy.bph077

Bitner, M. J. (1992). Servicescapes: The impact of physical surroundings on customers and employees. *Journal of Marketing*, *56*, 57–71.

Blackford, J. U., Buckholtz, J. W., Avery, S. N., & Zald, D. H. (2010). A unique role for the human amygdala in novelty detection. *NeuroImage*, *50*(3), 1188–1193.

Block, N. (1995). On a confusion about a function of consciousness. *Behavioral and Brain Sciences*, *18*(2), 227–247.

Blomstervik, I. H., & Olsen, S. O. (2022). Progress on novelty in tourism: An integration of personality, attitudinal and emotional theoretical foundations. *Tourism Management*, *93*, 104574.

Blumenthal, V. (2020). Consumer immersion in managed visitor attractions: The role of individual responses and antecedent factors. *Scandinavian Journal of Hospitality and Tourism*, *20*(1), 4–27. https://doi.org/10.1080/15022250.2020.1725624

Blythe, J. L., Armitage, D., Bennett, N. J., Silver, J. J., & Song, A. M. (2021). The politics of ocean governance transformations. *Frontiers in Marine Science*, *804*. https://doi.org/10.3389/fmars.2021.634718

Bodhi, B. (2011). What does mindfulness really mean? A canonical perspective. *Contemporary Buddhism*, *12*(1), 19–39. https://doi.org/10.1080/14639947.2011.564813

Bogicevic, V., Seo, S., Kandampully, J., Liu, S. Q., & Rudd, N. A. (2019). Virtual reality presence as a preamble of tourism experience: The role of mental imagery. *Tourism Management*, *74*, 55–64.

Boje, D. M. (2001). *Narrative methods for organizational & communication research*. SAGE Publications.

Bolls, P. D., Lang, A., & Potter, R. F. (2001). The effects of message valence and listener arousal on attention, memory, and facial muscular responses to radio advertisements. *Communication Research*, *28*(5), 627–651.

Bolls, P. D., Muehling, D. D., & Yoon, K. (2003). The effects of television commercial pacing on viewers' attention and memory. *Journal of Marketing Communications*, *9*(1), 17–28. https://doi.org/10.1080/1352726032000068032

Bongard, J., & Levin, M. (2021). Living things are not (20th century) machines: Updating mechanism metaphors in light of the modern science of machine behavior. *Frontiers in Ecology and Evolution*, *9*, 650726.

Bonifield, C., & Cole, C. (2007). Affective responses to service failure: Anger, regret, and retaliatory versus conciliatory responses. *Marketing Letters*, *18*(1), 85–99.

Bonner, E. T., & Friedman, H. L. (2011). A conceptual clarification of the experience of awe: An interpretative phenomenological analysis. *The Humanistic Psychologist*, *39*(3), 222–235. https://doi.org/10.1080/08873267.2011.593372

Bos, L., McCabe, S., & Johnson, S. (2015). Learning never goes on holiday: An exploration of social tourism as a context for experiential learning. *Current Issues in Tourism*, *18*(9), 859–875.

Bosangit, C., Hibbert, S., & McCabe, S. (2015). If I was going to die I should at least be having fun. *Annals of Tourism Research*, *55*, 1–14.

Boshoff, C. (2017). An assessment of consumers' subconscious responses to frontline employees' attractiveness in a service failure and recovery situation. *South African Journal of Economic and Management Sciences*, *20*(1), 1–13.

Boucsein, W. (2012). *Electrodermal activity* (2nd ed.). Springer Science & Business Media. https://doi.org/10.1007/978-1-4614-1126-0

Boulding, K. E. (1956). *The image: Knowledge in life and society*. University of Michigan Press.

Bowden, J., & Dagger, T. (2011). To delight or not to delight? An investigation of loyalty formation in the restaurant industry. *Journal of Hospitality Marketing & Management*, *20*(5), 501–524.

Bower, G. H., Monteiro, K. P., & Gilligan, S. G. (1978). Emotional mood as a context for learning and recall. *Journal of Verbal Learning and Verbal Behavior*, *17*(5), 573–585.

Braasch, B. (2008). Major concepts in tourism research–Memory. In *Leeds: Centre for tourism and cultural change*. Leeds Metropolitan University.

Bradley, M. M. (2009). Natural selective attention. *Orienting and Emotion Psychophysiology*, *46*(1), 1–11.

Bradley, M. M., & Lang, P. J. (2000). Measuring emotion: Behavior, feeling, and physiology. In R. D. R. Lane, L. Nadel, G. L. Ahern, J. Allen, & A. W. Kaszniak (Eds.), *Cognitive neuroscience of emotion* (pp. 242–276). Oxford University Press.

Bradshaw, C. J., Ehrlich, P. R., Beattie, A., Ceballos, G., Crist, E., Diamond, J., & Blumstein, D. T. (2021). Underestimating the challenges of avoiding a ghastly future. *Frontiers in Conservation Science*, *9*. https://doi.org/10.3389/fcosc.2020.615419

Braisby, N., & Gellatly, A. (2012). *Cognitive psychology* (Vol., 2nd ed.). OUP Oxford.

Brandon, P. (2016). Body and self: An entangled narrative. *Phenomenology and the Cognitive Sciences*, *15*(1), 67–83.

Brashears, M. E., & Quintane, E. (2015). The microstructures of network recall: How social networks are encoded and represented in human memory. *Social Networks*, *41*, 113–126.

Braun, K. A., Ellis, R., & Loftus, E. F. (2002). Make my memory: How advertising can change our memories of the past. *Psychology and Marketing*, *19*(1), 1–23.

Braun, K., Grinley, M., & Loftus, E. (2006). Tourist memory distortion. *Journal of Travel Research*, *44*(4), 360–367. https://doi.org/10.1177/0047287506286721

Braun-LaTour, K. A., Grinley, M. J., & Loftus, E. F. (2006). Tourist memory distortion. *Journal of Travel Research*, *44*(4), 360–367.

Braunstein, L. M., Gross, J. J., & Ochsner, K. N. (2017). Explicit and implicit emotion regulation: A multi-level framework. *Social Cognitive and Affective Neuroscience*, *12*(10), 1545–1557.

Braver, T. S., Krug, M. K., Chiew, K. S., Kool, W., Westbrook, J. A., Clement, N. J., … Group, M. (2014). Mechanisms of motivation-cognition interaction: Challenges and opportunities. *Cognitive, Affective, & Behavioral Neuroscience*, *14*(2), 443–472. https://doi.org/10.3758/s13415-014-0300-0

Breitsohl, J., & Garrod, B. (2016). Assessing tourists' cognitive, emotional, and behavioral reactions to an unethical destination incident. *Tourism Management*, *54*, 209–220.

Breunig, M. (2017). Experientially learning and teaching in a student-directed classroom. *Journal of Experiential Education*, *40*(3), 213–230.

Bricker, K. S., Donohoe, H. M., Becerra, L., & Nickerson, N. (2015). Theoretical perspectives on tourism—An introduction. In L. Becerra & N. Nickerson (Eds.), *Demystifying theories in tourism research* (pp. 1–6). CABI Publishing.

Briesemeister, B. B., Tamm, S., Heine, A., & Jacobs, A. M. (2013). Approach the good, withdraw from the bad—A review on frontal alpha asymmetry measures in applied psychological research. *Psychology*, *4*(3A), 261–267.

Briñol, P., & Petty, R. E. (2003). Overt head movements and persuasion: A self-validation analysis. *Journal of Personality and Social Psychology*, *84*(6), 1123–1139.

Briñol, P., Petty, R. E., Stavraki, M., Lamprinakos, G., Wagner, B., & Díaz, D. (2018). Affective and cognitive validation of thoughts: An appraisal perspective on anger, disgust, surprise, and awe. *Journal of Personality and Social Psychology*, *114*(5), 693. https://doi.org/10.1037/pspa0000118

Brochado, A., Stoleriu, O., & Lupu, C. (2021). Wine tourism: A multisensory experience. *Current Issues in Tourism*, *24*(5), 597–615.

Brockington, G., Gomes Moreira, A. P., Buso, M. S., Gomes da Silva, S., Altszyler, E., Fischer, R., & Moll, J. (2021). Storytelling increases oxytocin and positive emotions and decreases cortisol and pain in hospitalized children. *Proceedings of the National Academy of Sciences of the United States of America*, *118*(22). https://doi.org/10.1073pnas.2018409118

Brodie, R. J., Hollebeek, L. D., Jurić, B., & Ilić, A. (2011). Customer engagement conceptual domain, fundamental propositions, and implications for research. *Journal of Service Research*, *14*(3), 252–271.

Brosch, T., Scherer, K., Grandjean, D., & Sander, D. (2013). The impact of emotion on perception, attention, memory, and decision-making. *Swiss Medical Weekly*, *143*(1920), w13786–w13786.

Brown, J. (1958). Some tests of the decay theory of immediate memory. *The Quarterly Journal of Experimental Psychology*, *10*, 12–21. https://doi.org/10.1080/17470215808416249

Brown, N. A., Neel, R., & Sherman, R. A. (2015). Measuring the evolutionarily important goals of situations: Situational affordances for adaptive problems. *Evolutionary Psychology*, *13*(3), 1–15.

Brown, R., & Kulik, J. (1977). Flashbulb memories. *Cognition*, *5*(1), 73–99.

Brown, R., Lau, H., & LeDoux, J. E. (2019). Understanding the higher-order approach to consciousness. *Trends in Cognitive Sciences*, *23*(9), 754–768.

Bruner, J. (1986). *Actual minds, possible worlds*. Harvard University Press.

Bruner, J. (1990). *Acts of meaning*. Harvard UP.

Brunner-Sperdin, A., Peters, M., & Strobl, A. (2012). It is all about the emotional state: Managing tourists' experiences. *International Journal of Hospitality Management*, *31*(1), 23–30.

Buckley, R. (2020). Nature tourism and mental health: Parks, happiness, and causation. *Journal of Sustainable Tourism*, *28*(9), 1409–1424. https://doi.org/10.1080/09669582.2020.1742725

Buckley, R., & Akhoundogli, M. (2019). Tourism affordances as a research lens. *Journal of Travel Research*, *59*(7), 1331–1334. https://doi.org/10.1177/0047287519877241

Buckley, R. C., & Cooper, M. A. (2022). Tourism as a tool in nature-based mental health: Progress and prospects post-pandemic. *International Journal of Environmental Research and Public Health*, *19*(20), 13112. https://doi.org/10.3390/ijerph192013112

Buckner, R. L., & Carroll, D. C. (2007). Self-projection and the brain. *Trends in Cognitive Sciences*, *11*(2), 49–57.

Buda, D. M., d'Hauteserre, A. M., & Johnston, L. (2014). Feeling and tourism studies. *Annals of Tourism Research*, *46*, 102–114.

Bunzeck, N., Doeller, C. F., Dolan, R. J., & Duzel, E. (2012). Contextual interaction between novelty and reward processing within the mesolimbic system. *Human Brain Mapping*, *33*(6), 1309–1324.

Burns, A., Greene, B. R., McGrath, M. J., O'Shea, T. J., Kuris, B., Ayer, S. M., Stroiescu, F., & Cionca, V. (2010). Shimmer™–A wireless sensor platform for noninvasive biomedical research. *IEEE Sensors Journal*, *10*(9), 1527–1534.

Burton, S., Howlett, E., & Tangari, A. H. (2009). Food for thought: How will the nutrition labeling of quick service restaurant menu items influence consumers' product evaluations, purchase intentions, and choices? *Journal of Retailing*, *85*(3), 258–273.

Butler, R. (2015). The evolution of tourism and tourism research. *Tourism Recreation Research*, *40*(1), 16–27.

Buttafuoco, A., Pedale, T., Buchanan, T. W., & Santangelo, V. (2018). Only "efficient" emotional stimuli affect the content of working memory during free-recollection from natural scenes. *Cognitive Processing*, *19*(1), 125–132. https://doi.org/10.1007/s10339-017-0846-1

Buzinde, C. N. (2020). Theoretical linkages between well-being and tourism: The case of self-determination theory and spiritual tourism. *Annals of Tourism Research*, *83*.

Cacioppo, J. T., Cacioppo, S., & Petty, R. E. (2018). The neuroscience of persuasion: A review with an emphasis on issues and opportunities. *Social Neuroscience*, *13*(2), 129–172.

Cai, R. R., Lu, L., & Gursoy, D. (2018). Effect of disruptive customer behaviors on others' overall service experience: An appraisal theory perspective. *Tourism Management*, *69*, 330–334.

Calero, C., & Turner, L. W. (2020). Regional economic development and tourism: A literature review to highlight future directions for regional tourism research. *Tourism Economics*, *26*(1), 3–26.

Campos, A. C., Mendes, J., Do Valle, P. O., & Scott, N. (2017). Co-creating animal-based tourist experiences: Attention, involvement, and memorability. *Tourism Management*, *63*, 100–114.

Campos, A. C., Mendes, J., Oom do Valle, P., & Scott, N. (2014). Attention and memorability in tourist co-creation experiences: A qualitative approach. In *Paper presented at the TMS Algarve 2014: Management studies international conference,* Algarve.

Campos, A. C., Mendes, J., Oom do Valle, P., & Scott, N. (2016). Co-creation experiences: Attention and memorability. *Journal of Travel & Tourism Marketing*, *33*(9), 309–1336.

Campos, A. C., Mendes, J., Oom do Valle, P., & Scott, N. (2017). Attentive tourists: The importance of co-creative experiences. In N. Scott, J. Gao, & J. Ma (Eds.), *Tourism experience design* (pp. 93–109). CABI.

Campos, A. C., Mendes, J., Valle, P. O. D., & Scott, N. (2018). Co-creation of tourist experiences: A literature review. *Current Issues in Tourism*, *21*(4), 369–400.

Campos, A. C., Pinto, P., & Scott, N. (2020). Bottom-up factors of attention during the tourist experience: An empirical study. *Current Issues in Tourism*, *23*(24), 3111–3133.

Carbone, L. P. (1998). Total customer experience drives value. *Management Review*, *87*(7), 62–63.

Cardoso, L., Diasb, F., de Araújo, A. F., & Marquesa, M. I. A. (2019). A destination imagery processing model: Structural differences between dream and favourite destinations. *Annals of Tourism Research*, *74*, 81–94.

Carpenter, R. (2016). A review of instruments on cognitive appraisal of stress. *Archives of Psychiatric Nursing*, *30*(2), 271–279.

Carruthers, C., & Hood, C. D. (2011). Mindfulness and well-being: Implications for TR practice. *Therapeutic Recreation Journal*, *45*(3), 171–189. https://doi.org/10.1093/clipsy/bpg015

Carter, T. J., & Gilovich, T. (2010). The relative relativity of material and experiential purchases. *Journal of Personality and Social Psychology*, *98*(1), 146–159.

Carù, A., & Cova, B. (2007). Consumer immersion in an experiential context. In A. Carù & B. Cova (Eds.), *Consuming experience* (pp. 34–47). Routledge. https://doi.org/10.4324/9780203390498

Caruelle, D., Gustafsson, A., Shams, P., & Lervik-Olsen, L. (2019). The use of electrodermal activity (EDA) measurement to understand consumer emotions-A literature review and a call for action. *Journal of Business Research*, *104*, 146–160.

Carvalho, S. W., Samu, S., & Sivaramakrishnan, S. (2011). The effect of country-related brand associations and product attributes on attitude toward unfamiliar foreign brands: A schema congruity perspective. *Journal of International Consumer Marketing*, *23*(2), 135–150. https://doi.org/10.1080/08961530.2011.543055

Carver, C. S., & Connor-Smith, J. (2010). Personality and coping. *Annual Review of Psychology*, *61*, 679–704.

Casado-Aranda, L. A., Sánchez-Fernández, J., & Ibáñez-Zapata, J. Á. (2023). Evaluating communication effectiveness through eye tracking: Benefits, state of the art, and unresolved questions. *International Journal of Business Communication*, *60*(1), 24–61.

Castree, N., Adams, W. M., Barry, J., Brockington, D., Büscher, B., Corbera, E., … Wynne, B. (2014). Changing the intellectual climate. *Nature Climate Change*, *4*(9), 763–768. https://doi.org/10.1038/nclimate2339

Cavanagh, M. (2011). Students' experiences of active engagement through cooperative learning activities in lectures. *Active Learning in Higher Education*, *12*(1), 23–33.

Cavanagh, P. (2011). Visual cognition. *Vision Research*, *51*(13), 1538–1551.

Cave, J., & Dredge, D. (2020). Regenerative tourism needs diverse economic practices. *Tourism Geographies*, *22*(3), 503–513. https://doi.org/10.1080/14616688.2020.1768434

Čavlek, N. (2015). Experiential learning in tourism education. In *Tourism education: Global issues and trends* (pp. 101–113). Emerald Group Publishing Limited.

Chaitin, J. (2003). Narratives and story-telling. In G. Burgess & H. Burgess (Eds.), *Beyond intractability, conflict information consortium*. University of Colorado. http://www.beyondintractability.org/essay/narratives

Chalmers, D. J. (1998). The problems of consciousness. In H. Jasper, L. Descarries, V. Castellucci, & S. Rossignol (Eds.), *Consciousness: At the frontiers of neuroscience. Advances in neurology* (pp. 7–18). Lippincott-Raven Press.

Chandralal, L., & Valenzuela, F. R. (2013). Exploring memorable tourism experiences: Antecedents and behavioural outcomes. *Journal of Economics, Business and Management*, *1*(2), 177–181.

Chaney, D. W. (2013). An overview of the first use of the terms cognition and behavior. *Behavioral Sciences*, *3*(1), 143–153.

Chang, C. (2013). Imagery fluency and narrative advertising effects. *Journal of Advertising*, *42*(1), 54–68. https://doi.org/10.1080/00913367.2012.749087

Charatsari, C. (2014). Collecting and using autobiographical memories in rural social research: A step-by-step guide. *Journal of Agricultural & Food Information*, *15*(1), 19–41.

Chekalina, T., Fuchs, M., & Lexhagen, M. (2014). A value co-creation perspective on customer-based brand equity model for tourism destinations–A case from Sweden. *Matkailututkimus*, *10*(1), 8–24.

Chen, G., Bao, J., & Huang, S. (2014). Segmenting Chinese backpackers by travel motivations. *International Journal of Tourism Research*, *16*(4), 355–367.

Chen, H., & Rahman, I. (2018). Cultural tourism: An analysis of engagement, cultural contact, memorable tourism experience and destination loyalty. *Tourism Management Perspectives*, *26*, 153–163.

Chen, L. I.-L. (2022). Traversing the time of COVID-19 lockdown with mindful leisure. *Wakayama Tourism Review*, *2*, 30–33. https://doi.org/10.19002/10.19002.2.30

Chen, G., So, K. K. F., Hu, X., & Poomchaisuwan, M. (2022). Travel for affection: A stimulus-organism-response model of honeymoon tourism experiences. *Journal of Hospitality & Tourism Research*, *46*(6), 1187–1219.

Chen, Z., King, B., & Suntikul, W. (2020). Co-creation of value for cultural festivals: Behind the scenes in Macau. *Tourism Recreation Research*, *45*(4), 430–444.

Chen, T. (2015). The persuasive effectiveness of mini-films: Narrative transportation and fantasy proneness. *Journal of Consumer Behaviour*, *14*(1), 21–27.

Chen, L. I.-L., Scott, N., & Benckendorff, P. (2014). An exploration of mindfulness theories in eastern and western philosophies. In P. M. Chien (Ed.), *Proceedings of the 24th council for Australasian university tourism and hospitality education (CAUTHE) annual conference* (pp. 164–175). The University of Queensland.

Chen, L. I.-L., Scott, N., & Benckendorff, P. (2017). Mindful tourist experiences: A Buddhist perspective. *Annals of Tourism Research*, *64*, 1–12. https://doi.org/10.1016/j.annals.2017.01.013

Cheong, Y., & Kim, K. (2011). The interplay between advertising claims and product categories in food advertising: A schema congruity perspective. *Journal of Applied Communication Research*, *39*(1), 55–74.

Cherif, H., & Miled, B. (2013). Are brand communities influencing brands through co-creation? A cross-national example of the brand AXE: In France and in Tunisia. *International Business Research*, *6*(9), 14–29.

Cheung, S. Y., Mak, J. Y., & Dixon, A. W. (2016). Elite active sport tourists: Economic impacts and perceptions of destination image. *Event Management*, *20*(1), 99–108.

Chi, O. H., & Chi, C. G. (2022). Reminiscing other people's memories: Conceptualizing and measuring vicarious nostalgia evoked by heritage tourism. *Journal of Travel Research*, *61*(1), 33–49.

Chiew, K. S., & Braver, T. S. (2011). Positive affect versus reward: Emotional and motivational influences on cognitive control. *Frontiers in Psychology*, *2*, 279. https://doi.org/10.3389/fpsyg.2011.00279

Chirico, A. (2020). Awe. In V. P. Glăveanu (Ed.), *The Palgrave encyclopedia of the possible* (pp. 1–9). Springer International Publishing. https://doi.org/10.1007/978-3-319-98390-5_30-1

Chirico, A., Ferrise, F., Cordella, L., & Gaggioli, A. (2018). Designing awe in virtual reality: An experimental study. *Frontiers in Psychology*, *8*, 2351. https://doi.org/10.3389/fpsyg.2017.02351

Chirico, A., & Gaggioli, A. (2021). The potential role of awe for depression: Reassembling the puzzle. *Frontiers in Psychology*, *12*, 617715. https://doi.org/10.3389/fpsyg.2021.617715

Chirico, A., Yaden, D., Riva, G., & Gaggioli, A. (2016). The potential of virtual reality for the investigation of awe. *Frontiers in Psychology*, *7*, 1766. https://doi.org/10.3389/fpsyg.2016.01766

Chirico, A., & Yaden, D. B. (2018). Awe: A self-transcendent and sometimes transformative emotion. In H. C. Lench (Ed.), *The function of emotions* (pp. 221–233). Springer. https://doi.org/10.1007/978-3-319-77619-4_11

Choi, H., & Choi, H. C. (2019). Investigating tourists' fun-eliciting process toward tourism destination sites: An application of cognitive appraisal theory. *Journal of Travel Research*, *58*(5), 732–744.

Choi, H. M., Mohammad, A. A., & Kim, W. G. (2019). Understanding hotel frontline employees' emotional intelligence, emotional labor, job stress, coping strategies and burnout. *International Journal of Hospitality Management*, *82*, 199–208.

Christian, B., Miles, L., Fung, F., Best, S., & Macrae, N. (2013). The shape of things to come: Exploring goal-directed prospection. *Consciousness and Cognition*, *22*(2), 471–478.

Christianson, S. Å. (1992). Emotional stress and eyewitness memory: A critical review. *Psychological Bulletin*, *112*(2), 284.

Chronis, A. (2012). Between place and story: Gettysburg as tourism imaginary. *Annals of Tourism Research*, *39*(4), 1797–1816.

Chung, H. C., Chung, N., & Yang, S. B. (2017, August). An impact of tourism experience on the quality of life: The roles of autobiographical memory and SNS. In *Proceedings of the international conference on electronic commerce* (pp. 1–7).

Ciaramelli, E., Grady, C. L., & Moscovitch, M. (2008). Top-down and bottom-up attention to memory: A hypothesis (AtoM) on the role of the posterior parietal cortex in memory retrieval. *Neuropsychologia*, *46*(7), 1828–1851.

Cicerali, E. E., Cicerali, L. K., & Saldamlı, A. (2017). Linking psycho-environmental comfort factors to tourist satisfaction levels: Application of a psychology theory to tourism research. *Journal of Hospitality Marketing & Management*, *26*(7), 717–734.

Clark, A. (1998). *Being there: Putting brain, body, and world together again*. MIT Press.

Clark, A. (2013). Whatever next? Predictive brains, situated agents, and the future of cognitive science. *Behavioural Brain Science*, *36*(3), 181–204.

Clark, E., Mulgrew, K., Kannis-Dymand, L., Schaffer, V., & Hoberg, R. (2019). Theory of planned behavior: Predicting tourists' pro-environmental intentions after a humpback whale encounter. *Journal of Sustainable Tourism*, *27*(5), 649–667. https://doi.org/10.1080/09669582.2019.1603237

Clemente, S., Dolansky, E., Mantonakis, A., & White, K. (2014). The effects of perceived product-extrinsic cue incongruity on consumption experiences: The case of celebrity sponsorship. *Marketing Letters*, *25*(4), 373–384.

Clewett, D., & Davachi, L. (2017). The ebb and flow of experience determines the temporal structure of memory. *Current Opinion in Behavioral Sciences*, *17*, 186–193.

Cloninger, C. R., Przybeck, T. R., Svrakic, D. M., & Wetzel, R. D. (1994). *The temperament and character inventory (TCI): A guide to its development and use* (pp. 19–28). Center for Psychobiology of Personality, Washington University.

Cloninger, C. R., Svrakic, D. M., & Przybeck, T. R. (1993). A psychobiological model of temperament and character. *Archives of General Psychiatry*, *50*(12), 975–990.

Coenen, A., Fine, E., & Zayachkivska, O. (2014). Adolf Beck: A forgotten pioneer in electroencephalography. *Journal of the History of the Neurosciences*, *23*(3), 276–286.

Coghlan, A., Buckley, R., & Weaver, D. (2012). A framework for analysing awe in tourism experiences. *Annals of Tourism Research*, *39*(3), 1710–1714. https://doi.org/10.1016/j.annals.2012.03.007

Coghlan, A., & Filo, K. (2013). Using constant comparison method and qualitative data to understand participants' experiences at the nexus of tourism, sport and charity events. *Tourism Management*, *35*, 122–131.

Cohen, E. (1972). Towards a sociology of international tourism. *Social Research*, *39*(1), 164–182.

Cohen, E. (1979). A phenomenology of tourist experiences. *Sociology*, *13*(2), 179–201. https://doi.org/10.1177/003803857901300203

Cohen, J. B., & Areni, C. S. (1991). Affect and consumer behavior. In T. S. Robertson & H. H. Kassarjian (Eds.), *Handbook of consumer behavior* (pp. 188–240). Prentice-Hall.

Cohen, J. B., & Bernard, H. R. (2013). Evolutionary psychology and consumer behavior: A constructive critique. *Journal of Consumer Psychology*, *23*(3), 387–399. https://doi.org/10.1016/j.jcps.2013.03.006

Cohen, M. A., & Dennett, D. C. (2011). Consciousness cannot be separated from function. *Trends in Cognitive Sciences*, *15*(8), 358–364.

Cohen, R. A., Sparling-Cohen, Y. A., & O'Donnell, B. F. (1993). *The neuropsychology of attention*. Plenum Press.

Cohen, S. A., Prayag, G., & Moital, M. (2013). Consumer behaviour in tourism: Concepts, influences and opportunities. *Current Issues in Tourism*, *17*(10), 872–909. https://doi.org/10.1080/13683500.2013.850064

Cohen, S. A., & Cohen, E. (2019). New directions in the sociology of tourism. *Current Issues in Tourism*, *22*(2), 153–172.

Coles, N., March, D. S., Marmolejo-Ramos, F., Larsen, J. T., Arinze, N. C., Ndukaihe, I. L. G., Willis, M. L., Foroni, F., Reggev, N., Mokady, A., Forscher, P. S., Hunter, J. F., Kaminski, G., Yüvrük, E., Kapucu, A., Nagy, T., Hajdu, N., Tejada, J., Freitag, R. M. K., … Liuzza, M. T. (2022). A multi-lab test of the facial feedback hypothesis by the many smiles collaboration. *Nature Human Behaviour*, *6*(12), 1731–1742.

Collins, A. M., & Loftus, E. F. (1975). A spreading-activation theory of semantic processing. *Psychological Review*, *82*(6), 407–428.

Conway, M. A., & Pleydell-Pearce, C. W. (2000). The construction of autobiographical memories in the self-memory system. *Psychological Review*, *107*(2), 261–288.

Corbetta, M., & Shulman, G. L. (2002). Control of goal-directed and stimulus-driven attention in the brain. *Nature Reviews Neuroscience*, *3*(3), 201–215.

Cormick, C. (2019). Who doesn't love a good story?—What neuroscience tells about how we respond to narratives? *Journal of Communication Science*, *18*(5), 1–10. https://doi.org/10.22323/2.18050401

Correia, A., & Moital, M. (2009). Antecedents and consequences of prestige motivation in tourism: An expectancy-value motivation. In M. Kozak & A. Decrop (Eds.), *Handbook of tourist behavior* (pp. 34–50). Routledge.

Costall, A. (2006). 'Introspectionism' and the mythical origins of scientific psychology. *Consciousness and Cognition*, *15*(4), 634–654.

Cowan, N. (2008). What are the differences between long-term, short-term, and working memory? *Progress in Brain Research*, *169*, 323–338.

Cowan, K., & Ketron, S. (2019). A dual model of product involvement for effective virtual reality: The roles of imagination, co-creation, telepresence, and interactivity. *Journal of Business Research*, *100*, 483–492.

Craik, F. I., & Lockhart, R. S. (1972). Levels of processing: A framework for memory research. *Journal of Verbal Learning and Verbal Behavior*, *11*(6), 671–684.

Crompton, J. L. (1979). Motivations for pleasure vacations. *Annals of Tourism Research*, *6*(4), 408–424.

Crossley, M. L. (2000). Narrative psychology, trauma and the study of self/identity. *Theory & Psychology*, *10*(4), 527–546. https://doi.org/10.1177/0959354300104005

Crouch, G., Perdue, R., Timmermans, H., & Uysal, M. (2004). *Consumer psychology of tourism. Hospitality and leisure*. CABI Publishing.

Crouch, G. I. (2013). Homo sapiens on vacation: What can we learn from Darwin? *Journal of Travel Research*, *52*(5), 575–590. https://doi.org/10.1177/0047287512475219

Crowley, K. A. (2018). Memorability in narration: An overview of mnemonic features in oral and written tradition. *Research Journal Studies About Languages*, *32*, 77–93.

Csikszentmihalyi, M. (1990). *Flow: The psychology of optimal experience*. Harper & Row.

Custers, R., & Aarts, H. (2010). The unconscious will: How the pursuit of goals operates outside of conscious awareness. *Science*, *329*, 47–50.

D'Argembeau, A., & Mathy, A. (2011). Tracking the construction of episodic future thoughts. *Journal of Experimental Psychology: General*, *140*(2), 258–271.

D'Esposito, M., & Postle, B. R. (2015). The cognitive neuroscience of working memory. *Annual Review of Psychology*, *66*(1), 115–142.

Da Silva deMatos, N. M., de Sa, E. S., & de Oliveira Duarte, P. A. (2021). A review and extension of the flow experience concept. Insights and directions for tourism research. *Tourism Management Perspectives*, *38*, 100802.

Dahlin, B. (2001). Critique of the schema concept. *Scandinavian Journal of Educational Research*, *45*(3), 287–300.

Dalakas, V. (2006). The effect of cognitive appraisals on emotional responses during service encounters. *Services Marketing Quarterly*, *27*(1), 23–41. https://doi.org/10.1300/J396v27n01_02

Damasio, A. (1998). Investigating the biology of consciousness. *Philosophical Transactions of the Royal Society of London – Series B: Biological Sciences*, *353*(1377), 1879–1882.

Damasio, A. R. (1995). Toward a neurobiology of emotion and feeling: Operational concepts and hypotheses. *The Neuroscientist*, *1*(1), 19–25.

Damasio, A., & Carvalho, G. B. (2013). The nature of feelings: Evolutionary and neurobiological origins. *Nature Reviews Neuroscience*, *14*(2), 143–152. https://doi.org/10.1038/nrn3403

Damasio, A. R. (2003). *Looking for Spinoza: Joy, sorrow, and the feeling brain*. Houghton Mifflin Harcourt.

Damasio, A. R. (2004). Emotions and feelings. In *Feelings and emotions: The Amsterdam symposium* (Vol. 5, pp. 49–57). Cambridge University Press.

Dancygier, B. (2011). *The language of stories: A cognitive approach*. Cambridge University Press.

Dann, G. (1977). Anomie, ego-enhancement, and tourism. *Annals of Tourism Research*, *4*, 184–194.

Dann, G. (1981). Tourist motivation. An appraisal. *Annals of Tourism Research*, *8*(2), 187–219.

Dann, G. (2005). The theoretical state of the art in the sociology and anthropology of tourism. *Tourism Analysis*, *10*(1), 3–15.

Dann, G., & Jacobsen, J. K. S. (2003). Tourism smellscapes. *Tourism Geographies*, *5*(1), 3–25.

Darcy, S., Cameron, B., & Pegg, S. (2010). Accessible tourism and sustainability: A discussion and case study. *Journal of Sustainable Tourism*, *18*(4), 515–537. https://doi.org/10.1080/09669581003690668

Davenport, T. H., & Beck, J. C. (2001). *The attention economy: Understanding the new currency of business*. Harvard Business Press.

Davis, J. R., & Gold, G. J. (2011). An examination of emotional empathy, attributions of stability, and the link between perceived remorse and forgiveness. *Personality and Individual Differences*, *50*(3), 392–397.

Dawson, M. E., Schell, A. M., & Filion, D. L. (2007). The electrodermal system. In J. T. Cacioppo & G. G. Berntson (Eds.), *Handbook of psychophysiology* (3rd ed., pp. 159–181). Cambridge University Press.

Dayan, P., Kakade, S., & Montague, P. R. (2000). Learning and selective attention. *Nature Neuroscience*, *3*(11), 1218–1223.

De Bloom, J., Ritter, S., Kühnel, J., Reinders, J., & Geurts, S. (2014). Vacation from work: A 'ticket to creativity'? The effects of recreational travel on cognitive flexibility and originality. *Tourism Management*, *44*(0), 164–171.

De Lange, F. P., Jensen, O., Bauer, M., & Toni, I. (2008). Interactions between posterior gamma and frontal alpha/beta oscillations during imagined actions. *Frontiers in Human Neuroscience*, *2*(7), 1–12.

De Rojas, C., & Camarero, C. (2008). Visitors' experience, mood, and satisfaction in a heritage context: Evidence from an interpretation center. *Tourism Management*, *29*(3), 525–537.

De Sousa, A. (2013). Towards an integrative theory of consciousness: Part 1 (neurobiological and cognitive models). *Mens Sana Monographs*, *11*(1), 100–150.

De Vignemont, F. (2011). Embodiment, ownership and disownership. *Consciousness and Cognition*, *20*(1), 82–93. https://doi.org/10.1016/j.concog.2010.09.004

De Vries, P. C. (2016). Paul Ricoeur's hermeneutics and the discourse of mark 13: Appropriating the apocalyptic. In *Studies in the thought of Paul Ricoeur*. Lexington Books.

De Vries, R., Jager, G., Tijssen, I., & Zandstra, E. H. (2018). Shopping for products in a virtual world: Why haptics and visuals are equally important in shaping consumer perceptions and attitudes. *Food Quality and Preference*, *66*, 64–75.

Deci, E. L. (1975). *Intrinsic motivation*. Plenum Press.

Deci, E. L., & Ryan, R. M. (1985). The general causality orientations scale: Self-determination in personality. *Journal of Research in Personality*, *19*(2), 109–134.

Deci, E. L., & Ryan, R. M. (2000). The "what" and "why" of goal pursuits: Human needs and the self-determination of behavior. *Psychological Inquiry*, *11*(4), 227–268.

Deci, E. L., & Ryan, R. M. (2015). Self-determination theory. In J. D. Wright (Ed.), *International encyclopedia of the social & behavioral sciences* (pp. 486–491). Elsevier. https://doi.org/10.1016/B978-0-08-097086-8.26036-4

Decrop, A., & Snelders, D. (2004). Planning the summer vacation–An adaptable process. *Annals of Tourism Research*, *31*(4), 1008–1030.

Dekhili, S., & Achabou, M. A. (2014). Towards greater understanding of ecolabel effects: The role of country of origin. *Journal of Applied Business Research*, *30*(2), 433–438.

Dekhili, S., & Hallem, Y. (2020). An examination of the relationship between co-creation and well-being: An application in the case of tourism. *Journal of Travel & Tourism Marketing*, *37*(1), 33–47.

Delgadillo, Y., & Escalas, J. E. (2004). Narrative word-of-mouth communication: Exploring memory and attitude effects of consumer storytelling. *Advances in Consumer Research*, *31*, 186–192.

Dellaert, B. G. C., Arentze, T. A., & Horeni, O. (2014). Tourists' mental representations of complex travel decision problems. *Journal of Travel Research*, *53*(1), 3–11. https://doi.org/10.1177/0047287513506297

Demblon, J., & D'Argembeau, A. (2014). The organization of prospective thinking: Evidence of event clusters in freely generated future thoughts. *Consciousness and Cognition*, *24*(Suppl. C), 75–83.

Deng, X., Unnava, H. R., & Lee, H. (2019). "Too true to be good?" when virtual reality decreases interest in actual reality. *Journal of Business Research*, *100*, 561–570.

Denny, B. T., & Ochsner, K. N. (2014). Behavioral effects of longitudinal training in cognitive reappraisal. *Emotion*, *14*, 425–433.

Deshpande, R. (1983). 'Paradigms lost': On theory and method in research in marketing. *Journal of Marketing*, *47*(4), 101–110.

Desmet, P., & Hekkert, P. (2007). Framework of product experience. *International Journal of Design*, *1*(1), 57–66.

Dewe, P. (1991). Primary appraisal, secondary appraisal, and coping: Their role in stressful work encounters. *Journal of Occupational Psychology*, *64*(4), 331–351.

Dewe, P. (2000). Measures of coping with stress at work: A review and critique. In P. Dewe, M. Leiter, & T. Cox (Eds.), *Coping, health, and organizations* (pp. 3–30). Taylor & Francis.

Dias, C., & Dias, V. (2019). Memorable tourism experience design an effective destination marketing tool. *Anatolia*, *30*(4), 626–628.

Diefendorff, J. M., & Gosserand, R. H. (2003). Understanding the emotional labor process: A control theory perspective. *Journal of Organizational Behavior*, *24*(8), 945–959.

Dietrich, A. (2004). The cognitive neuroscience of creativity. *Psychonomic Bulletin & Review*, *11*(6), 1011–1026.

Dijksterhuis, A. (2004). Think different: The merits of unconscious thought in preference development and decision making. *Journal of Personality and Social Psychology*, *87*(5), 586–598.

Dijksterhuis, A., & Aarts, H. (2010). Goals, attention, and (un) consciousness. *Annual Review of Psychology*, *61*, 467–490.

Dimberg, U., Thunberg, M., & Grunedal, S. (2002). Facial reactions to emotional stimuli: Automatically controlled emotional responses. *Cognition & Emotion*, *16*(4), 449–471.

Diţoiu, M. C., & Căruntu, A. L. (2014). Sensory experiences regarding five-dimensional brand destination. *Procedia-Social and Behavioral Sciences*, *109*, 301–306.

Dlodlo, N. (2018). The influence of product presentation on hotel website revisit intentions: The mediating role of mental imagery. *International Journal of Social Sciences and Humanity Studies*, *10*(1), 49–65.

Do, T.-P. (2011). *Mindfulness-based stress reduction (MBSR): A qualitative study of MBSR with older adults*. Ph.D. Dissertation/Thesis, Alliant International University.

Dolan, R. J. (2002). Emotion, cognition, and behavior. *Science*, *298*(5596), 1191–1194.

Dolnicar, S. (2020). Designing for more environmentally friendly tourism. *Annals of Tourism Research*, *84*, 102933. https://doi.org/10.1016/j.annals.2020.102933

Dominey, P. F. (2021). Narrative event segmentation in the cortical reservoir. *PLoS Computational Biology*, *17*(10), e1008993. https://doi.org/10.1371/journal.pcbi.1008993

Dong, R., & Ni, S. G. (2020). Openness to experience, extraversion, and subjective well-being among Chinese college students: The mediating role of dispositional awe. *Psychological Reports*, *123*(3), 903–928. https://doi.org/10.1177/0033294119826884

Donovan, R. J., & Rossiter, J. R. (1982). Store atmosphere: An environmental psychology approach. *Journal of Retailing*, *58*(1), 34–57.

Dooling, D. J., & Lachman, R. (1971). Effects of comprehension on retention of prose. *Journal of Experimental Psychology*, *88*(2), 216–222.

Dowray, S., Swartz, J. J., Braxton, D., & Viera, A. J. (2013). Potential effect of physical activity based menu labels on the calorie content of selected fast food meals. *Appetite*, *62*, 173–181.

Drichoutis, A. C., Nayga, R. M., & Lazaridis, P. (2012). Food away from home expenditures and obesity among older Europeans: Are there gender differences? *Empirical Economics*, *42*(3), 1051–1078.

Dubé, L., & Menon, K. (2000). Multiple roles of consumption emotions in post-purchase satisfaction with extended service transactions. *International Journal of Service Industry Management*, *11*(3), 287–304. https://doi.org/10.1108/09564230010340788

Duchowski, A. T. (2017). *Eye tracking methodology: Theory and practice*. Springer.

Duerden, M. D., Lundberg, N. R., Ward, P., Taniguchi, S. T., Hill, B., Widmer, M. A., & Zabriskie, R. (2018). From ordinary to extraordinary: A framework of experience types. *Journal of Leisure Research*, *49*(3–5), 196–216.

Duhachek, A. (2005). Coping: A multidimensional, hierarchical framework of responses to stressful consumption episodes. *Journal of Consumer Research*, *32*(1), 41–53.

Duhigg, C. (2013). *The power of habit: Why we do what we do and how to change*. Random House.

Düzel, E., Bunzeck, N., Guitart-Masip, M., & Düzel, S. (2010). NOvelty-related motivation of anticipation and exploration by dopamine (NOMAD): Implications for healthy aging. *Neuroscience & Biobehavioral Reviews*, *34*(5), 660–669.

Eacott, M. J., & Easton, A. (2012). Remembering the past and thinking about the future: Is it really about time? *Learning and Motivation*, *43*, 200–208.

Eagly, A. H., Wood, W., & Chaiken, S. (1978). Causal inferences about communicators and their effect on opinion change. *Journal of Personality and Social Psychology*, *36*(4), 424–435.

Ebbinghaus, H. (1885). Memory: A contribution to experimental psychology. Classics in the history of psychology. *Annals of Neurosciences*, *20*(4), 155–156.

Ebner, N. C., Riediger, M., & Lindenberger, U. (2009). Schema reliance for developmental goals increases from early to late adulthood: Improvement for the young, loss prevention for the old. *Psychology and Aging*, *24*(2), 310–323. https://doi.org/10.1037/a0015430

Eccles, J. S., Adler, T. F., Futterman, R., Goff, S. B., Kaczala, C. M., Meece, J., & Midgley, C. (1983). Expectancies, values, and academic behaviors. In J. T. Spence (Ed.), *Achievement and achievement motives* (pp. 75–146). W. H. Freeman.

Eccles, J. S., & Wigfield, A. (2002). Motivational beliefs, values, and goals. *Annual Review of Psychology*, *51*(1), 109–132.

Eckstein, M. K., Guerra-Carrillo, B., Singley, A. T. M., & Bunge, S. A. (2017). Beyond eye gaze: What else can eye tracking reveal about cognition and cognitive development? *Developmental Cognitive Neuroscience*, *25*, 69–91.

Ehlers, A., & Clark, D. M. (2000). A cognitive model of posttraumatic stress disorder. *Behaviour Research and Therapy*, *38*(4), 319–345. https://doi.org/10.1016/S0005-7967(99)00123-0

Eichenbaum, H., & Cohen, N. J. (2004). *From conditioning to conscious recollection: Memory systems of the brain*. Oxford Psychology Series. https://doi.org/10.1093/acprof:oso/9780195178043.003.0014

Eiser, J. R., & Eiser, J. R. (1986). *Social psychology: Attitudes, cognition and social behaviour*. Cambridge University Press.

Ek, R., Larsen, J., Hornskov, S. B., & Mansfeldt, O. K. (2008). A dynamic framework of tourist experiences: Space-time and performances in the experience economy. *Scandinavian Journal of Hospitality and Tourism*, *8*(2), 122–140.

Ekman, P. (1992). An argument for basic emotions. *Cognition & Emotion*, *6*(3–4), 169–200.

Ekman, P., Friesen, W. V., & Ellsworth, P. (1972). *Emotion in the human face: Guidelines for research and an integration of findings*. Pergamon.

Elder, R. S., & Krishna, A. (2022). A review of sensory imagery for consumer psychology. *Journal of Consumer Psychology*, *32*(2), 293–315.

Ellis, G. D., Freeman, P. A., Jamal, T., & Jiang, J. (2018). A theory of structured experience. *Annals of Leisure Research*, *22*(1), 97–118.

Ellis, A., Park, E., Kim, S., & Yeoman, I. (2018). What is food tourism? *Tourism Management*, *68*, 250–263. https://doi.org/10.1016/j.tourman.2018.03.025

Ellis, G. D., Jamal, T., & Jiang, J. (2015). Fast-thinking and slow-thinking: A process approach to understand situated tourist experiences. In *Travel and tourism research association: Advancing tourism research globally* (p. 36). https://scholarworks.umass.edu/ttra/ttra2015/Academic_Papers_Oral/36

Ellsworth, P. C., & Scherer, K. R. (2003). Appraisal processes in emotion. In R. J. Davidson, K. R. Sherer, & H. H. Goldsmith (Eds.), *Handbook of affective sciences* (pp. 572–595). Oxford University Press.

Ellsworth, P. C., & Smith, C. A. (1988a). Shades of joy: Patterns of appraisal differentiating pleasant emotions. *Cognition & Emotion*, *2*(4), 301–331. https://doi.org/10.1080/02699938808412702

Ellsworth, P. C., & Smith, C. A. (1988b). From appraisal to emotion: Differences among unpleasant feelings. *Motivation and Emotion*, *12*(3), 271–302.

Engel, A. K., Fries, P., & Singer, W. (2001). Dynamic predictions: Oscillations and synchrony in top–down processing. *Nature Reviews Neuroscience*, *2*(10), 704–716.

English, T., Lee, I. A., John, O. P., & Gross, J. J. (2017). Emotion regulation strategy selection in daily life: The role of social context and goals. *Motivation and Emotion*, *41*(2), 230–242.

Epstein, S. (1994). Integration of the cognitive and the psychodynamic unconscious. *American Psychologist*, *49*(8), 709–724.

Eraqi, M. I. (2011). Co-creation and the new marketing mix as an innovative approach for enhancing tourism industry competitiveness in Egypt. *International Journal of Services and Operations Management*, *8*(1), 76–91.

Erasmus, A. C., Boshoff, E., & Rousseau, G. G. (2002). The potential of using script theory in consumer behaviour research. *Journal of Family Ecology and Consumer Sciences*, *30*, 1–9.

Ernst, A., Philippe, F. L., & d'Argembeau, A. (2018). Wanting or having to: The role of goal self-concordance in episodic future thinking. *Consciousness and Cognition*, *66*, 26–39.

Escalas, J. (2004). Imagine yourself in the product: Mental simulation, narrative transportation, and persuasion. *Journal of Advertising*, *33*(2), 37–48.

Escalas, J. (2006). Self-referencing and persuasion: Narrative transportation versus analytical elaboration. *Journal of Consumer Research*, *33*(4), 421–429. https://doi.org/10.1086/510216

Escalas, J., & Luce, M. (2004). Understanding the effects of process-focused versus outcome-focused thought in response to advertising. *Journal of Consumer Research*, *31*(2), 274–285.

Evans, J. S. B. (2003). In two minds: Dual-process accounts of reasoning. *Trends in Cognitive Sciences*, *7*(10), 454–459.

Evans, J. S. B. (2008). Dual-processing accounts of reasoning, judgment, and social cognition. *Annual Review of Psychology*, *59*, 255–278.

Escalas, J. E. (2007). Self-referencing and persuasion: Narrative transportation versus analytical elaboration. *Journal of Consumer Research*, *33*(4), 421–429.

Everett, S. (2008). Beyond the visual gaze? The pursuit of an embodied experience through food tourism. *Tourist Studies*, *8*(3), 337–358.

Everett, S. (2019). Theoretical turns through tourism taste-scapes: The evolution of food tourism research. *Research in Hospitality Management*, *9*(1), 3–12.

Falk, J. H., Ballantyne, R., Packer, J., & Benckendorff, P. (2012). Travel and learning: A neglected tourism research area. *Annals of Tourism Research*, *39*(2), 908–927. https://doi.org/10.1016/j.annals.2011.11.016

Farkic, J., Isailovic, G., & Taylor, S. (2021). Forest bathing as a mindful tourism practice. *Annals of Tourism Research Empirical Insights*, *2*(2), 100028.

Farmaki, A. (2021). Memory and forgetfulness in tourism crisis research. *Tourism Management*, *83*, 104210. https://doi.org/10.1016/j.tourman.2020.104210

Fazey, I., Schäpke, N., Caniglia, G., Hodgson, A., Kendrick, I., Lyon, C., Page, G., Patterson, J., Riedy, C., Strasser, T., Verveen, S., Adams, D., Goldstein, B., Klaes, M., Leicester, G., Linyard, A., McCurdy, A., Ryan, P., Sharpe, B., … Young, H. R. (2020). Transforming knowledge systems for life on earth: Visions of future systems and how to get there. *Energy Research & Social Science*, *70*, 101724.

Feighey, W. (2003). Negative image? Developing the visual in tourism research. *Current Issues in Tourism*, *6*(1), 76–85.

Fennis, B., Das, E., & Fransen, M. (2012). Print advertising: Vivid content. *Journal of Business Research*, *65*(6), 861–864.

Fernandez Velasco, P., & Loev, S. (2020). Affective experience in the predictive mind: A review and new integrative account. *Synthese*, *198*(11), 10847–10882.

Ferreri, L., Aucouturier, J.-J., Muthalib, M., Bigand, E., & Bugaiska, A. (2013). Music improves verbal memory encoding while decreasing prefrontal cortex activity: An fNIRS study. *Frontiers in Microbiology*, *7*, 779. https://doi.org/10.3389/fnhum.2013.00779

Fesenmaier, D. R., & Xiang, Z. (2017). Introduction to tourism design and design science in tourism. In D. R. Fesenmaier & Z. Xiang (Eds.), *Design science in tourism* (pp. 3–16). Springer International Publishing Switzerland.

Fiebelkorn, I. C., & Kastner, S. (2020). Functional specialization in the attention network. *Annual Review of Psychology*, *71*, 221–249.

Filep, S., Cao, D., Jiang, M., & DeLacy, T. (2013). Savouring tourist experiences after a holiday. *Leisure/Loisir*, *37*(3), 191–203.

Filep, S., & Laing, J. (2019). Trends and directions in tourism and positive psychology. *Journal of Travel Research*, *58*(3), 343–354.

Finlayson, M. A., & Corman, S. R. (2013). The military interest in narrative. *Sprache und Datenverarbeitung*, *37*(1–2), 173–191.

Fiore, A., Kim, J., & Lee, H.-H. (2005). Effect of image interactivity technology on consumer responses toward the online retailer. *Journal of Interactive Marketing*, *19*(3), 38–53.

Fisher, E. B., Thorpe, C. T., McEvoy DeVellis, B., & DeVellis, R. F. (2007). Healthy coping, negative emotions, and diabetes management. *The Diabetes Educator*, *33*(6), 1080–1103.

Fiske, S. T., & Linville, P. W. (1980). What does the schema concept buy us? *Personality and Social Psychology Bulletin*, *6*(4), 543–557.

Fiske, S. T., & Neuberg, S. L. (1990). A continuum of impression formation, from category-based to individuating processes: Influences of information and motivation on attention and interpretation. In *Advances in experimental social psychology* (Vol. 23, pp. 1–74). Elsevier.

Fivush, R. (2011). The development of autobiographical memory. *Annual Review of Psychology*, *62*(1), 559–582.

Flavián, C., Ibáñez-Sánchez, S., & Orús, C. (2019a). Integrating virtual reality devices into the body: Effects of technological embodiment on customer engagement and behavioral intentions toward the destination. *Journal of Travel & Tourism Marketing*, *36*(7), 847–863.

Flavián, C., Ibáñez-Sánchez, S., & Orús, C. (2019b). The impact of virtual, augmented, and mixed reality technologies on the customer experience. *Journal of Business Research*, *100*, 547–560.

Flavián, C., Ibáñez-Sánchez, S., & Orús, C. (2021). Impacts of technological embodiment through virtual reality on potential guests' emotions and engagement. *Journal of Hospitality Marketing & Management*, *30*(1), 1–20.

Florek, M., & Insch, A. (2011). When fit matters: Leveraging destination and event image congruence. *Journal of Hospitality Marketing & Management*, *20*, 265–286.

Floyd, M. F. (1997). Pleasure, arousal, and dominance: Exploring affective determinants of recreation satisfaction. *Leisure Sciences*, *19*(2), 83–96.

Fodness, D. (1994). Measuring tourist motivation. *Annals of Tourism Research*, *21*(3), 555–581.

Folkman, S. (2013). Stress, coping, and hope. In *Psychological aspects of cancer* (pp. 119–121). Springer.

Folkman, S., Lazarus, R. S., Dunkel-Schetter, C., DeLongis, A., & Gruen, R. J. (1986). Dynamics of a stressful encounter: Cognitive appraisal, coping, and encounter outcomes. *Journal of Personality and Social Psychology*, *50*(5), 992–1003.

Font, X., English, R., Gkritzali, A., & Tian, W. S. (2021). Value co-creation in sustainable tourism: A service-dominant logic approach. *Tourism Management*, *82*, 104200.

Förster, J., Marguc, J., & Gillebaart, M. (2010). Novelty categorization theory. *Social and Personality Psychology Compass*, *4*(9), 736–755. https://doi.org/10.1111/j.1751-9004.2010.00289.x

Foster, J. J., & Awh, E. (2019). The role of alpha oscillations in spatial attention: Limited evidence for a suppression account. *Current Opinion in Psychology*, *29*, 34–40.

Foster, M. I., & Keane, M. T. (2015). Why some surprises are more surprising than others: Surprise as a metacognitive sense of explanatory difficulty. *Cognitive Psychology*, *81*, 74–116.

Foxe, J. J., & Snyder, A. C. (2011). The role of alpha-band brain oscillations as a sensory suppression mechanism during selective attention. *Frontiers in Psychology*, *2*, 154–154.

Francken, J. C., & Slors, M. (2018). Neuroscience and everyday life: Facing the translation problem. *Brain and Cognition*, *120*, 67–74.

Frankland, P. W., Josselyn, S. A., & Köhler, S. (2019). The neurobiological foundation of memory retrieval. *Nature Neuroscience*, *22*(10), 1576–1585.

Fredrickson, B. L. (2000). Extracting meaning from past affective experiences: The importance of peaks, ends, and specific emotions. *Cognition & Emotion*, *14*(4), 577–606.

Fredrickson, B., & Levenson, R. W. (1998). Positive emotions speed recovery from the cardiovascular sequelae of negative emotions. *Cognition & Emotion*, *12*(2), 191–220.

Fredrickson, B. L. (2001). The role of positive emotions in positive psychology: The broaden-and-build theory of positive emotions. *American Psychologist*, *56*(3), 218–226.

Fredrickson, B. L. (2004). Gratitude like other positive emotions: Broaders and builds. In R. Emmons & M. E. McCullough (Eds.), *The psychology of gratitude* (pp. 100–119). Oxford University Press.

Freeman, L. C., Freeman, S. C., & Michaelson, A. G. (1988). On human social intelligence. *Journal of Social and Biological Structures*, *11*(4), 415–425.

Freeman, L. C., Romney, A. K., & Freeman, S. C. (1987). Cognitive structure and informant accuracy. *American Anthropologist*, *89*(2), 310–325.

Fridgen, J. (1984). Environmental psychology and tourism. *Annals of Tourism Research*, *11*(1), 19–39.

Friedman, D., Cycowicz, Y. M., & Gaeta, H. (2001). The novelty P3: An event-related brain potential (ERP) sign of the brain's evaluation of novelty. *Neuroscience & Biobehavioral Reviews*, *25*(4), 355–373.

Friedman, D., Goldman, R., Stern, Y., & Brown, T. R. (2009). The brain's orienting response: An event-related functional magnetic resonance imaging investigation. *Human Brain Mapping*, *30*(4), 1144–1154.

Frijda, N. H. (1986). *The emotions*. Cambridge University Press.

Frijda, N. H. (1987). Emotion, cognitive structure, and action tendency. *Cognition & Emotion*, *1*(2), 115–143.

Frijda, N. H. (1993). The place of appraisal in emotion. *Cognition & Emotion*, *7*(3–4), 357–387.

Frijda, N. H., Kuipers, P., & Ter Schure, E. (1989). Relations among emotion, appraisal, and emotional action readiness. *Journal of Personality and Social Psychology*, *57*(2), 212–228.

Frith, C. D., & Frith, U. (1999). Interacting minds: A biological basis. *Science*, *286*(5445), 1692–1695. https://doi.org/10.1126/science.286.5445.1692

Fronda, G., Cassioli, F., Sebastiani, R., Galeone, A. B., & Balconi, M. (2021). Paint it green: A neuroscientific approach to hotel sustainability and ecological tourism. *Environment, Development and Sustainability*, *23*(10), 15513–15528.

Gabbott, M., Tsarenko, Y., & Mok, W. H. (2011). Emotional intelligence as a moderator of coping strategies and service outcomes in circumstances of service failure. *Journal of Service Research*, *14*(2), 234–248.

Gaggioli, A. (2016). Transformative experience design. In A. Gaggioli, G. Ferscha, G. Riva, S. Dunne, & I. Viaud-Delmon (Eds.), *Human computer confluence transforming* (pp. 97–121). De Gruyter.

Gallo, D. A., & Wheeler, M. E. (2013). Episodic memory. In D. Reisberg (Ed.), *The Oxford handbook of cognitive psychology* (pp. 189–205). Oxford University Press.

Gallo, I. (2013). *The processing and evaluation of experiences*. Ph.D., University of California.

Gao, J., & Kerstetter, D. L. (2018). From sad to happy to happier: Emotion regulation strategies used during a vacation. *Annals of Tourism Research*, *69*, 1–14.

Gao, L., Scott, N., & Ding, P. (2016). Attributes, theme and value of a visit to Zhouzhuang, China. *Journal of Destination Marketing & Management*, *5*, 239–248.

Gao, L., Scott, N., Ding, P., & Cooper, C. (2012). Tourist experience development: Designed attributes, perceived experiences, and customer value. *Strategic Marketing in Tourism Services*, *14*, 216–230.

Garcia-Prieto, P., & Scherer, K. R. (2016). Connecting social identity theory and cognitive appraisal theory of emotions. In *Social identities* (pp. 189–208). Psychology Press.

Garland, E. L., Fredrickson, B., Kring, A. M., Johnson, D. P., Meyer, P. S., & Penn, D. L. (2010). Upward spirals of positive emotions counter downward spirals of negativity: Insights from the broaden-and-build theory and affective neuroscience on the treatment of emotion dysfunctions and deficits in psychopathology. *Clinical Psychology Review*, *30*(7), 849–864.

Garnham, A. (2019). Cognitivism. In *The Routledge companion to philosophy of psychology* (pp. 99–110). Routledge.

Gartner, W. C. (1993). Image formation process. In M. Uysal & D. R. Fesenmaier (Eds.), *Communication and channel systems in tourism marketing* (pp. 191–215). Haworth.

Gauthier, C. J., & Fan, A. P. (2019). BOLD signal physiology: Models and applications. *NeuroImage*, *187*, 116–127.

Gawronski, B., & Bodenhausen, G. V. (2006). Associative and propositional processes in evaluation: An integrative review of implicit and explicit attitude change. *Psychological Bulletin*, *132*(5), 692–731.

Gawronski, B., & Creighton, L. A. (2013). Dual process theories. In D. E. Carlston (Ed.), *The Oxford handbook of social cognition* (pp. 282–312). Oxford University Press.

Gazzaniga, M. S., Ivry, R. B., & Mangun, G. R. (2014). *Cognitive neuroscience: The biology of the mind* (4th ed.). Norton Loose Leaf.

Gehart, D. R. (2012). Mindfulness in therapy. In *Mindfulness and acceptance in couple and family therapy* (pp. 3–19). Springer US.

Gendron, M., & Barrett, L. F. (2009). Reconstructing the past: A century of ideas about emotion in psychology. *Emotion Review*, *1*(4), 316–339.

Gentile, C., Spiller, N., & Noci, G. (2007). How to sustain the customer experience: An overview of experience components that co-create value with the customer. *European Management Journal*, *25*(5), 395–410.

Gerrig, R. (1993). *Experience narrative worlds*. Yale University Press.

Gerring, J. (1999). What makes a concept good? A criterial framework for understanding concept formation in the social sciences. *Polity*, *31*(3), 357–393.

Gethin, R. (2011). On some definitions of mindfulness. *Contemporary Buddhism*, *12*(1), 263–279.

Ghosh, T., & Sarkar, A. (2016). "To feel a place of heaven": Examining the role of sensory reference cues and capacity for imagination in destination marketing. *Journal of Travel & Tourism Marketing*, *33*, 25–37.

Ghosh, V. E., & Gilboa, A. (2014). What is a memory schema? A historical perspective on current neuroscience literature. *Neuropsychologia*, *53*, 104–114.

Gibson, C. (2010). Geographies of tourism: (Un)ethical encounters. *Progress in Human Geography*, *34*(4), 521–527. https://doi.org/10.1177/030913250934868

Gibson, J. J. (2015). *The ecological approach to visual perception* (Classic ed.). Psychology Press. (Original work published 1979).

Gigerenzer, G. (2020). How to explain behavior? *Topics in Cognitive Science*, *12*(4), 1363–1381.

Giglio, V. J., Luiz, O. J., Chadwick, N. E., & Ferreira, C. E. (2018). Using an educational video-briefing to mitigate the ecological impacts of scuba diving. *Journal of Sustainable Tourism*, *26*(5), 782–797. https://doi.org/10.1080/09669582.2017.1408636

Gilbert, D. (2006). *Stumbling on happiness*. Vintage Books.

Gilbert, D., & Wilson, T. D. (2007). Prospection: Experiencing the future. *Science*, *317*(5843), 1351–1354.

Gilboa, A., & Marlatte, H. (2017). Neurobiology of schemas and schema-mediated memory. *Trends in Cognitive Sciences*, *21*(8), 618–631.

Gleaves, D. H., Smith, S. M., Butler, L. D., & Spiegel, D. (2004). False and recovered memories in the laboratory and clinic: A review of experimental and clinical evidence. *Clinical Psychology: Science and Practice*, *11*(1), 3–28.

Glenberg, A. M. (1997). What memory is for: Creating meaning in the service of action. *Behavioral and Brain Sciences*, *20*(1), 41–50.

Glenberg, A. M. (2010). Embodiment as a unifying perspective for psychology. *Wiley Interdisciplinary Reviews: Cognitive Science*, *1*(4), 586–596.

Glenberg, A. M., Witt, J. K., & Metcalfe, J. (2013). From the revolution to embodiment: 25 years of cognitive psychology. *Perspectives on Psychological Science*, *8*(5), 573–585.

Gnoth, J. (1997). Tourism motivation and expectation formation. *Annals of Tourism Research*, *24*(2), 283–304.

Godovykh, M., & Tasci, A. D. (2021). Emotions, feelings, and moods in tourism and hospitality research: Conceptual and methodological differences. *Tourism and Hospitality Research*, *22*(2), 247–253.

Gold, C., Fachner, J., & Erkkilä, J. (2013). Validity and reliability of electroencephalographic frontal alpha asymmetry and frontal midline theta as biomarkers for depression. *Scandinavian Journal of Psychology*, *54*(2), 118–126.

Goldinger, S. D., Papesh, M. H., Barnhart, A. S., Hansen, W. A., & Hout, M. C. (2016). The poverty of embodied cognition. *Psychonomic Bulletin & Review*, *23*, 959–978.

Goldstein, E. B. (2010). *Sensation and perception* (8th ed.). Wadsworth.

Gollwitzer, P. M., & Oettingen, G. (2012). Goal pursuit. In R. M. Ryan (Ed.), *The Oxford handbook of human motivation* (pp. 208–231). Oxford University Press.

Goolaup, S., Solér, C., & Nunkoo, R. (2018). Developing a theory of surprise from travelers' extraordinary food experiences. *Journal of Travel Research*, *57*(2), 218–231. https://doi.org/10.1177/0047287517691154

Goossens, C. (2000). Tourism information and pleasure motivation. *Annals of Tourism Research*, *27*(2), 301–321.

Goossens, L., Pauwels, P., Rudzka-Ostyn, B., Simon-Vandenbergen, A.-M., & Vanparys, J. (1995). *By word of mouth: Metaphor, metonymy and linguistic action in a cognitive perspective*. John Benjamins Publishing.

Gordon, R., Ciorciari, J., & van Laer, T. (2018). Using EEG to examine the role of attention, emotion, working memory, and imagination in narrative-transportation. *European Journal of Marketing*, *52*(1/2), 92–117.

Gössling, S., Cohen, S. A., & Hibbert, J. F. (2018). Tourism as connectedness. *Current Issues in Tourism*, *21*(14), 1586–1600. https://doi.org/10.1080/13683500.2016.1157142

Gössling, S., Scott, D., & Hall, C. M. (2020). Pandemics, tourism, and global change: A rapid assessment of COVID-19. *Journal of Sustainable Tourism*, *29*(1), 1–20.

Goussinsky, R. (2015). Customer aggression felt anger and emotional deviance: The moderating role of job autonomy. *International Journal of Quality and Service Sciences*, *7*(1), 50–71.

Govers, R., Go, F., & Kumar, K. (2007). Promoting tourism destination image. *Journal of Travel Research*, *46*(1), 15–23.

Govers, R., Go, F. M., & Kumar, K. (2007). Promoting tourism destination image. *Journal of Travel Research*, *46*(1), 15–23. https://doi.org/10.1177/0047287507302374

Grabe, M. E., Lang, A., Zhou, S., & Bolls, P. D. (2000). Cognitive access to negatively arousing news: An experimental investigation of the knowledge gap. *Communication Research*, *27*(1), 3–26. https://doi.org/10.1177/009365000027001001

Graesser, A., Olde, B., & Klettke, B. (2002). How does the mind construct and represent stories? In M. Green, J. Strange, & T. Brock (Eds.), *Narrative impact: Social and cognitive foundations* (pp. 229–262). Lawrence Erlbaum Associates.

Graham, D. J., & Jeffery, R. W. (2011). Location, location, location: Eye-tracking evidence that consumers preferentially view prominently positioned nutrition information. *Journal of the American Dietetic Association*, *111*(11), 1704–1711.

Graham, S., & Weiner, B. (1996). Theories and principles of motivation. In D. C. Berliner & R. C. Calfee (Eds.), *Handbook of educational psychology* (Vol. 4, pp. 63–84). Macmillan Library Reference.

Grandey, A. A., & Gabriel, A. S. (2015). Emotional labor at a crossroads: Where do we go from here? *Annual Review of Organizational Psychology and Organizational Behavior*, *2*(1), 323–349.

Grant, A. M., & Shin, J. (2012). Work motivation: Directing, energizing, and maintaining effort (and research). In R. M. Ryan (Ed.), *The Oxford handbook of human motivation* (pp. 505–519). Oxford University Press.

Grappi, S., & Romani, S. (2012). Emotions that drive consumers away from brands: Measuring negative emotions toward brands and their behavioral effects. *International Journal of Research in Marketing*, *29*(1), 55–67. https://doi.org/10.1016/j.ijresmar.2011.07.001

Great Barrier Reef Marine Park Authority (GBRMPA). (2017). *Annual report 2017–18*. https://hdl.handle.net/11017/3403

Greenwald, A. G., & Banaji, M. R. (1995). Implicit social cognition: Attitudes, self-esteem, and stereotypes. *Psychological Review*, *102*(1), 4–27.

Greenwald, A. G., McGhee, D. E., & Schwartz, J. L. (1998). Measuring individual differences in implicit cognition: The implicit association test. *Journal of Personality and Social Psychology*, *74*(6), 1464–1480.

Grénman, M., & Räikkönen, J. (2015). Well-being and wellness tourism–Same, same but different? Conceptual discussions and empirical evidence. *Matkailututkimus (The Finnish Journal of Tourism Research)*, *11*(1), 7–25.

Gretzel, U., Fesenmaier, D. R., Formica, S., & O'Leary, J. T. (2006). Searching for the future: Challenges faced by destination marketing organizations. *Journal of Travel Research*, *45*(2), 116–126.

Grissemann, U. S., & Stokburger-Sauer, N. E. (2012). Customer co-creation of travel services: The role of company support and customer satisfaction with the co-creation performance. *Tourism Management*, *33*(6), 1483–1492.

Groeppe-Klein, A. (2005). Arousal and consumer in-store behavior. *Brain Research Bulletin*, *67*(5), 428–437.

Groeppel-Klein, A., & Baun, D. (2001). The role of customers' arousal for retail store results from an experimental pilot study using electrodermal activity as indicator. *ACR North American Advances*, *28*, 412–419.

Grohs, R., & Reisinger, H. (2005). Image transfer in sports sponsorships: An assessment of moderating effects. *International Journal of Sports Marketing & Sponsorship*, *7*(1), 36–42.

Groome, D., & Eysenck, M. (2015). *An introduction to applied cognitive psychology* (Vol, 2nd ed.). Psychology Press.

Gross, J. J. (1998). The emerging field of emotion regulation: An integrative review. *Review of General Psychology*, *2*(3), 271–299.

Gross, J. J. (2014). Emotion regulation: Conceptual and empirical foundations. In J. J. Gross (Ed.), *Handbook of emotion regulation* (pp. 3–20). The Guilford Press.

Gross, J. J. (2015). Emotion regulation: Current status and future prospects. *Psychological Inquiry*, *26*(1), 1–26.

Gross, J. J., & Barrett, L. F. (2011). Emotion generation and emotion regulation: One or two depends on your point of view. *Emotion Review*, *3*(1), 8–16.

Gross, J. J., & John, O. P. (2003). Individual differences in two emotion regulation processes: Implications for affect, relationships, and wellbeing. *Journal of Personality and Social Psychology*, *85*(2), 348.

Gross, J. J., & Thompson, R. A. (2007). Emotion regulation: Conceptual foundations. In J. J. Gross (Ed.), *Handbook of emotion regulation* (pp. 3–24). Guilford Press.

Gross, S., & Huber, D. (2020). Tourist information boards on motorways: Perceptions, memory effects and decision-making support. *Journal of Qualitative Research in Tourism*, *1*(1), 28–50.

Grossenbacher, P. G., & Lovelace, C. T. (2001). Mechanisms of synesthesia: Cognitive and physiological constraints. *Trends in Cognitive Sciences*, *5*(1), 36–41.

Guan, F., Chen, J., Chen, O., Liu, L., & Zha, Y. (2019). Awe and prosocial tendency. *Current Psychology*, *38*(4), 1033–1041. https://doi.org/10.1007/s12144-019-00244-7

Gunn, C. A. (1972). *Vacationscape: Developing tourist areas*. Bureau of Business Research, University of Texas.

Guo, S., Zhao, N., Zhang, J., Xue, T., Liu, P., Xu, S., & Xu, D. (2017). Landscape visual quality assessment based on eye movement: College student eye-tracking experiments on tourism landscape pictures. *Resources Science*, *39*(06), 1137–1147.

Gutman, J. (1997). Means–End chains as goal hierarchies. *Psychology and Marketing*, *14*(6), 545–560.

Guttentag, D., Smith, S., Potwarka, L., & Havitz, M. (2018). Why tourists choose Airbnb: A motivation-based segmentation study. *Journal of Travel Research*, *57*(3), 342–359. https://doi.org/10.1177/0047287517696980

Gwinner, K. P., & Eaton, J. (1999). Building brand image through event sponsorship: The role of image transfer. *Journal of Advertising*, *28*(4), 47–57.

Gyimóthy, S., & Mykletun, R. J. (2004). Play in adventure tourism: The case of Arctic trekking. *Annals of Tourism Research*, *31*(4), 855–878.

Gyurak, A., Gross, J. J., & Etkin, A. (2011). Explicit and implicit emotion regulation: A dual-process framework. *Cognition & Emotion*, *25*(3), 400–412.

Ha, S., Huang, R., & Park, J.-S. (2019). Persuasive brand messages in social media: A mental imagery processing perspective. *Journal of Retailing and Consumer Services*, *48*, 41–49.

Habib, R., & Lepage, M. (2000). Novelty assessment in the brain. In E. Tulving (Ed.), *Memory, consciousness, and the brain: The Tallinn conference* (pp. 265–277). Psychology Press.

Hadinejad, A., Moyle, B. D., Kralj, A., & Scott, N. (2019). Physiological and self-report methods to the measurement of emotion in tourism. *Tourism Recreation Research*, *44*(4), 466–478.

Hajcak, G., Weinberg, A., MacNamara, A., & Foti, D. (2012). ERPs and the study of emotion. In *Oxford handbook of event-related potential components* (pp. 440–510). Oxford University Press.

Halkias, G. (2015). Mental representation of brands: A schema-based approach to consumers' organization of market knowledge. *The Journal of Product and Brand Management*, *24*(5), 438–448.

Halkias, G., & Kokkinaki, F. (2014). The degree of ad–brand incongruity and the distinction between schema-driven and stimulus-driven attitudes. *Journal of Advertising*, *43*(4), 397–409.

Halverson, J. R. (2011). *Why story is not narrative, centre for strategic communications*. https://csc.asu.edu/2011/12/08/why-story-is-not-narrative/. Accessed on January 10, 2022.

Han, S., Lerner, J. S., & Keltner, D. (2007). Feelings and consumer decision making: The appraisal-tendency framework. *Journal of Consumer Psychology*, *17*(3), 158–168.

Hanna, P., Wijesinghe, S., Paliatsos, I., Walker, C., Adams, M., & Kimbu, A. (2019). Active engagement with nature: Outdoor adventure tourism, sustainability, and wellbeing. *Journal of Sustainable Tourism*, *27*(9), 1355–1373. https://doi.org/10.1080/09669582.2019.1621883

Hannon, B., & Daneman, M. (2007). Prospective memory: The relative effects of encoding, retrieval, and the match between encoding and retrieval. *Memory*, *15*(5), 572–604.

Hansen, A. H., & Mossberg, L. (2017). Tour guides' performance and tourists' immersion: Facilitating consumer immersion by performing a guide plus role. *Scandinavian Journal of Hospitality Tourism*, *17*(3), 259–278. https://doi.org/10.1080/15022250.2016.1162347

Hansenne, M., Pinto, E., Pitchot, W., Reggers, J., Scantamburlo, G., Moor, M., & Ansseau, M. (2002). Further evidence on the relationship between dopamine and novelty seeking: A neuroendocrine study. *Personality and Individual Differences*, *33*(6), 967–977.

Hanslmayr, S., Gross, J., Klimesch, W., & Shapiro, K. L. (2011). The role of alpha oscillations in temporal attention. *Brain Research Reviews*, *67*(1–2), 331–343.

Harmon-Jones, E. (2003). Clarifying the emotive functions of asymmetrical frontal cortical activity. *Psychophysiology*, *40*(6), 838–848.

Harmon-Kizer, T. R. (2017). The effects of schema congruity on consumer response to celebrity advertising. *Journal of Marketing Communications*, *23*(2), 162–175.

Harré, R., & Parrott, W. G. (Eds.). (1996). *The emotions: Social, cultural and biological dimensions*. SAGE.

Hassenzahl, M. (2010). Experience design: Technology for all the right reasons. *Synthesis Lectures on Human-Centered Informatics*, *3*(1), 1–95.

Hassenzahl, M., Eckoldt, K., Diefenbach, S., Laschke, M., Lenz, E., & Kim, J. (2013). Designing moments of meaning and pleasure. Experience design and happiness. *International Journal of Design*, *7*(3), 21–31.

Hastie, R., & Kumar, P. A. (1979). Person memory: Personality traits as organizing principles in memory for behaviors. *Journal of Personality and Social Psychology*, *37*(1), 25.

Havitz, M. E., & Dimanche, F. (1997). Leisure involvement revisited: Conceptual conundrums and measurement advances. *Journal of Leisure Research*, *29*(3), 245–278.

Havlena, W., & Holbrook, M. (1986). The varieties of consumption experience: Comparing two typologies of emotion in consumer behaviour. *Journal of Consumer Research*, *13*(3), 394–404.

Hayes, D., & MacLeod, N. (2007). Packaging places: Designing heritage trails using an experience economy perspective to maximize visitor engagement. *Journal of Vacation Marketing*, *13*(1), 45–58.

Hayes-Roth, B., & Hayes-Roth, F. (1979). A cognitive model of planning. *Cognitive Science*, *3*(4), 275–310.

He, M., Liu, B., & Li, Y. (2022). Recovery experience of wellness tourism and place attachment: Insights from feelings-as-information theory. *International Journal of Contemporary Hospitality Management*, *34*(8), 2934–2952.

He, Z., Wu, L., & Li, X. R. (2018). When art meets tech: The role of augmented reality in enhancing museum experiences and purchase intentions. *Tourism Management*, *68*, 127–139.

Heath, C., Larrick, R. P., & Wu, G. (1999). Goals as reference points. *Cognitive Psychology*, *38*, 79–109.

Heckler, S. E., & Childers, T. L. (1992). The role of expectancy and relevancy in memory for verbal and visual information: What is incongruency? *Journal of Consumer Research*, *18*(4), 475–492.

Heinen, S., & Sommer, R. (2009). *Narratology in the age of cross-disciplinary narrative research*. De Gruyter. https://doi.org/10.1515/9783110222432

Hendee, J. C., Catton, W. R., Jr., Marlow, L. D., & Brockman, C. F. (1968). *Wilderness users in the Pacific Northwest, their characteristics, values and management preferences*. Research Paper NNW-61. U.S. Department of Agriculture, Forest Service, Pacific Northwest Forest and Range Experiment Station.

Henderson, H. A., Pine, D. S., & Fox, N. A. (2015). Behavioral inhibition and developmental risk: A dual-processing perspective. *Neuropsychopharmacology*, *40*(1), 207–224.

Henke, K. (2010). A model for memory systems based on processing modes rather than consciousness. *Nature Reviews Neuroscience*, *11*(7), 523–532.

Herman, D. (2009). Cognitive approaches to narrative analysis. In G. Brône & J. Vandaele (Eds.), *Cognitive poetics: Goals, gains and gaps* (pp. 79–124). De Gruyter Mouton.

Herman, D. (2013). Cognitive narratology (revised version; uploaded 22 September 2013). In P. Hühn, J. Pier, W. Schmid, & J. Schönert (Eds.), *The living handbook of narratology*. Hamburg University. http://www.lhn.uni-hamburg.de/article/cognitive-narratology-revised-version-uploaded-22-september-2013

Herman, J. L., & Tetrick, L. E. (2009). Problem-focused versus emotion-focused coping strategies and repatriation adjustment. *Human Resource Management*, *48*(1), 69–88.

Hernández-Méndez, J., & Muñoz-Leiva, F. (2015). What type of online advertising is most effective for etourism 2.0? An eye tracking study based on the characteristics of tourists. *Computers in Human Behavior*, *50*, 618–625.

Hertzman, E., Anderson, D., & Rowley, S. (2008). Edutainment heritage tourist attractions: A portrait of visitors' experiences at Storyeum. *Museum Management and Curatorship*, *23*(2), 155–175.

Hewett, R., Liefooghe, A., Visockaite, G., & Roongrerngsuke, S. (2018). Bullying at work: Cognitive appraisal of negative acts, coping, wellbeing, and performance. *Journal of Occupational Health Psychology*, *23*(1), 71–84.

Hicks, J. R., & Stewart, W. P. (2020). Learning from wildlife-inspired awe. *The Journal of Environmental Education*, *51*(1), 44–54. https://doi.org/10.1080/00958964.2019.1594148

Hidalgo, M. C., & Hernandez, B. (2001). Place attachment: Conceptual and empirical questions. *Journal of Environmental Psychology*, *21*(3), 273–281. https://doi.org/10.1006/jevp.2001.0221

Higgins, E. T., & Scholer, A. A. (2009). Engaging the consumer: The science and art of the value creation process. *Journal of Consumer Psychology*, *19*(2), 100–114.

Higgins-Desbiolles, F. (2020). The "war over tourism": Challenges to sustainable tourism in the tourism academy after COVID-19. *Journal of Sustainable Tourism*, *29*(4), 551–569. https://doi.org/10.1080/09669582.2020.1803334

Hill, J., Curtin, S., & Gough, G. (2014). Understanding tourist encounters with nature: A thematic framework. *Tourism Geographies*, *16*(1), 68–87. https://doi.org/10.1080/14616688.2013.851265

Hillman, W. (2007). Travel authenticated? Postcards, tourist brochures, and travel photography. *Tourism Analysis*, *12*(3), 135–148.

Hilton, J. L., Klein, J. G., & Von Hippel, W. (1991). Attention allocation and impression formation. *Personality and Social Psychology Bulletin*, *17*(5), 548–559.

Hinde, S. J., Smith, T. J., & Gilchrist, I. D. (2018). Does narrative drive dynamic attention to a prolonged stimulus? *Cognitive Research: Principles and Implications*, *3*, 1–12. https://doi.org/10.1186/s41235-018-0140-5

Hirschman, E. C. (1985). Dual consciousness and altered states: Implications for consumer research. *Journal of Business Research*, *13*(3), 223–234.

Hirschman, E. C., & Holbrook, M. B. (1982). Hedonic consumption: Emerging concepts, methods and propositions. *Journal of Marketing*, *46*(3), 92–101.

Hirsh, J. B., DeYoung, C. G., & Peterson, J. B. (2009). Metatraits of the big five differentially predict engagement and restraint of behavior. *Journal of Personality*, *77*(4), 1085–1102.

Hirst, I. S. (2003). Perspectives of mindfulness. *Journal of Psychiatric and Mental Health Nursing*, *10*(3), 359–366. https://doi.org/10.1046/j.1365-2850.2003.00610.x

Hirst, W., Yamashiro, J. K., & Coman, A. (2018). Collective memory from a psychological perspective. *Trends in Cognitive Sciences*, *22*(5), 438–451. https://doi.org/10.1016/j.tics.2018.02.010

Hjalager, A. M., & Konu, H. (2011). Co-branding and co-creation in wellness tourism: The role of cosmeceuticals. *Journal of Hospitality Marketing & Management*, *20*(8), 879–901.

Hoang, C. C. (2020). Negative emotions and coping behaviors of passenger in the airline industry, Vietnam. *The Journal of Asian Finance, Economics, and Business*, *7*(10), 865–874.

Hoberg, R., Kannis-Dymand, L., Mulgrew, K., Schaffer, V., & Clark, E. (2021). Humpback whale encounters: Encouraging pro-environmental behaviours. *Current Issues in Tourism*, *24*(13), 1918–1929. https://doi.org/10.1080/13683500.2020.1808597

Hobson, N. M., Schroeder, J., Risen, J. L., Xygalatas, D., & Inzlicht, M. (2018). The psychology of rituals: An integrative review and process-based framework. *Personality and Social Psychology Review*, *22*(3), 260–284.

Hoemann, K., Gendron, M., & Barrett, L. F. (2017). Mixed emotions in the predictive brain. *Current Opinion in Behavioral Sciences*, *15*, 51–57.

Holbrook, M. B. (2006). Consumption experience, customer value, and subjective personal introspection: An illustrative photographic essay. *Journal of Business Research*, *59*(6), 714–725.

Holbrook, M. B., & Hirschman, E. C. (1982). The experiential aspects of consumption: Consumer fantasies, feelings, and fun. *Journal of Consumer Research*, *9*(2), 132–140.

Hommel, B. (2022). Goaliath: A theory of goal-directed behavior. *Psychological Research*, *86*(4), 1054–1077. https://doi.org/10.1007/s00426-021-01563-w

Hopp, H., Rohrmann, S., & Hodapp, V. (2012). Suppression of negative and expression of positive emotions: Divergent effects of emotional display rules in a hostile service interaction. *European Journal of Work & Organizational Psychology*, *21*(1), 84–105.

Hosany, S. (2012). Appraisal determinants of tourist emotional responses. *Journal of Travel Research*, *51*(3), 303–314.

Hosany, S., & Gilbert, D. (2010). Measuring tourists' emotional experiences toward Hedonic holiday destinations. *Journal of Travel Research*, *49*(4), 513–526.

Hosany, S., Hunter-Jones, P., & McCabe, S. (2020). Emotions in tourist experiences: Advancing our conceptual, methodological and empirical understanding. *Journal of Destination Marketing & Management*, *16*. https://doi.org/10.1016/j.jdmm.2020.100444

Hosany, S., Martin, D., & Woodside, A. G. (2020). Emotions in tourism: Theoretical designs, measurements, analytics, and interpretations. *Journal of Travel Research*, *60*(7), 1391–1407.

Hosany, S., Martin, D., & Woodside, A. G. (2021). Emotions in tourism: Theoretical designs, measurements, analytics, and interpretations. *Journal of Travel Research*, *60*(7), 1391–1407. https://doi.org/10.1177/0047287520937079

Hosany, S., & Prayag, G. (2013). Patterns of tourists' emotional responses, satisfaction, and intention to recommend. *Journal of Business Research*, *66*(6), 730–737.

Hosany, S., Prayag, G., Deesilatham, S., Caušević, S., & Odeh, K. (2015). Measuring tourists' emotional experiences: Further validation of the destination emotion scale. *Journal of Travel Research*, *54*(4), 482–495.

Hosany, S., Sthapit, E., & Björk, P. (2022). Memorable tourism experience: A review and research agenda. *Psychology and Marketing*, *39*(8), 1467–1486. https://doi.org/10.1002/mar.21665

Hosseini, S., Cortes Macias, R., & Almeida Garcia, F. (2021). Memorable tourism experience research: A systematic review of the literature. *Tourism Recreation Research*, 1–15.

Howell, A. J., Dopko, R. L., Passmore, H.-A., & Buro, K. (2011). Nature connectedness: Associations with well-being and mindfulness. *Personality and Individual Differences*, *51*(2), 166–171. https://doi.org/10.1016/j.paid.2011.03.037

Hoyt, E. (2001). *Whale watching 2001: Worldwide tourism numbers, expenditures, and expanding socioeconomic benefits*. A special report from the International Fund for Animal Welfare. http://cetaceanhabitat.org/pdf_bin/hoyt_ww_2001_report.pdf

Hsieh, Y.-C., Chiu, H.-C., Tang, Y.-C., & Lin, W.-Y. (2018). Does raising value co creation increase all customers' happiness? *Journal of Business Ethics*, *152*(4), 1053 1067.

Hsu, C. H. C., Cai, L. A., & Li, M. (2010). Expectation, motivation, and attitude: A tourist behavioral model. *Journal of Travel Research*, *49*(3), 282–296.

Hsu, L., & Chen, Y.-J. (2020). Neuromarketing, subliminal advertising, and hotel selection: An EEG study. *Australasian Marketing Journal*, *28*(4), 200–208.

Hsu, T.-H., & Chiang, C.-Y. (2011). Script comparisons during service encounters in fast-food chains. *Tourism and Hospitality Research*, *11*(1), 19–29. https://doi.org/10.1057/thr.2010.19

Huang, Y., Scott, N., Ding, P., & Cheng, D. (2012). Impression of Liusanjie: Effect of mood on experience and satisfaction. *International Journal of Tourism Research*, *14*(1), 91–102.

Huang, Y., Yaple, Z. A., & Yu, R. (2020). Goal-oriented and habitual decisions: Neural signatures of model-based and model-free learning. *NeuroImage*, *215*, 116834.

Hubel, D. H., & Wiesel, T. N. (1962). Receptive fields, binocular interaction, and functional architecture in the cat's visual cortex. *The Journal of Physiology*, *160*(1), 106–154.

Hudson, J. A., Fivush, R., & Kuebli, J. (1992). Scripts and episodes: The development of event memory. *Applied Cognitive Psychology*, *6*(6), 483–505.

Hudson, S., & Brent Ritchie, J. R. (2009). Branding a memorable destination experience. The case of 'Brand Canada'. *International Journal of Tourism Research*, *11*(2), 217–228. https://doi.org/10.1002/jtr.720

Huffman, C., & Houston, M. J. (1993). Goal-oriented experiences and the development of knowledge. *Journal of Consumer Research*, *20*, 190–207.

Hughes, K., & Moscardo, G. (2019). ICT and the future of tourist management. *Journal of Tourism Futures*, *5*(3), 228–240. https://doi.org/10.1108/jtf-12-2018-0072

Hull, C. L. (1943). *Principles of behavior: An introduction to behavior theory*. Appleton-Century-Crofts.

Hunt, E. (1989). Cognitive science: Definition, status, and questions. *Annual Review of Psychology*, *40*(1), 603–629. https://doi.org/10.1146/annurev.ps.40.020189.003131

Hunt, J. D. (1975). Image as a factor in tourism development. *Journal of Travel Research*, *13*(3), 1–7.

Hunzaker, M. F. (2016). Cultural sentiments and schema-consistency bias in information transmission. *American Sociological Review*, *81*(6), 1223–1250.

Hunzaker, M. F., & Valentino, L. (2019). Mapping cultural schemas: From theory to method. *American Sociological Review*, *84*(5), 950–981.

Hur, S., Lim, H., & Lyu, J. (2020). "I" or "she/he"? The effects of visual perspective on consumers' evaluation of brands' social media marketing: From imagery fluency perspective. *Journal of Global Fashion Marketing*, *11*(1), 1–17.

Hurley, R. A., Hutcherson, D. E., Tonkin, C. E., Dailey, S. B., & Rice, J. C. (2015). Measuring physiological arousal towards packaging: Tracking electrodermal activity within the consumer shopping environment. *Journal of Applied Packaging Research*, *7*(3), 76–90.

Huta, V. (2015). Eudaimonia and Hedonia: Their complementary functions in life and how they can be pursued in practice. In S. Joseph (Ed.), *Positive psychology in practice: Promoting human flourishing in work, health, education and everyday life* (2nd ed.). Wiley.

Hwang, J., & Lee, J. (2018). A strategy for enhancing senior tourists' well-being perception: Focusing on the experience economy. *Journal of Travel & Tourism Marketing*, *36*(3), 314–329.

Hwang, J., & Seo, S. (2016). A critical review of research on customer experience management: Theoretical, methodological, and cultural perspectives. *International Journal of Contemporary Hospitality Management*, *28*(10), 2218–2246.

Hyun, M., & O'Keefe, R. (2012). Virtual destination image: Testing a telepresence model. *Journal of Business Research*, *65*(1), 29–35.

Iani, F. (2021). Embodied cognition: So flexible as to be "disembodied"? *Consciousness and Cognition*, *88*, 103075. https://doi.org/10.1016/j.concog.2021.103075

Ihamäki, P. (2012). Geocachers: The creative tourism experience. *Journal of Hospitality and Tourism Technology*, *3*(3), 152–175.

Ind, N., Coates, N., & Lerman, K. (2020). The gift of co-creation: What motivates customers to participate. *Journal of Brand Management*, *27*(2), 181–194.

Io, M. U. (2015). Exploring the impact of past memories on tourist experiences of homeland visit: The case of the Chinese immigrants in Macao. *Journal of Vacation Marketing*, *21*(2), 191–204.

Iran-Nejad, A. (1989). Associative and nonassociative schema theories of learning. *Bulletin of the Psychonomic Society*, *27*(1), 1–4.

Iriki, A., Tanaka, M., & Iwamura, Y. (1996). Coding of modified body schema during tool use by macaque postcentral neurones. *NeuroReport*, *7*(14), 2325–2330.

Iso-Ahola, S. E. (1982). Towards a social psychological theory of tourism motivation: A rejoinder. *Annals of Tourism Research*, *9*(2), 256–261.

Ives, C. D., Abson, D. J., von Wehrden, H., Dorninger, C., Klaniecki, K., & Fischer, J. (2018). Reconnecting with nature for sustainability. *Sustainability Science*, *13*, 1389–1397. https://doi.org/10.1007/s11625-018-0542-9

Iwasaki, Y., & Mannell, R. C. (2000). Hierarchical dimensions of leisure stress coping. *Leisure Sciences*, *22*(3), 163–181.

Izard, C. E. (1971). *The face of emotion*. Appleton-Century-Crofts.

Izard, C. E. (2008). Emotion theory and research: Highlights, unanswered questions, and emerging issues. *Annual Review of Psychology*, *60*(1), 1–25. https://doi.org/10.1146/annurev.psych.60.110707.163539

Izard, E. E. (1977). *Human emotions*. Plenum.

Jacob, G. A., & Arntz, A. (2013). Schema therapy for personality disorders–A review. *International Journal of Cognitive Therapy*, *6*(2), 171–185. https://doi.org/10.1521/ijct.2013.6.2.171

Jacobs, M. H., Fehres, P., & Campbell, M. (2012). Measuring emotions toward wildlife: A review of generic methods and instruments. *Human Dimensions of Wildlife*, *17*(4), 233–247.

Jacoby, J. (2002). Stimulus-organism-response reconsidered: An evolutionary step in modeling (consumer) behavior. *Journal of Consumer Psychology*, *12*(1), 51–57.

Jafari, J., & Ritchie, J. B. (1981). Toward a framework for tourism education: Problems and prospects. *Annals of Tourism Research*, *8*(1), 13–34.

James, W. (1981/1890). *The principles of psychology*. Harvard University Press.

Jang, S. S., & Namkung, Y. (2009). Perceived quality, emotions, and behavioral intentions: Application of an extended Mehrabian–Russell model to restaurants. *Journal of Business Research*, *62*(4), 451–460.

Järv, R. (2010). Fairy tales and tourist trips. *Fabula*, *51*(3–4), 281–294.

Järvi, H., Keränen, J., Ritala, P., & Vilko, J. (2020). Value co-destruction in hotel services: Exploring the misalignment of cognitive scripts among customers and providers. *Tourism Management*, *77*, 104030.

Jennings, G. (2001). *Tourism research*. John Wiley and Sons.

Jensen, M. T. (2016). Tourism research and audio methods. *Annals of Tourism Research*, *56*, 158–160.

Jensen, O., Bonnefond, M., & VanRullen, R. (2012). An oscillatory mechanism for prioritizing salient unattended stimuli. *Trends in Cognitive Sciences*, *16*(4), 200–206.

Ji, M., Wong, I. A., Eves, A., & Leong, A. M. W. (2018). A multilevel investigation of China's regional economic conditions on co-creation of dining experience and outcomes. *International Journal of Contemporary Hospitality Management*, *30*(4), 2132–2152.

Jia, Y., Ouyang, J., & Guo, Q. (2021). When rich pictorial information backfires: The interactive effects of pictures and psychological distance on evaluations of tourism products. *Tourism Management*, *85*, 104315.

Jiang, Y., Li, S., Huang, J., & Scott, N. (2020). Worry and anger from flight delay: Antecedents and consequences. *International Journal of Tourism Research*, *22*(3), 289–302.

Jiang, Y., Ramkissoon, H., & Mavondo, F. (2016). Destination marketing and visitor experiences: The development of a conceptual framework. *Journal of Hospitality Marketing & Management*, *25*(6), 653–675.

Jiménez-Barreto, J., & Campo-Martínez, S. (2018). Destination website quality, users' attitudes, and the willingness to participate in online co-creation experiences. *European Journal of Management and Business Economics*, *27*(1), 26–41.

Jobe, J. B. (2003). Cognitive psychology and self-reports: Models and methods. *Quality of Life Research*, *12*(3), 219–227.

Johnson, A., & Stewart, D. W. (2005). A reappraisal of the role of emotion in consumer behaviour: Traditional and contemporary approaches. In N. Malhotra (Ed.). *Review of marketing research* (Vol. 1, pp. 3–34). M. E. Sharpe.

Johnson, M. K., Foley, M. A., Suengas, A. G., & Raye, C. L. (1988). Phenomenal characteristics of memories for perceived and imagined autobiographical events. *Journal of Experimental Psychology: General*, *117*(4), 371.

Johnson, Z. S., Tian, Y., & Lee, S. (2016). Country-of-origin fit: When does a discrepancy between brand origin and country of manufacture reduce consumers' product evaluations? *Journal of Brand Management*, *23*(4), 403–418.

Johnson-Laird, P. N. (2010). Mental models and human reasoning. *Proceedings of the National Academy of Sciences*, *107*(43), 18243–18250.

Jordan, E. J., & Prayag, G. (2022). Residents' cognitive appraisals, emotions, and coping strategies at local dark tourism sites. *Journal of Travel Research*, *61*(4), 887–902.

Jordan, E. J., Vogt, C. A., & DeShon, R. P. (2015). A stress and coping framework for understanding resident responses to tourism development. *Tourism Management*, *48*, 500–512.

Jorgenson, J., Nickerson, N., Dalenberg, D., Angle, J., Metcalf, E., & Freimund, W. (2019). Measuring visitor experiences: Creating and testing the tourism autobiographical memory scale. *Journal of Travel Research*, *58*(4), 566–578.

Josiassen, A., Assaf, A. G., Woo, L., & Kock, F. (2016). The imagery–Image duality model: An integrative review and advocating for improved delimitation of concepts. *Journal of Travel Research*, *55*(6), 789–803.

Joye, Y., & Dewitte, S. (2016). Up speeds you down. Awe-evoking monumental buildings trigger behavioral and perceived freezing. *Journal of Environmental Psychology*, *47*, 112–125. http://doi.org/10.1016/j.jenvp.2016.05.001

Jun, S. H., & Vogt, C. (2013). Travel information processing applying a dual-process model. *Annals of Tourism Research*, *40*, 191–212.

Jungmann, S. M., & Witthöft, M. (2020). Health anxiety, cyberchondria, and coping in the current COVID-19 pandemic: Which factors are related to coronavirus anxiety? *Journal of Anxiety Disorders*, *73*, 102239.

Kabat-Zinn, J. (2002). Meditation is about paying attention. *Reflections: The SoL Journal*, *3*(3), 68–71. https://doi.org/10.1162/152417302317363949

Kabat-Zinn, J. (2003a). Mindfulness-based interventions in context: Past, present, and future. *Clinical Psychology: Science and Practice*, *10*(2), 144–156. https://doi.org/10.1093/clipsy.bpg016

Kabat-Zinn, J. (2003b). Mindfulness-based stress reduction (MBSR). *Constructivism in the Human Sciences*, *8*(2), 73–107.

Kahneman, D. (2002). *Maps of bounded rationality*. Nobel Prize Committee.

Kahneman, D. (2011). *Thinking, fast and slow*. Macmillan.

Kahneman, D., & Tversky, A. (1979). Prospect theory: An analysis of decisions under risk. *Econometrica*, *47*(2), 263–291.

Kahneman, D., & Tversky, A. (2013). Prospect theory: An analysis of decision under risk. In *Handbook of the fundamentals of financial decision making: Part I* (pp. 99–127).

Kaidesoja, T., Sarkia, M., & Hyyryläinen, M. (2019). Arguments for the cognitive social sciences. *Journal for the Theory of Social Behaviour*, *49*(4), 480–498.

Kandampully, J., Mok, C., & Sparks, B. A. (2001). *Service quality management in hospitality, tourism, and leisure*. Haworth Hospitality Press.

Kane, J., Van Boven, L., & McGraw, A. P. (2012). Prototypical prospection: Future events are more prototypically represented and simulated than past events. *European Journal of Social Psychology*, *42*(3), 354–362.

Kang, C., & Whittingham, K. (2010). Mindfulness: A dialogue between Buddhism and clinical psychology. *Mindfulness*, *1*, 161–173. https://doi.org/10.1007/s12671-010-0018-1

Kang, E., Scott, N., Lee, T., & Ballantyne, R. (2012). Benefits from visiting a 'dark tourism' site: The case of the Jeju April 3rd Peace Park, Korea. *Tourism Management*, *33*(2), 257–265.

Kang, H. (2020). Impact of VR on impulsive desire for a destination. *Journal of Hospitality and Tourism Management*, *42*, 244–255.

Kang, M. (2009). *Effects of podcast tours on tourists' experiences in a national park*. Ph.D. Dissertation/Thesis, Texas A&M University.

Kankhuni, Z., & Ngwira, C. (2022). Overland tourists' natural soundscape perceptions: Influences on experience, satisfaction, and electronic word-of-mouth. *Tourism Recreation Research*, *47*(5–6), 591–607. https://doi.org/10.1080/02508281.2021.1878653

Kara, B. (2013). Landscape design and cognitive psychology. *Procedia-Social and Behavioral Sciences*, *82*, 288–291.

Kastner, S., & Pinsk, M. A. (2004). Visual attention as a multilevel selection process. *Cognitive, Affective, & Behavioral Neuroscience*, *4*(4), 483–500.

Katsuki, F., & Constantinidis, C. (2014). Bottom-up and top-down attention: Different processes and overlapping neural systems. *The Neuroscientist*, *20*(5), 509–521.

Katz, D. (1960). The functional approach to the study of attitude. *Social Cognition*, *11*, 163–204.

Kawasaki, A., & Uehara, I. (2020). Cultural life scripts of Japanese adolescents. *Applied Cognitive Psychology*, *34*(2), 357–371.

Ke, J., & Yoon, J. (2020). Design for breathtaking experiences: An exploration of design strategies to evoke awe in human–Product interactions. *Multimodal Technologies and Interaction*, *4*(4), 82. https://doi.org/10.3390/mti4040082

Keller, K. L. (1993). Conceptualizing, measuring, and managing customer-based brand equity. *Journal of Marketing*, *57*(1), 1–22.

Kelly, J. E., III. (2015). Computing, cognition, and the future of knowing: How humans and machines are forging a new age of understanding. In *IBM research and solutions portfolio* (pp. 2–11).

Keltner, D., & Haidt, J. (2003). Approaching awe, a moral, spiritual, and aesthetic emotion. *Cognition & Emotion*, *17*(2), 297–314. https://doi.org/10.1080/02699930302297

Keltner, D., & Kring, A. M. (1998). Emotion, social function, and psychopathology. *Review of General Psychology*, *2*(3), 320–342.

Kemp, R., & McConachie, B. (2018). *The Routledge companion to theatre, performance and cognitive science*. Routledge.

Kensinger, E. A., & Ford, J. H. (2020). Retrieval of emotional events from memory. *Annual Review of Psychology*, *71*, 251–272. https://doi.org/10.1146/annurev-psych-010419-051123

Kesteren, M. T. R. V., Ruiter, D. J., Fernandez, G. S. E., & Henson, R. N. (2012). How schema and novelty augment memory formation. *Trends in Neurosciences*, *35*(4), 211–219.

Kever, A., Pollatos, O., Vermeulen, N., & Grynberg, D. (2015). Interoceptive sensitivity facilitates both antecedent-and response-focused emotion regulation strategies. *Personality and Individual Differences*, *87*, 20–23.

Kidd, D., & Castano, E. (2017). Different stories: How levels of familiarity with literary and genre fiction relate to mentalizing. *Psychology of Aesthetics, Creativity, and the Arts*, *11*(4), 474–486. https://doi.org/10.1037/aca0000069

Killeen, P. R. (1982). Incentive theory. In *Paper presented at the 1981 Nebraska symposium on motivation: Response structure and organization, Lincoln.*

Kim, D.-Y., Hwang, Y.-H., & Fesenmaier, D. R. (2005). Modeling tourism advertising effectiveness. *Journal of Travel Research*, *44*(1), 42–49.

Kim, E., Tang, L. R., Meusel, C., & Gupta, M. (2018). Optimization of menu-labeling formats to drive healthy dining: An eye tracking study. *International Journal of Hospitality Management*, *70*, 37–48.

Kim, H., & Chen, J. S. (2016). Destination image formation process: A holistic model. *Journal of Vacation Marketing*, *22*(2), 154–166.

Kim, H. C., & Jeong, J. Y. (2016). Effect of celebrity endorsement in marketing of musicals: Poster versus social networking site. *Social Behavior and Personality: International Journal*, *44*(8), 1243–1253.

Kim, J., & Fesenmaier, D. R. (2015). Measuring emotions in real time: Implications for tourism experience design. *Journal of Travel Research*, *54*(4), 419–429.

Kim, J., & Fesenmaier, D. R. (2017a). Sharing tourism experiences: The posttrip experience. *Journal of Travel Research*, *56*(1), 28–40.

Kim, J., & Fesenmaier, D. R. (2017b). Tourism experience and tourism design. In D. R. Fesenmaier & Z. Xiang (Eds.), *Design science in tourism: Foundations of destination management* (pp. 17–29). Springer International Publishing.

Kim, J., Giroux, M., Park, J., & Lee, J. C. (2021). An evolutionary perspective in tourism: The role of socioeconomic status on extremeness aversion in travel decision making. *Journal of Travel Research*. Online.

Kim, J., Kang, J. H., & Kim, Y. K. (2014). Impact of mega sport events on destination image and country image. *Sport Marketing Quarterly*, *23*(3), 161–175.

Kim, J., & Seo, Y. (2018). An evolutionary perspective on risk taking in tourism. *Journal of Travel Research*, *58*(8), 1235–1248.

Kim, J.-H. (2010). Determining the factors affecting the memorable nature of travel experiences. *Journal of Travel & Tourism Marketing*, *27*(8), 780–796.

Kim, J. H. (2014). The antecedents of memorable tourism experiences: The development of a scale to measure the destination attributes associated with memorable experiences. *Tourism Management*, *44*, 34–45.

Kim, J. H. (2018). The impact of memorable tourism experiences on loyalty behaviors: The mediating effects of destination image and satisfaction. *Journal of Travel Research*, *57*(7), 856–870. https://doi.org/10.1177/0047287517721369

Kim, J. H., Guo, J., & Wang, Y. (2021). Tourists' negative emotions: Antecedents and consequences. *Current Issues in Tourism*, 1–19.

Kim, J. H., & Jang, S. (2016). Memory retrieval of cultural event experiences: Examining internal and external influences. *Journal of Travel Research*, *55*(3), 322–339.

Kim, J. H., Ritchie, B. J. R., & McCormick, B. (2012). Development of a scale to measure memorable tourism experiences. *Journal of Travel Research*, *51*(1), 12–25.

Kim, J.-H., Ritchie, J. R., & Tung, W. (2010). The effect of memorable experience on behavioral intentions in tourism: A structural equation modeling approach. *Tourism Analysis*, *15*(6), 637–648.

Kim, J. H., Shim, Y., Choi, I., & Choi, E. (2021). The role of coping strategies in maintaining wellbeing during the COVID-19 outbreak in South Korea. *Social Psychological and Personality Science*, 1948550621990595.

Kim, J. H., & Youn, H. (2017). How to design and deliver stories about tourism destinations. *Journal of Travel Research*, *56*(6), 808–820.

Kim, J. S., Lee, T. J., & Kim, N. J. (2021). What motivates people to visit an unknown tourist destination? Applying an extended model of goal-directed behavior. *International Journal of Tourism Research*, *23*(1), 13–25.

Kim, M. J., & Hall, M. (2019). A hedonic motivation model in virtual reality tourism: Comparing visitors and non-visitors. *International Journal of Information Management*, *46*, 236–249.

Kim, M. S., & Duda, J. L. (2003). The coping process: Cognitive appraisals of stress, coping strategies, and coping effectiveness. *The Sport Psychologist*, *17*(4), 406–425.

Kim, N. Y., & Park, J. W. (2016). A study on the impact of airline service delays on emotional reactions and customer behavior. *Journal of Air Transport Management*, *57*, 19–25.

Kim, S. (2012). The impact of tv drama attributes on touristic experiences at film tourism destinations. *Tourism Analysis*, *17*, 573–585. https://doi.org/10.3727/108354212X13485873913804

Kim, S., & Jun, J. (2016). The impact of event advertising on attitudes and visit intentions. *Journal of Hospitality and Tourism Management*, *29*, 1–8.

Kim, S.-B., Kim, D.-Y., & Bolls, P. (2014). Tourist mental-imagery processing: Attention and arousal. *Annals of Tourism Research*, *45*, 63–76.

Kim, Y., Ribeiro, M. A., & Li, G. (2022a). Tourism memory characteristics scale: Development and validation. *Journal of Travel Research*, *61*(6), 1308–1326.

Kim, Y., Ribeiro, M. A., & Li, G. (2022b). Tourism memory, mood repair and behavioral intention. *Annals of Tourism Research*, *93*, 103369.

Kim-Spoon, J., Deater-Deckard, K., Calkins, S. D., King-Casas, B., & Bell, M. A. (2019). Commonality between executive functioning and effortful control related to adjustment. *Journal of Applied Developmental Psychology*, *60*, 47–55.

Kirillova, K. (2019). Travel beautifully: The role of aesthetics in experience design. In P. Wassler, V. Michael, & P. Dieter (Eds.). *Atmospheric turn in culture and tourism: Place, design and process impacts on customer behaviour, marketing and branding* (Vol. 16, pp. 153–163). Emerald Publishing Limited.

Kirillova, K., Fu, X., Lehto, X., & Cai, L. (2014). What makes a destination beautiful? Dimensions of tourist aesthetic judgment. *Tourism Management*, *42*, 282–293.

Kirillova, K., & Lehto, X. (2016). Aesthetic and restorative qualities of vacation destinations: How are they related? *Tourism Analysis*, *21*(5), 513–527.

Kirillova, K., Lehto, X., & Cai, L. (2017a). Tourism and existential transformation: An empirical investigation. *Journal of Travel Research*, *56*(5), 638–650. https://doi.org/10.1177/0047287516650277

Kirillova, K., Lehto, X., & Cai, L. (2017b). What triggers transformative tourism experiences? *Tourism Recreation Research*, *42*(4), 498–511. https://doi.org/10.1080/02508281.2017.1342349

Kirsch, L. P., Urgesi, C., & Cross, E. S. (2016). Shaping and reshaping the aesthetic brain: Emerging perspectives on the neurobiology of embodied aesthetics. *Neuroscience & Biobehavioral Reviews*, *62*, 56–68.

Kisfalvi, V., & Oliver, D. (2015). Creating and maintaining a safe space in experiential learning. *Journal of Management Education*, *39*(6), 713–740. http://doi.org/10.1016/j.neubiorev.2015.12.005

Kitson, A., Buie, E., Stepanova, E. R., Chirico, A., Riecke, B. E., & Gaggioli, A. (2019, May). Transformative experience design: Designing with interactive technologies to support transformative experiences. In *Extended abstracts of the 2019 CHI conference on human factors in computing systems. Paper presented at the 2019 CHI conference on human factors in computing systems*, Glasgow, Scotland. https://doi.org/10.1145/3290607.3311762

Klaaren, K. J., Hodges, S. D., & Wilson, T. D. (1994). The role of affective expectations in subjective experience and decision-making. *Social Cognition*, *12*(2), 77–101.

Klebba, J. M. (1985). Physiological measures of research: A review of brain activity, electrodermal response, pupil dilation, and voice analysis methods and studies. *Current Issues and Research in Advertising*, *8*(1), 53–76.

Klein, S. B., Cosmides, L., Tooby, J., & Chance, S. (2002). Decisions and the evolution of memory: Multiple systems, multiple functions. *Psychological Review*, *109*(2), 306–329.

Klenosky, D. B. (2002). The "pull" of tourism destinations: A means-end investigation. *Journal of Travel Research*, *40*(4), 385–395.

Klimesch, W., Sauseng, P., & Hanslmayr, S. (2007). EEG alpha oscillations: The inhibition–timing hypothesis. *Brain Research Reviews*, *53*(1), 63–88.

KLM Royal Dutch Airlines (Producer). (2010). *KLM surprise*. [Video].

Knobloch, U., Robertson, K., & Aitken, R. (2017). Experience, emotion, and Eudaimonia: A consideration of tourist experiences and well-being. *Journal of Travel Research*, *56*(5), 651–662. https://doi.org/10.1177/0047287516650937

Ko, L. W., Komarov, O., Hairston, W. D., Jung, T. P., & Lin, C. T. (2017). Sustained attention in real classroom settings: An EEG study. *Frontiers in Human Neuroscience*, *11*, 388.

Koch, C., & Tsuchiya, N. (2007). Attention and consciousness: Two distinct brain processes. *Trends in Cognitive Sciences*, *11*(1), 16–22.

Kock, F., Josiassen, A., & Assaf, A. G. (2018). On the origin of tourist behavior. *Annals of Tourism Research*, *73*, 180–183.

Kock, F., & Ringberg, T. (2019). Embodied cognition effects on tourist behavior. *Annals of Tourism Research*, *78*, 102725. https://doi.org/10.1016/j.annals.2019.05.002

Koehler, J. J. (1993). The influence of prior beliefs on scientific judgments of evidence quality. *Organizational Behavior and Human Decision Processes*, *56*(1), 28–55.

Kolb, D. A. (1984). *Experiential learning: Experience as the source of learning and development*. Pearson Education.

Kong, S., Huang, Z., Scott, N., Zhang, Z. A., & Shen, Z. (2019). Web advertisement effectiveness evaluation: Attention and memory. *Journal of Vacation Marketing*, *25*(1), 130–146.

Körner, A., Topolinski, S., & Strack, F. (2015). Routes to embodiment. *Frontiers in Psychology*, *6*, 940. https://doi.org/10.3389/fpsyg.2015.00940

Korteling, J. E., Brouwer, A. M., & Toet, A. (2018). A neural network framework for cognitive bias. *Frontiers in Psychology*, *9*, 1561. https://doi.org/10.3389/fpsyg.2018.01561

Koseoglu, M. A., Mehraliyev, F., & Xiao, H. (2019). Intellectual connections in tourism studies. *Annals of Tourism Research*, *79*, 102760.

Kotera, Y., Lyons, M., Vione, K. C., & Norton, B. (2021). Effect of nature walks on depression and anxiety: A systematic review. *Sustainability*, *13*(7), 4015. https://doi.org/10.3390/su13074015

Kottler, J. A. (2002). Transformative travel: International counselling in action. *International Journal for the Advancement of Counselling*, *24*(4), 207–210.

Krapp, A. (1999). Interest, motivation, and learning: An educational-psychological perspective. *European Journal of Psychology of Education*, *14*(1), 23–40.

Kreibig, S. D. (2010). Autonomic nervous system activity in emotion: A review. *Biological Psychology*, *84*(3), 394–421. https://doi.org/10.1016/j.biopsycho.2010.03.010

Kreziak, D., & Frochot, I. (2011). Construction de l'expérience touristique: Les stratégies des touristes en stations de sport d'hiver [Co-construction of the tourism experience: Tourists strategies in ski resorts]. *Décisions Marketing*, *64*, 23–34.

Krishna, A. (2012). An integrative review of sensory marketing: Engaging the senses to affect perception, judgment, and behavior. *Journal of Consumer Psychology*, *22*(3), 332–351.

Krishna, A., Elder, R. S., & Caldara, C. (2010). Feminine to smell but masculine to touch? Multisensory congruence and its effect on the aesthetic experience. *Journal of Consumer Psychology*, *20*(4), 410–418.

Krishna, A., Morrin, M., & Sayin, E. (2013). Smellizing cookies and salivating: A focus on olfactory imagery. *Journal of Consumer Research*, *41*(1), 18–34.

Krishna, A., & Schwarz, N. (2014). Sensory marketing, embodiment, and grounded cognition: A review and introduction. *Journal of Consumer Psychology*, *24*(2), 159–168.

Kristjánsson, Á., & Egeth, H. (2020). How feature integration theory integrated cognitive psychology, neurophysiology, and psychophysics. *Attention, Perception, & Psychophysics*, *82*(1), 7–23. https://doi.org/10.3758/s13414-019-01803-7

Kroeber-Riel, W. (1979). Activation research: Psychobiological approaches in consumer research. *Journal of Consumer Research*, *5*(4), 240–250.

Kruger, S., Sirgy, M. J., Lee, D.-J., & Yu, G. (2015). Does life satisfaction of tourists increase if they set travel goals that have high positive valence? *Tourism Analysis*, *20*(2), 173–188.

Kruglanski, A. W., Shah, J. Y., Fishbach, A., Friedman, R., Chun, W. Y., & Sleeth-Keppler, D. (2002). A theory of goal systems. *Advances in Experimental Social Psychology*, *34*, 331–378.

Kruglanski, A. W., Shah, J. Y., Fishbach, A., Friedman, R., Chun, W. Y., & Sleeth-Keppler, D. (2018). A theory of goal systems. In R. Persaud (Ed.), *The motivated mind* (pp. 207–250). Routledge.

Krugman, H. E. (1965). The impact of television advertising: Learning without involvement. *Public Opinion Quarterly*, *29*(3), 349–356.

Kuan, T.-F. (2012). Cognitive operations in Buddhist meditation: Interface with Western psychology. *Contemporary Buddhism*, *13*(1), 35–60. https://doi.org/10.1080/14639947.2012.669281

Kucukergin, K. G., Kucukergin, F. N., & Dedeoglu, B. B. (2020). An overview of the destination physical servicescape with SOR paradigm: The importance of prestige sensitivity. *Asia Pacific Journal of Tourism Research*, *25*(5), 473–488.

Kukkonen, K. (2014). Plot. In P. Hühn, J. Pier, W. Schmid, & J. Schönert (Eds.), *The living handbook of narratology*. Hamburg University. http://www.lhn.uni-hamburg.de/article/plot

Kukkonen, K., & Caracciolo, M. (2014). Introduction: What is the "second generation". *STYLE*. https://doi.org/10.5325/style.48.3.261

Kumar, A., & Oliver, R. L. (1997). Special session summary: Cognitive appraisals, consumer emotions, and consumer response. *Advances in Consumer Research*, *24*, 17–18.

Kurzban, R., Duckworth, A., Kable, J. W., & Myers, J. (2013). An opportunity cost model of subjective effort and task performance. *Behavioral and Brain Sciences*, *36*(6), 661–679. https://doi.org/10.1017/S0140525X12003196

Kyle, G., & Chick, G. (2004). Enduring leisure involvement: The importance of personal relationships. *Leisure Studies*, *23*(3), 243–266.

Laborde, S., Mosley, E., & Thayer, J. F. (2017). Heart rate variability and cardiac vagal tone in psychophysiological research–Recommendations for experiment planning, data analysis, and data reporting. *Frontiers in Psychology*, *8*, 213–213.

Labrague, L. J., McEnroe-Petitte, D. M., Gloe, D., Thomas, L., Papathanasiou, I. V., & Tsaras, K. (2017). A literature review on stress and coping strategies in nursing students. *Journal of Mental Health*, *26*(5), 471–480.

Lachman, R., Lachman, J. L., & Butterfield, E. C. (1979). *Cognitive psychology and information processing: An introduction*. Erlbaum.

Lachman, R., Lachman, J. L., & Butterfield, E. C. (2015). *Cognitive psychology and information processing: An introduction*. Psychology Press.

Lackey, N. Q., Tysor, D. A., McNay, G. D., Joyner, L., Baker, K. H., & Hodge, C. (2021). Mental health benefits of nature-based recreation: A systematic review. *Annals of Leisure Research*, *24*(3), 379–393. https://doi.org/10.1080/11745398.2019.1655459

Laing, J. H., & Crouch, G. I. (2009). Myth, adventure, and fantasy at the frontier: Metaphors and imagery behind an extraordinary travel experience. *International Journal of Tourism Research*, *11*(2), 127–141.

Lakoff, G., & Johnson, M. (2008). *Metaphors we live by*. University of Chicago press.

Lakoff, G., Johnson, M., & Sowa, J. F. (1999). Review of philosophy in the flesh: The embodied mind and its challenge to western thought. *Computational Linguistics*, *25*(4), 631–634.

Lakshmanan, A., & Krishnan, S. (2009). How does imagery in interactive consumption lead to false memory? A reconstructive memory perspective. *Journal of Consumer Psychology*, *19*(3), 451–462.

Landauer, T. K. (1986). How much do people remember? Some estimates of the quantity of learned information in long-term memory. *Cognitive Science*, *10*(4), 477–493.

Lang, P. J. (1977). Imagery in therapy: An information processing analysis of fear. *Behavior Therapy*, *8*(5), 862–886.

Lang, P. J. (1979). A bio-informational theory of emotional imagery. *Psychophysiology*, *16*(6), 495–512.

Lang, P. J., & Bradley, M. M. (2013). Appetitive and defensive motivation: Goal-directed or goal-determined? *Emotion Review*, *5*(3), 230–234.

Lang, P. J., Bradley, M. M., & Cuthbert, B. N. (2005). *International affective picture system (IAPS): Technical manual and affective ratings*. NIMH, Center for the Study of Emotion & Attention.

Langer, E. J. (1992). Matters of mind: Mindfulness/mindlessness in perspective. *Consciousness and Cognition*, *1*(4), 289–305.

Larsen, S. (2007). Aspects of a psychology of the tourist experience. *Scandinavian Journal of Hospitality and Tourism*, *7*(1), 7–18.

Larsen, S., & Jenssen, D. (2004). The school trip: Travelling with, not to or from. *Scandinavian Journal of Hospitality and Tourism*, *4*(1), 43–57.

Larsson, L., Nyström, M., Andersson, R., & Stridh, M. (2015). Detection of fixations and smooth pursuit movements in high-speed eye-tracking data. *Biomedical Signal Processing and Control*, *18*, 145–152.

Lau, H., & Rosenthal, D. (2011). Empirical support for higher-order theories of conscious awareness. *Trends in Cognitive Sciences*, *15*(8), 365–373.

Lazarus, R. S. (1974). Psychological stress and coping in adaptation and illness. *The International Journal of Psychiatry in Medicine*, *5*(4), 321–333.

Lazarus, R. S. (1984). *On the primacy of cognition*. University of California.

Lazarus, R. S. (1991). Goal congruent (positive) and problematic emotions. In R. S. Lazarus (Ed.), *Emotion and adaptation*. Oxford University Press.

Lazarus, R. S. (1991a). *Emotion and adaptation*. Oxford University Press.

Lazarus, R. S. (1991b). Progress on a cognitive-motivational-relational theory of emotion. *American Psychologist*, *46*(8), 819–834.

Lazarus, R. S. (1991c). Cognition and motivation in emotion. *American Psychologist*, *46*(4), 352–367.

Lazarus, R. S. (1993). Coping theory and research: Past, present, and future. *Psychosomatic Medicine*, *55*, 234–247.

Lazarus, R. S. (1998). *Fifty years of research and theory by RS Lazarus: An analysis of historical and perennial issues*. Lawrence Erlbaum.

Lazarus, R. S. (2001). Relational meaning and discrete emotions. In K. R. Scherer & A. Schorr (Eds.), *Appraisal processes in emotion: Theory, methods, research* (pp. 1–19). Oxford University Press.

Lazarus, R. S., & Folkman, S. (1984). *Stress, appraisal, and coping*. Springer Publishing Company.

Lazarus, R. S., & Smith, C. A. (1988). Knowledge and appraisal in the cognition—Emotion relationship. *Cognition & Emotion*, *2*(4), 281–300.

Le, D., Hadinejad, A., Moyle, B., Ma, J., & Scott, N. (2020). A review of eye-tracking methods in tourism research. *Eye Tracking in Tourism*, 13–27.

Le, D., Pratt, M., Wang, Y., Scott, N., & Lohmann, G. (2020). How to win the consumer's heart? Exploring appraisal determinants of consumer pre-consumption emotions. *International Journal of Hospitality Management*, *88*, 102542.

Le, D., Scott, N., Becken, S., & Connolly, R. M. (2019). Tourists' aesthetic assessment of environmental changes, linking conservation planning to sustainable tourism development. *Journal of Sustainable Tourism*, *27*(10), 1477–1494.

Le, D., Scott, N., & Lohmann, G. (2019). Applying experiential marketing in selling tourism dreams. *Journal of Travel & Tourism Marketing*, *36*(2), 220–235. https://doi.org/10.1080/10548408.2018.1526158

Le, D., Scott, N., & Wang, Y. (2021). Impact of prior knowledge and psychological distance on tourist imagination of a promoted tourism event. *Journal of Hospitality and Tourism Management*, *49*, 101–111.

Lean, G. L. (2012). Transformative travel: A mobilities perspective. *Tourist Studies*, *12*(2), 151–172. https://doi.org/10.1177/1468797612454624

LeDoux, J. E., Michel, M., & Lau, H. (2020). A little history goes a long way toward understanding why we study consciousness the way we do today. *Proceedings of the National Academy of Sciences USA*, *117*(13), 6976–6984.

Lee, A. Y. (2001). The mere exposure effect: An uncertainty reduction explanation revisited. *Personality and Social Psychology Bulletin*, *27*(10), 1255–1266.

Lee, J. K., Ahn, T., & Lee, K.-Y. (2016). Exploring the impact of country-of-origin fit and team identification in sports brand evaluation. *European Sport Management Quarterly*, *16*(4), 413–432.

Lee, K.-Y. (2012). Consumer processing of virtual experience in e-commerce: A test of an integrated framework. *Computers in Human Behavior*, *28*(6), 2134–2142.

Lee, K. H., & Kim, D. Y. (2017). Explicit and implicit image cognitions toward destination: Application of the single-target implicit association test (ST-IAT). *Journal of Destination Marketing & Management*, *6*(4), 396–406.

Lee, S. (2018Lee, K.-Y. (2012)). Investigating antecedents and outcome of telepresence on a hotel's website. *International Journal of Contemporary Hospitality Management*, *30*(2), 757–775.

Lee, S. A., & Jeong, M. (2014). Enhancing online brand experiences: An application of congruity theory. *International Journal of Hospitality Management*, *40*, 49–58.

Lee, S. A., & Oh, H. (2017). Sharing travel stories and behavioral outcomes: A case of travel. *Tourism Management*, *62*, 147–158.

Lee, T. H., & Crompton, J. (1992). Measuring novelty seeking in tourism. *Annals of Tourism Research*, *19*(4), 732–751.

Lee, W. J., & Gretzel, U. (2012). Designing persuasive destination websites: A mental imagery processing perspective. *Tourism Management*, *33*(5), 1270–1280. https://doi.org/10.1016/j.tourman.2011.10.012

Lehner, E., & D'Argembeau, A. (2016). The role of personal goals in autonoetic experience when imagining future events. *Consciousness and Cognition*, *42*, 267–276. https://doi.org/10.1016/j.concog.2016.04.002

Lehto, X. Y. (2013). Assessing the perceived restorative qualities of vacation destinations. *Journal of Travel Research*, *52*(3), 325–339.

Lehto, X. Y., Lee, G., & Ismail, J. (2014). Measuring congruence of affective images of destinations and their slogans. *International Journal of Tourism Research*, *16*(3), 250–260.

Lei, M., Chen, W., Wu, J., Zhang, Y., & Lv, Y. (2022). Neurophysiological measures in hospitality and tourism: Review, critique, and research agenda. *Journal of Hospitality & Tourism Research*, 10963480221091117.

León, C. (2016). An architecture of narrative memory. *Biologically Inspired Cognitive Architectures*, *16*, 19–33. https://doi.org/10.1016/j.bica.2016.04.002

Lerner, J. S., & Keltner, D. (2000). Beyond valence: Toward a model of emotion-specific influences on judgement and choice. *Cognition & Emotion*, *14*(4), 473–493. https://doi.org/10.1080/026999300402763

Leschziner, V., & Brett, G. (2021). Have schemas been good to think with? In *Sociological forum* (Vol. 36, pp. 1207–1228).

Levenson, R. W. (2003). Blood, sweat, and fears: The autonomic architecture of emotion. *Annals of the New York Academy of Sciences*, *1000*(1), 348–366.

Levin, A. M., Davis, J. C., & Levin, I. (1996). Theoretical and empirical linkages between consumers' responses to different branding strategies. *Advances in Consumer Research*, *23*, 296–300.

Levine, L. J. (1997). Reconstructing memory for emotions. *Journal of Experimental Psychology: General*, *126*(2), 165–177.

Levine, L. J., Prohaska, V., Burgess, S. L., Rice, J. A., & Laulhere, T. M. (2001). Remembering past emotions: The role of current appraisals. *Cognition & Emotion*, *15*(4), 393–417.

Levy, N. (2014). Consciousness, implicit attitudes, and moral responsibility. *Noûs*, *48*(1), 21–40.

Li, M., Chen, Y., Wang, J., & Liu, T. (2020). Children's attention toward cartoon executed photos. *Annals of Tourism Research*, *80*, 102799.

Li, J., Zhang, Z., Jing, F., Gao, J., Ma, J., Shao, G., & Scott, N. (2020). An evaluation of urban green space in Shanghai, China, using eye tracking. *Urban Forestry and Urban Greening*, *56*, 126903.

Li, Q., Huang, Z. J., & Christianson, K. (2016). Visual attention toward tourism photographs with text: An eye-tracking study. *Tourism Management*, *54*, 243–258.

Li, S., Jiang, Y., Cheng, B., & Scott, N. (2021). The effect of flight delay on customer loyalty intention: The moderating role of emotion regulation. *Journal of Hospitality and Tourism Management*, *47*, 72–83.

Li, S., Lyu, T., Chen, M., & Zhang, P. (2022). The prospects of using EEG in tourism and hospitality research. *Journal of Hospitality & Tourism Research*, *46*(1), 189–211.

Li, S., Scott, N., & Walters, G. (2014). Potential methods for measuring emotion in Chinese tourism advertising. In *Paper presented at the G20 first east-west dialogue on tourism and the Chinese Dream, Griffith University Gold Coast*.

Li, S., Scott, N., & Walters, G. (2015). Current and potential methods for measuring emotion in tourism experiences: A review. *Current Issues in Tourism*, *18*(9), 805–827. https://doi.org/10.1080/13683500.2014.975679

Li, Y., & Dong, X. (2019). Expression of port context in tourism landscape design based on visual methods analysis—Taking Qinhuangdao Port as an example. *Vacation Tour*, *3*, 53–55.

Li, X., Meng, F., Uysal, M., & Mihalik, B. (2013). Understanding China's long-haul outbound travel market: An overlapped segmentation approach. *Journal of Business Research*, *66*(6), 786–793. https://doi.org/10.1016/j.jbusres.2011.09.019

Li, X., & Stepchenkova, S. (2012). Chinese outbound tourists' destination image of America: Part I. *Journal of Travel Research*, *51*(3), 250–266. https://doi.org/10.1177/0047287511410349

Li, Y., Liu, B., Chen, P., & Huan, T. C. (2021). Tourism service providers' physical attractiveness and customers' service quality evaluation: Is warmth or competence more important? *Tourism Review*, *76*(6), 1260–1278.

Li, Y., Liu, B., & Huan, T. C. T. (2019). Renewal or not? Consumer response to a renewed corporate social responsibility strategy: Evidence from the coffee shop industry. *Tourism Management*, *72*, 170–179.

Li, Y., Liu, B., & Xie, L. (2022). Celebrity endorsement in international destination marketing: Evidence from eye-tracking techniques and laboratory experiments. *Journal of Business Research*, *150*, 553–566.

Li, Y., Liu, B., Zhang, R., & Huan, T. C. (2020). News information and tour guide occupational stigma: Insights from the stereotype content model. *Tourism Management Perspectives*, *35*, 100711.

Li, Y., Song, Y., & Fang, S. (2022). Time perception and tourist behavioral decision when travelling. *Current Issues in Tourism*, 1 7.

Li, Y., Zhang, C., & Fang, S. (2022). Can beauty save service failures? The role of recovery employees' physical attractiveness in the tourism industry. *Journal of Business Research*, *141*, 100–110.

Li, S., Sung, B., Lin, Y., & Mitas, O. (2022). Electrodermal activity measure: A methodological review. *Annals of Tourism Research*, *96*, 103460.

Liang, C., Ho, P., Yen, C., Chen, C., Kuo, S., Huang, C., & Ma, K. (2017). The relationship between the striatal dopamine transporter and novelty seeking and cognitive flexibility in opioid dependence. *Progress in Neuro-Psychopharmacology and Biological Psychiatry*, *74*, 36–42.

Liberman, N., Trope, Y., & Wakslak, C. (2007). Construal level theory and consumer behavior. *Journal of Consumer Psychology*, *17*(2), 113–117.

Lien, N.-H., & Chen, Y.-L. (2013). Narrative ads: The effect of argument strength and story format. *Journal of Business Research*, *66*(4), 516–522.

Lim, Y., & Weaver, P. A. (2014). Customer-based brand equity for a destination: The effect of destination image on preference for products associated with a destination brand. *International Journal of Tourism Research*, *16*(3), 223–231.

Lin, P. M. C., Tung, V. W. S., Qiu Zhang, H., & Gu, Q. (2018). Tourist experience on memorable hospitality services. *Journal of China Tourism Research*, *14*(2), 123–145.

Lin, Y., Kerstetter, D., Nawijn, J., & Mitas, O. (2014). Changes in emotions and their interactions with personality in a vacation context. *Tourism Management*, *40*, 416–424.

Lindell, A. K., & Lindell, K. L. (2014). Beauty captures the attention of the beholder. *Journal of Cognitive Psychology*, *26*(7), 768–780.

Lindenberg, S., & Steg, L. (2007). Normative, gain and hedonic goal frames guiding environmental behavior. *Journal of Social Issues*, *63*(1), 117–137. https://doi.org/10.1111/j.1540-4560.2007.00499.x

Lindquist, K. A., Gendron, M., Oosterwijk, S., & Barrett, L. F. (2013). Do people essentialize emotions? Individual differences in emotion essentialism and emotional experience. *Emotion*, *13*(4), 629–644. https://doi.org/10.1037/a0032283

Lindquist, K. A., Wager, T. D., Kober, H., Bliss-Moreau, E., & Barrett, L. F. (2012). The brain basis of emotion: A meta-analytic review. *Behavioral and Brain Sciences*, *35*(3), 121–143.

Line, N. D., Hanks, L., & Zhang, L. (2016). Sustainability communication: The effect of message construals on consumers' attitudes towards green restaurants. *International Journal of Hospitality Management*, *57*, 143–151.

Lisman, J. E., & Grace, A. A. (2005). The hippocampal-VTA loop: Controlling the entry of information into long-term memory. *Neuron*, *46*(5), 703–713.

Liu, A., Wang, X. L., Liu, F., Yao, C., & Deng, Z. (2018). Soundscape and its influence on tourist satisfaction. *The Service Industries Journal*, *38*(3–4), 164–181.

Liu, Y., Hultman, M., Eisingerich, A. B., & Wei, X. (2020). How does brand loyalty interact with tourism destination? Exploring the effect of brand loyalty on place attachment. *Annals of Tourism Research*, *81*, 102879. https://doi.org/10.1016/j.annals.2020.102879

Liu, B., & Li, Y. (2021). What you see is what you get: A research review and prospects of tourism eye-tracking studies. *Tourism and Hospitality Prospects*, *5*(1), 76–102.

Liu, B., & Li, Y. (2022). Teddy-bear effect in service recovery. *Annals of Tourism Research*, *94*, 103400.

Liu, B., Li, Y., Kralj, A., Moyle, B., & He, M. (2022a). Inspiration and wellness tourism: The role of cognitive appraisal. *Journal of Travel & Tourism Marketing*, *39*(2), 173–187.

Liu, B., Moyle, B., & Kralj, A. (2022b). When and how sharing tourism experiences on social media backfires: TMSP model of sharing driven outcomes. *Journal of Hospitality and Tourism Management*, *53*, 155–159.

Liu, D., & Upchurch, R. S. (2020). A glimpse into energy tourism via application of eye-tracking technology. *Journal of Leisure Research*, *51*(2), 230–244.

Liu, A., Wang, X. L., Liu, F., Yao, C., & Deng, Z. (2018). Soundscape and its influence on tourist satisfaction. *Service Industries Journal*, *38*(3–4), 164–181. https://doi.org/10.1080/02642069.2017.1382479

Livingstone, K. M., & Isaacowitz, D. M. (2015). Situation selection and modification for emotion regulation in younger and older adults. *Social Psychological and Personality Science*, *6*(8), 904–910.

Loewenstein, G. F., Weber, E. U., Hsee, C. K., & Welch, N. (2001). Risk as feelings. *Psychological Bulletin*, *127*(2), 267–286.

Loftus, E. F. (2005). Planting misinformation in the human mind: A 30-year investigation of the malleability of memory. *Learning & Memory*, *12*(4), 361–366.

Logothetis, N. K. (2008). What we can do and what we cannot do with fMRI. *Nature*, *453*(7197), 869–878.

Lohse, M., & Overgaard, M. (2019). Emotional priming depends on the degree of conscious experience. *Neuropsychologia*, *128*, 96–102.

Loken, B. (2006). Consumer psychology: Categorization, inferences, affect, and persuasion. *Annual Review of Psychology*, *57*, 453–485.

Loo, P. T., Khoo-Lattimore, C., & Boo, H. C. (2021). How should I respond to a complaining customer? A model of cognitive-emotive-behavioral from the perspective of restaurant service employees. *International Journal of Hospitality Management*, *95*. https://doi.org/10.1016/j.ijhm.2021.102882

Lorini, E., & Castelfranchi, C. (2007). The cognitive structure of surprise: Looking for basic principles. *Topoi*, *26*(1), 133–149.

Loureiro, S. M. C. (2014). The role of the rural tourism experience economy in place attachment and behavioral intentions. *International Journal of Hospitality Management*, *40*, 1–9.

Loureiro, S. M. C., Stylos, N., & Bellou, V. (2021). Destination atmospheric cues as key influencers of tourists' word-of-mouth communication: Tourist visitation at two mediterranean capital cities. *Tourism Recreation Research*, *46*(1), 85–108.

Louvet, B., Gaudreau, P., Menaut, A., Genty, J., & Deneuve, P. (2007). Longitudinal patterns of stability and change in coping across three competitions: A latent class growth analysis. *Journal of Sport & Exercise Psychology*, *29*(1), 100–117.

Løvoll, H. S. (2019). The inner feeling of glacier hiking: An exploratory study of "immersion" as it relates to flow, hedonia and eudaimonia. *Scandinavian Journal of Hospitality and Tourism*, *19*(3), 300–316. https://doi.org/10.1080/15022250.2019.1581084

Lu, D., Liu, Y., Lai, I., & Yang, L. (2017). Awe: An important emotional experience in sustainable tourism. *Sustainability*, *9*(12), 2189. https://doi.org/10.3390/su9122189

Luchman, J. N., & González-Morales, M. G. (2013). Demands, control, and support: A meta-analytic review of work characteristics interrelationships. *Journal of Occupational Health Psychology*, *18*(1), 37.

Luck, S. J. (2014). *An introduction to the event-related potential technique*. MIT press.

Luck, S. J., & Kappenman, E. S. (2011). *The Oxford handbook of event-related potential components*. Oxford University Press.

Lugmayr, A., Sutinen, E., Suhonen, J., Sedano, C. I., Hlavacs, H., & Montero, C. S. (2016). Serious storytelling – A first definition and review. *Multimedia Tools and Applications*, *76*(76), 15707–15733.

Luna-Cortés, G. (2021). Measuring tourists' stress in Colombia. *Tourism*, *69*(1), 32–46. https://doi.org/10.37741/t.69.1.3

Lundquist, D. (2014). *Management of dusky dolphin tourism at Kaikoura, New Zealand* (pp. 337–351). Cambridge University Press. ISBN: 9780521195973.

Lustenberger, C., Boyle, M. R., Foulser, A. A., Mellin, J. M., & Fröhlich, F. (2015). Functional role of frontal alpha oscillations in creativity. *Cortex*, *67*, 74–82.

Lv, X., Li, H., & Xia, L. (2020). Effects of haptic cues on consumers' online hotel booking decisions: The mediating role of mental imagery. *Tourism Management*, *77*, 104025. https://doi.org/10.1016/j.tourman.2019.104025

Lv, X., Lu, R., Xu, S., Sun, J., & Yang, Y. (2022). Exploring visual embodiment effect in dark tourism: The influence of visual darkness on dark experience. *Tourism Management*, *89*, 104438.

Lwin, M. O., Morrin, M., Chong, C. S. T., & Goh, S. X. (2016). Odor semantics and visual cues: What we smell impacts where we look, what we remember, and what we want to buy. *Journal of Behavioral Decision Making*, *29*(2–3), 336–350.

Ma, J., Gao, J., & Scott, N. (2017). Introduction: Creating memorable experiences. In N. Scott, J. Gao, & J. Ma (Eds.), *Visitor experience design* (pp. 3–12). CABI.

Ma, J., Gao, J., Scott, N., & Ding, P. (2013). Customer delight derived from theme park experiences: The antecedents of delight based on cognitive appraisal theory. *Annals of Tourism Research*, *42*, 359–381.

Ma, J., Scott, N., Gao, J., & Ding, P. (2017). Delighted or satisfied? Positive emotional responses derived from theme park experiences. *Journal of Travel & Tourism Marketing*, *34*(1), 1–19.

Ma, Q. G., Hu, L. F., Pei, G. X., Ren, P. Y., & Ge, P. (2014). Applying neuroscience to tourism management: A primary exploration of neurotourism. *Applied Mechanics and Materials*, *670*, 1637–1640.

Ma, S., Zhao, X., Gong, Y., & Wengel, Y. (2021). Proposing "healing tourism" as a post-COVID-19 tourism product. *Anatolia, 32*(1), 136–139. https://doi.org/10.1080/13032917.2020.1808490

Macaulay, R., Lee, K., Johnson, K., & Williams, K. (2022). Mindful engagement, psychological restoration, and connection with nature in constrained nature experiences. *Landscape and Urban Planning, 217*, 104263. https://doi.org/10.1016/j.landurbplan.2021.104263

MacCannell, D. (1973). Staged authenticity: Arrangements of social space in tourist settings. *American Journal of Sociology, 79*(3), 589–603.

MacCannell, D. (1976). *The tourist: A new theory of the leisure class*. Schocken Books.

MacInnes, S., Grün, B., & Dolnicar, S. (2022). Habit drives sustainable tourist behaviour. *Annals of Tourism Research, 92*, 103329. https://doi.org/10.1016/j.annals.2021.103329

MacInnis, D., & Price, L. (1987). The role of imagery in information processing: Review and extensions. *Journal of Consumer Research, 13*(4), 473–491.

MacInnis, D. J., & Price, L. L. (1987). The role of imagery in information processing: Review and extensions. *Journal of Consumer Research, 13*(4), 473–491.

MacLeod, C. M. (1991). Half a century of research on the Stroop effect: An integrative review. *Psychological Bulletin, 109*(2), 163–203.

Macmillan Education. (n.d.). *Sadness*. Macmillan English Dictionary. https://www.macmillandictionary.com/thesaurus-category/british/feeling-sad-or-unhappy

Maddux, J. E. (1999). Expectancies and the social–Cognitive perspective: Basic principles, processes, and variables. In I. Kirsch (Ed.), *How expectancies shape experience* (pp. 17–39). American Psychological Association.

Madsen, J., & Parra, L. C. (2021). Similar cognitive processing synchronizes brains, hearts, and eyes. *bioRxiv*. [Preprint]. https://doi.org/10.1101/2021.09.16.460722

Magnini, V. P., & Dallinger, I. (2018). Consumer information overload and the need to prompt script deviations. *Journal of Quality Assurance in Hospitality & Tourism, 19*(3), 285–297. https://doi.org/10.1080/1528008x.2016.1230038

Makransky, G., Borre-Gude, S., & Mayer, R. E. (2019). Motivational and cognitive benefits of training in immersive virtual reality based on multiple assessments. *Journal of Computer Assisted Learning, 35*(6), 691–707. https://doi.org/10.1111/jcal.12375

Mandler, G. (1982). The structure of value: Accounting for taste. In M. S. Clark & S. T. Fiske (Eds.), *Affect and cognition: The 17th annual carnegie symposium* (pp. 3–36). Lawrence Erlbaum Associates.

Mandler, J. M. (1992). How to build a baby: II. Conceptual primitives. *Psychological Review, 99*(4), 587–604.

Manisaligil, A., & Haksoz, C. (2016). Days of future past: Implications of mental time travel on organizational behavior. *Academy of Management Proceedings, 2016*(1), 13074. Briarcliff Manor, NY 10510: Academy of Management.

Mannell, R. C., & Iso-Ahola, S. E. (1987). Psychological nature of leisure and tourism experience. *Annals of Tourism Research, 14*(3), 314–331.

Mano, H., & Oliver, R. L. (1993). Assessing the dimensionality and structure of the consumption experience: Evaluation, feeling, and satisfaction. *Journal of Consumer Research, 20*, 451–466.

Manstead, A. S. R., Tetlock, P. E., & Manstead, T. (1989). Cognitive appraisals and emotional experience: Further evidence. *Cognition & Emotion, 3*(3), 225–239.

Manthiou, A., Kang, J., Chiang, L., & Tang, L. (2016). Investigating the effects of memorable experiences: An extended model of script theory. *Journal of Travel & Tourism Marketing, 33*(3), 362–379. https://doi.org/10.1080/10548408.2015.1064055

Manthiou, A., Kang, J., & Hyun, S. S. (2017). An integration of cognitive appraisal theory and script theory in the luxury cruise sector: The bridging role of recollection and storytelling. *Journal of Travel & Tourism Marketing, 34*(8), 1071–1088.

Mar, R. A. (2018). Stories and the promotion of social cognition. *Current Directions in Psychological Science, 27*(4), 257–262. https://doi.org/10.1177/0963721417749654

Mari, M., & Poggesi, S. (2013). Servicescape cues and customer behavior: A systematic literature review and research agenda. *Service Industries Journal, 33*(2), 171–199.

Marschall, S. (2012). Tourism and memory. *Annals of Tourism Research, 39*(4), 2216–2219.

Martinez-Conde, S., Alexander, R. G., Blum, D., Britton, N., Lipska, B. K., Quirk, G. J., & Macknik, S. L. (2019). The storytelling brain: How neuroscience stories help bridge the gap between research and society. *Journal of Neuroscience*, *39*(42), 8285–8290. https://doi.org/10.1523/jneurosci.1180-19.2019

Mason, O., & Hargreaves, I. (2001). A qualitative study of mindfulness-based cognitive therapy for depression. *British Journal of Medical Psychology*, *74*(2), 197–212.

Massara, F., Liu, S. S., & Melara, R. D. (2010). Adapting to a retail environment: Modeling consumer–Environment interactions. *Journal of Business Research*, *63*(7), 673–681.

Massaro, D. W., & Cowan, N. (1993). Information processing models: Microscopes of the mind. *Annual Review of Psychology*, *44*(1), 383–425.

Mather, E. (2013). Novelty, attention, and challenges for developmental psychology. *Frontiers in Psychology*, *4*, 491.

Mathisen, L. (2013). Staging natural environments: A performance perspective. In *Advances in hospitality and leisure* (Vol. 9, pp. 163–183). Emerald Group Publishing Limited.

Mattila, A. S., & Ro, H. (2008). Discrete negative emotions and customer dissatisfaction responses in a casual restaurant setting. *Journal of Hospitality & Tourism Research*, *32*(1), 89–107.

Mattila, A. S., & Wirtz, J. (2001). Congruency of scent and music as a driver of in-store evaluations and behavior. *Journal of Retailing*, *77*(2), 273–289.

Mauss, I. B., & Robinson, M. D. (2009). Measures of emotion: A review. *Cognition & Emotion*, *23*(2), 209–237. https://doi.org/10.1080/02699930802204677

Mayer, F. S., Frantz, C. M., Bruehlman-Senecal, E., & Dolliver, K. (2009). Why is nature beneficial? The role of connectedness to nature. *Environment and Behavior*, *41*(5), 607–643. https://doi.org/10.1177/0013916508319745

Mayer, J. D., & Salovey, P. (1997). What is emotional intelligence? In P. Salovey & D. J. Sluyter (Eds.), *Emotional development and emotional intelligence: Educational implications* (pp. 3–34). Basic Books.

Mayr, E., Knipfer, K., & Wessel, D. (2009). In-sights into mobile learning: An exploration of mobile eye tracking methodology for learning in museums. In *Researching mobile learning: Frameworks, tools, and research designs* (pp. 189–204).

McCabe, A. S. (2000). Tourism motivation process. *Annals of Tourism Research*, *27*(4), 1049–1051. https://doi.org/10.1016/s0160-7383(99)00130-9

McCabe, S., & Foster, C. (2006). The role and function of narrative in tourist interaction. *Journal of Tourism and Cultural Change*, *4*, 194–215. https://doi.org/10.2167/jtcc071.0

McCabe, S., Li, C., & Chen, Z. (2016). Time for a radical reappraisal of tourist decision making? Toward a new conceptual model. *Journal of Travel Research*, *55*(1), 3–15.

McCabe, S., & Stokoe, E. (2010). Have you been away? Holiday talk in everyday interaction. *Annals of Tourism Research*, *37*(4), 1117–1140.

McCarroll, C. J., & Cosentino, E. (2020). Rewarding one's future self: Psychological connectedness, episodic prospection, and a puzzle about perspective. *Review of Philosophy and Psychology*, *11*(2), 449–467.

McCartney, G., & Chen, Y. (2020). Co-creation tourism in an ancient Chinese town. *Journal of China Tourism Research*, *16*(2), 159–182.

McCloskey, M., Wible, C. G., & Cohen, N. J. (1988). Is there a special flashbulb-memory mechanism? *Journal of Experimental Psychology: General*, *117*(2), 171–181.

McCormick, K. (2016). Celebrity endorsements: Influence of a product-endorser match on Millennials attitudes and purchase intentions. *Journal of Retailing and Consumer Services*, *32*, 39–45.

McGaugh, J. L. (2000). Memory–A century of consolidation. *Science*, *287*(5451), 248–251.

McGroarty, B., Ellis, S., Chapman, J., Eadon-Clarke, P., Panek, R., & Raphael, R. (2020). *Global wellnes trends report—The future of wellness 2020*. Global Wellness Summit.

McKercher, B., Okumus, F., & Okumus, B. (2008). Food tourism as a viable market segment: It's all how you cook the numbers. *Journal of Travel & Tourism Marketing*, *25*(2), 137–148.

McLeay, F., Lichy, J., & Major, B. (2019). Co-creation of the ski-chalet community experiencescape. *Tourism Management*, *74*, 413–424.

Meesters, C., Merckelbach, H., Muris, P., & Wessel, I. (2000). Autobiographical memory and trauma in adolescents. *Journal of Behavior Therapy and Experimental Psychiatry*, *31*(1), 29–39.

Mehmetoglu, M., & Engen, M. (2011). Pine and Gilmore's concept of experience economy and its dimensions: An empirical examination in tourism. *Journal of Quality Assurance in Hospitality & Tourism*, *12*(4), 237–255.

Mehrabian, A., & Russell, J. A. (1974). *An approach to environmental psychology*. MIT Press.

Mei, X. Y., Hågensen, A. M. S., & Kristiansen, H. S. (2020). Storytelling through experiencescape: Creating unique stories and extraordinary experiences in farm tourism. *Tourism and Hospitality Research*, *20*(1), 93–104. https://doi.org/10.1177/1467358418813410

Meidenbauer, K. L., Stenfors, C. U., Bratman, G. N., Gross, J. J., Schertz, K. E., Choe, K. W., & Berman, M. G. (2020). The affective benefits of nature exposure: What's nature got to do with it? *Journal of Environmental Psychology*, *72*, 101498. https://doi.org/10.1111/spc3.12630

Meier, L. L., Semmer, N. K., Elfering, A., & Jacobshagen, N. (2008). The double meaning of control: Three-way interactions between internal resources, job control, and stressors at work. *Journal of Occupational Health Psychology*, *13*(3), 244.

Meißner, M., Pfeiffer, J., Pfeiffer, T., & Oppewal, H. (2019). Combining virtual reality and mobile eye tracking to provide a naturalistic experimental environment for shopper research. *Journal of Business Research*, *100*, 445–458.

Mele, C., Spena, T. R., Kaartemo, V., & Marzullo, M. L. (2021). Smart nudging: How cognitive technologies enable choice architectures for value co-creation. *Journal of Business Research*, *129*, 949–960.

Mennella, R., Patron, E., & Palomba, D. (2017). Frontal alpha asymmetry neurofeedback for the reduction of negative affect and anxiety. *Behaviour Research and Therapy*, *92*, 32–40.

Merchant, J. (2016). The image schema and innate archetypes: Theoretical and clinical implications. *Journal of Analytical Psychology*, *61*(1), 63–78.

Messmer, D. J., & Johnson, R. R. (1993). Inquiry conversion and travel advertising effectiveness. *Journal of Travel Research*, *31*(4), 14–21. https://doi.org/10.1177/004728759303100403

Meyer, W.-U., Reisenzein, R., & Schützwohl, A. (1997). Toward a process analysis of emotions: The case of surprise. *Motivation and Emotion*, *21*(3), 251–274.

Meuter, N. (2013). Narration in various disciplines. In P. Hühn, J. Pier, W. Schmid, & J. Schönert (Eds.), *The living handbook of narratology*. Hamburg University. http://www.lhn.uni-hamburg.de/article/narration-various-disciplines

Meyers-Levy, J., & Tybout, A. M. (1989). Schema congruity as a basis for product evaluation. *Journal of Consumer Research*, *16*(1), 39–54.

Miao, L., Lehto, X., & Wei, W. (2014). The hedonic value of hospitality consumption: Evidence from spring break experiences. *Journal of Hospitality Marketing & Management*, *23*(2), 99–121.

Miao, L., Mattila, A. S., & Mount, D. (2011). Other consumers in service encounters: A script theoretical perspective. *International Journal of Hospitality Management*, *30*(4), 933–941. https://doi.org/10.1016/j.ijhm.2011.01.012

Michaelidou, N., & Dibb, S. (2008). Consumer involvement: A new perspective. *The Marketing Review*, *8*(1), 83–99.

Michalkó, G., Irimiás, A., & Timothy, D. J. (2015). Disappointment in tourism: Perspectives on tourism destination management. *Tourism Management Perspectives*, *16*, 85–91.

Michopoulou, E., Al-Qasmi, I., & Melpignano, C. (2021). Co-creating value in desert tourism experiences. *Tourism Planning & Development*, *18*(2), 245–265.

Miller, D., Hadjimarcou, J., & Miciak, A. (2000). A scale for measuring advertisement-evoked mental imagery. *Journal of Marketing Communications*, *6*(1), 1–20.

Miller, E. K., & Wallis, J. D. (2009). Executive function and higher-order cognition: Definition and neural substrates. *Encyclopedia of Neuroscience*, *4*, 99–104. https://doi.org/10.1016/B978-008045046-9.00418-6

Miller, G. A. (2003). The cognitive revolution: A historical perspective. *Trends in Cognitive Sciences*, *7*(3), 141–144.

Milstein, N., & Gordon, I. (2020). Validating measures of electrodermal activity and heart rate variability derived from the empatica E4 utilized in research settings that involve interactive dyadic states. *Frontiers in Behavioral Neuroscience*, *14*, 148–148.

Ming-Zher, P., Swenson, N. C., & Picard, R. W. (2010). A wearable sensor for unobtrusive, long-term assessment of electrodermal activity. *IEEE Transactions on Biomedical Engineering*, *57*(5), 1243–1252. https://doi.org/10.1109/TBME.2009.2038487

Minkiewicz, J., Evans, J., & Bridson, K. (2014). How do consumers co-create their experiences? An exploration in the heritage sector. *Journal of Marketing Management*, *30*(1–2), 30–59.

Mischel, W. (2008). The toothbrush problem. *Association for Psychological Science Observer*, *21*, 11.

Mitas, O., & Bastiaansen, M. (2018). Novelty: A mechanism of tourists' enjoyment. *Annals of Tourism Research*, *72*, 98–108.

Mitas, O., Cuenen, R., Bastiaansen, M., Chick, G., & van den Dungen, E. (2020). The war from both sides: How Dutch and German visitors experience an exhibit of Second World War stories. *International Journal of the Sociology of Leisure*, *3*(3), 277–303.

Mitas, O., Ward, P., Praat, A., & Altan, M. (2021). The sandwich and the beating heart 7. In *Experiences Research Summit, Breda.*

Mitchell, T. R., Thompson, L., Peterson, E., & Cronk, R. (1997). Temporal adjustments in the evaluation of events: The "rosy view". *Journal of Experimental Social Psychology*, *33*(4), 421–448.

Miyake, A., & Friedman, N. P. (2012). The nature and organization of individual differences in executive functions: Four general conclusions. *Current Directions in Psychological Science*, *21*(1), 8–14.

Miyake, A., Friedman, N. P., Emerson, M. J., Witzki, A. H., Howerter, A., & Wager, T. D. (2000). The unity and diversity of executive functions and their contributions to complex 'frontal lobe' tasks: A latent variable analysis. *Cognitive Psychology*, *41*(1), 49–100.

Mograbi, G. J. C. (2011). Neural basis of decision-making and assessment: Issues on testability and philosophical relevance. *Mens Sana Monographs*, *9*(1), 251–259.

Mok, H. M. K. (1990). A quasi-experimental measure of the effectiveness of destinational advertising: Some evidence from Hawaii. *Journal of Travel Research*, *29*(1), 30–34. https://doi.org/10.1177/004728759002900107

Mondschein, A., & Moga, S. T. (2019). New directions in cognitive-environmental research. *Journal of the American Planning Association*, *84*(3–4), 263–275.

Monsell, S., & Driver, J. (2000). *Control of cognitive processes: Attention and performance XVIII* (Vol. 18). MIT Press.

Moorhouse, N., tom Dieck, M., & Jung, T. (2017). Augmented reality to enhance the learning experience in cultural heritage tourism: An experiential learning cycle perspective. *eReview of Tourism Research*, *8*.

Moors, A., Ellsworth, P. C., Scherer, K. R., & Frijda, N. H. (2013). Appraisal theories of emotion: State of the art and future development. *Emotion Review*, *5*(2), 119–124. https://doi.org/10.1177/1754073912468165

Mordue, T., Moss, O., & Johnston, L. (2020). The impacts of onshore windfarms on a UK rural tourism landscape: Objective evidence, local opposition, and national politics. *Journal of Sustainable Tourism*, *28*(11), 1882–1904.

Morris, J. D., Woo, C., Geason, J. A., & Kim, J. (2002). The power of affect: Predicting intention. *Journal of Advertising Research*, *42*(3), 7–18.

Moscardo, G. (1996). Mindful visitors: Heritage and Tourism. *Annals of Tourism Research*, *23*(2), 376–397.

Moscardo, G. (2010). The shaping of tourist experience: The importance of stories and themes. In M. Morgan, P. Lugosi, & J. R. B. Ritchie (Eds.), *The tourism and leisure experience: Consumer and managerial perspectives* (pp. 43–58). Channel.

Moscardo, G. (2017). Stories as a tourist experience design tool. In D. Fesenmaier & Z. Xiang (Eds.), *Design science in tourism* (pp. 97–124). Springer International Publishing.

Moscardo, G. (2018). Tourist experience design, quality services and experiences in hospitality and tourism. In *Bridging tourism theory and practice* (Vol. 9, pp. 93–107). Emerald Publishing Limited.

Moscardo, G. (2020). Stories and design in tourism. *Annals of Tourism Research*, *83*, 102950. https://doi.org/10.1016/j.annals.2020.1029

Moscardo, G. (2021). The story turns in tourism: Forces and futures. *Journal of Tourism Futures*, *7*(2), 168–173. https://doi.org/10.1108/JTF-11-2019-0131

Moscardo, G., & Pearce, P. L. (1986). Visitor centres and environmental interpretation: An exploration of the relationships among visitor enjoyment, understanding and mindfulness. *Journal of Environmental Psychology*, *6*(2), 89–108. https://doi.org/10.1016/s0272-4944(86)80011-1

Mossberg, L. (2007). A marketing approach to the tourist experience. *Scandinavian Journal of Hospitality and Tourism*, *7*(1), 59–74.

Mossberg, L. (2008). Extraordinary experiences through storytelling. *Scandinavian Journal of Hospitality and Tourism*, *8*(3), 195–210.

Mossberg, L., Therkelsen, A., Huijbens, E. H., Björk, P., & Olsson, A. K. (2010). *Storytelling and destination development*. Nordic Innovation Centre.

Moyle, B. D., Moyle, C. L., Bec, A., & Scott, N. (2019). The next frontier in tourism emotion research. *Current Issues in Tourism*, *22*(12), 1393–1399.

Munro, G. D., & Ditto, P. H. (1997). Biased assimilation, attitude polarization, and affect in reactions to stereotype-relevant scientific information. *Personality and Social Psychology Bulletin*, *23*(6), 636–653.

Murayama, K., Elliot, A. J., & Friedman, R. (2012). Achievement goals. In R. M. Ryan (Ed.), *The Oxford handbook of human motivation* (pp. 191–207). Oxford University Press.

Murray, H. (1938). *Exploration and personality*. Oxford University Press.

Murray, E. A., Wise, S. P., & Graham, K. S. (2017). *The evolution of memory systems: Ancestors, anatomy, and adaptations*. Oxford University Press.

Murray, E. J. (1964). *Motivation and emotion*. Prentice-Hall.

Muschalla, B., & Schönborn, F. (2021). Induction of false beliefs and false memories in laboratory studies—A systematic review. *Clinical Psychology & Psychotherapy*, *28*(5), 1194–1209.

Nairne, J. S., & Pandeirada, J. N. (2016). Adaptive memory: The evolutionary significance of survival processing. *Perspectives on Psychological Science*, *11*(4), 496–511.

Nardini, G., & Lutz, R. (2018). How mental simulation evokes negative affective misforecasting of hedonic experiences. *Journal of Consumer Marketing*, *35*(6), 633–643.

Nave, K., Deane, G., Miller, M., & Clark, A. (2020). Wilding the predictive brain. *Wiley Interdiscipilinary Reviews of Cognitive Science*, *11*(6), e1542.

Nawijn, J., & Biran, A. (2019). Negative emotions in tourism: A meaningful analysis. *Current Issues in Tourism*, *22*(19), 2386–2398.

Nawijn, J., Mitas, O., Lin, Y., & Kerstetter, D. (2013). How do we feel on vacation? A closer look at how emotions change over the course of a trip. *Journal of Travel Research*, *52*(2), 265–274.

Negrete, A. (2021). Remembering rhythm and rhyme: Memorability of narratives for science communication. *Geoscience Communication*, *4*, 1–9. https://doi.org/10.5194/gc-4-1-2021

Nelson, S., & Spence, B. (2020). Narrative time. *Oxford Research Encyclopedia of Literature*. https://oxfordre.com/literature/view/10.1093/acrefore/9780190201098.001.0001/acrefore-9780190201098-e-1076. Accessed on January 11, 2022.

Nelson, V. (2014). Tourist identities in narratives of unexpected adventure in Madeira. *International Journal of Tourism Research*, *17*(6), 537–544. https://doi.org/10.1002/jtr.2018

Neuhofer, B., & Buhalis, D. (2014). Experience, co-creation, and technology: Issues, challenges and trends for technology enhanced tourism experiences. In *The Routledge handbook of tourism marketing* (pp. 124–139).

Neuhofer, B., Buhalis, D., & Ladkin, A. (2013). Experiences, co-creation, and technology: A conceptual approach to enhance tourism experiences. In *Tourism and global change: On the edge of something big* (pp. 546–555).

Neuhofer, B., Celuch, K., & To, T. L. (2020). Experience design and the dimensions of transformative festival experiences. *International Journal of Contemporary Hospitality Management*, *32*(9), 2881–2901.

Neuper, C., & Klimesch, W. (Eds.). (2006). *Event-related dynamics of brain oscillations*. Elsevier.

Newman, G. E. (2019). The psychology of authenticity. *Review of General Psychology*, *23*(1), 8–18.

Neys, W., & Pennycook, G. (2019). Logic, fast and slow: Advances in dual-process theorizing. *Current Directions in Psychological Science*, *28*(5), 503–509.

Neys, W. D. (2006). Dual processing in reasoning: Two systems but one reasoner. *Psychological Science*, *17*(5), 428–433.

Ngan, H. F. B., & Yu, C. E. (2019). To smile or not to smile–An eye-tracking study on service recovery. *Current Issues in Tourism*, *22*(19), 2327–2332.

Nhất Hạnh, T. C., & Laity, A. (2009). *The blooming of a lotus: Guided meditation for achieving the miracle of mindfulness*. Beacon Press.

Nickerson, R. S., & Adams, M. J. (1979). Long-term memory for a common object. *Cognitive Psychology*, *11*(3), 287–307.

Nielsen, M. B., & Knardahl, S. (2014). Coping strategies: A prospective study of patterns, stability, and relationships with psychological distress. *Scandinavian Journal of Psychology*, *55*(2), 142–150.

Nishida, H. (1999). A cognitive approach to intercultural communication based on schema theory. *International Journal of Intercultural Relations*, *23*(5), 753–777.

Nomura, K., & Akai, S. (2012). Empathy with fictional stories: Reconsideration of the fantasy scale of the interpersonal reactivity index. *Psychological Reports*, *110*(1), 304–314. https://doi.org/10.2466/02.07.09.11.pr0.110.1.304-314

Noordewier, M. K., & Breugelmans, S. M. (2013). On the valence of surprise. *Cognition & Emotion*, *27*(7), 1326–1334.

Nordbø, I., & Prebensen, N. K. (2015). Hiking as mental and physical experience. In *Advances in hospitality and leisure* (Vol. 11, pp. 169–186). Emerald Group Publishing Limited.

Norman, D. A. (1980). Twelve issues for cognitive science. *Cognitive Science*, *4*(1), 1–32.

Noronha, C., Guan, J., & Sio, S. H. I. (2021). Accounting for gaming in the time of plague: COVID-19 in Macau. *Sustainability Accounting, Management and Policy Journal*, *12*(5), 943–964.

Nosek, B. A., & Banaji, M. R. (2009). Implicit attitude. In *Oxford companion to consciousness* (pp. 84–85).

Novick, J. M., Bunting, M. F., Engle, R. W., & Dougherty, M. R. (Eds.). (2019). *Cognitive and working memory training: Perspectives from psychology, neuroscience, and human development*. Oxford University Press.

Nunez, P. L. (2000). Toward a quantitative description of large-scale neocortical dynamic function and EEG. *Behavioral and Brain Sciences*, *23*(3), 371–398.

Núñez, R., Allen, M., Gao, R., Miller Rigoli, C., Relaford-Doyle, J., & Semenuks, A. (2019). What happened to cognitive science? *Nature Human Behaviour*, *3*(8), 782–791.

Nyer, P. U. (1997). A study of the relationships between cognitive appraisals and consumption emotions. *Journal of the Academy of Marketing Science*, *25*(4), 296–304.

Obeidat, Z. M. I., Xiao, S. H., Iyer, G. R., & Nicholson, M. (2017). Consumer revenge using the internet and social media: An examination of the role of service failure types and cognitive appraisal processes. *Psychology and Marketing*, *34*(4), 496–515.

Oberauer, K. (2019). Working memory and attention—A conceptual analysis and review. *Journal of Cognition*, *2*(1), 36. https://doi.org/10.5334/joc.58

O'Connor, S., Campbell, R., Cortez, H., & Knowles, T. (2009). Whale watching worldwide: Tourism numbers, expenditures and expanding economic benefits. In *A special report from the international fund for animal welfare, Yarmouth MA, USA* (pp. 228). prepared by Economists at Large.

O'Dell, T., & Billing, P. (Eds.). (2005). *Experiencescapes; Tourism, culture and economy*. Business School Press.

O'Donnell, A. (2015). Contemplative pedagogy and mindfulness: Developing creative attention in an age of distraction. *Journal of Philosophy of Education*, *49*(2), 187–202.

Oettingen, G., & Reininger, K. M. (2016). The power of prospection: Mental contrasting and behavior change. *Social and Personality Psychology Compass*, *10*(11), 591–604.

Oh, H., Assaf, A. G., & Baloglu, S. (2016). Motivations and goals of slow tourism. *Journal of Travel Research*, *55*(2), 205–219. https://doi.org/10.1177/0047287514546228

Ohme, R., Reykowska, D., Wiener, D., & Choromanska, A. (2010). Application of frontal EEG asymmetry to advertising research. *Journal of Economic Psychology*, *31*(5), 785–793.

Okhuysen, G., & Bonardi, J. P. (2011). The challenges of building theory by combining lenses. *Academy of Management Review*, *36*(1), 6–11.

Oliver, R. (1993). Cognitive, affective, and attribute bases of the satisfaction response. *Journal of Consumer Research*, *20*, 418–430.

Olofsson, J. K., Nordin, S., Sequeira, H., & Polich, J. (2008). Affective picture processing: An integrative review of ERP findings. *Biological Psychology*, *77*(3), 247–265.

Omdahl, B. L. (2014). *Cognitive appraisal, emotion, and empathy*. Psychology Press.

Ooi, C. S. (2005). A theory of tourism experiences: The management of attention. In T. O'Dell & P. Billing (Eds.), *Experiencescapes: Tourism, culture and economy* (pp. 51–68). Copenhagen Business School Press.

Orams, M. B. (2000). Tourists getting close to whales, is it what whale-watching is all about? *Tourism Management*, *21*(6), 561–569. https://doi.org/10.1016/S0261-5177(00)00006-6

Orquin, J. L., & Loose, S. M. (2013). Attention and choice: A review on eye movements in decision making. *Acta Psychologica*, *144*(1), 190–206.

Ortony, A. E. (1993). *Metaphor and thought*. Cambridge University Press.

Ortony, A., & Clore, G. (2015). Can an appraisal model be compatible with psychological constructionism. In L. F. Barrett & J. A. Russell (Eds.), *The psychological construction of emotion* (pp. 305–333). Guildford Press.

Ortony, A., Clore, G. L., & Collins, A. (1988). *The cognitive structure of emotions*. Cambridge University Press.

Ortony, A., & Turner, T. J. (1990). What's basic about basic emotions? *Psychological Review*, *97*(3), 315.

Orús, C., Ibáñez-Sánchez, S., & Flavián, C. (2021). Enhancing the customer experience with virtual and augmented reality: The impact of content and device type. *International Journal of Hospitality Management*, *98*, 103019.

Otto, R. (1958). *The idea of the holy*. Oxford University Press.

Otto, J. E., & Ritchie, J. B. (1996). The service experience in tourism. *Tourism Management*, *17*(3), 165–174. https://doi.org/10.1016/0261-5177(96)00003-9

Overmars, S., & Poels, K. (2015). How product representation shapes virtual experiences and repatronage intentions: The role of mental imagery processing and experiential value. *International Review of Retail Distribution & Consumer Research*, *25*(3), 236–259. https://doi.org/10.1080/09593969.2014.988279

Pagel, J. F., Blagrove, M., Levin, R., Stickgold, B., & White, S. (2001). Definitions of dream: A paradigm for comparing field descriptive specific studies of dream. *Dreaming*, *11*(4), 195–202.

Palombo, D. J., Te, A. A., Checknita, K. J., & Madan, C. R. (2021). Exploring the facets of emotional episodic memory: Remembering "what," "when," and "which". *Psychological Science*, *32*(7), 1104–1114.

Pan, B., Zhang, L., & Law, R. (2013). The complex matter of online hotel choice. *Cornell Hospitality Quarterly*, *54*(1), 74–83.

Pan, S. (2011). The role of TV commercial visuals in forming memorable and impressive destination images. *Journal of Travel Research*, *50*(2), 171–185. https://doi.org/10.1177/0047287509355325

Pan, S., & Ryan, C. (2009). Tourism sense-making: The role of the senses and travel journalism. *Journal of Travel & Tourism Marketing*, *26*(7), 625–639.

Papies, E. K., Best, M., Gelibter, E., & Barsalou, L. W. (2017). The role of simulations in consumer experiences and behavior: Insights from the grounded cognition theory of desire. *Journal of the Association for Consumer Research*, *2*(4), 402–418. https://doi.org/10.1086/693110

Parisi, A. (2022). *The ecological implications of the metaverse*. [MOOC]. Coursera. https://www.coursera.org/learn/what-is-the-metaverse/lecture/osfCL?t=141

Park, S., Nicolau, J. L., & Fesenmaier, D. R. (2013). Assessing advertising in a hierarchical decision model. *Annals of Tourism Research*, *40*(1), 260–282. https://doi.org/10.1016/j.annals.2012.09.009

Park, D.-B., & Yoon, Y.-S. (2009). Segmentation by motivation in rural tourism: A Korean case study. *Tourism Management*, *30*(1), 99–108.

Park, M., & Yoo, J. (2020). Effects of perceived interactivity of augmented reality on consumer responses: A mental imagery perspective. *Journal of Retailing and Consumer Services*, *52*, 101912.

Parrinello, G. L. (2012). Tourism and neuroscience: A preliminary approach. *Tourismos: An International Multidisciplinary Journal of Tourism*, *7*(2), 39–54.

Pavlov, Y. G., & Kotchoubey, B. (2022). Oscillatory brain activity and maintenance of verbal and visual working memory: A systematic review. *Psychophysiology*, *59*(5), e13735.

Pawaskar, P., & Goel, M. (2014). A conceptual model: Multisensory marketing and destination branding. *Procedia Economics and Finance*, *11*, 255–267.

Payne, A. F., Storbacka, K., & Frow, P. (2008). Managing the co-creation of value. *Journal of the Academy of Marketing Science*, *36*(1), 83–96.

Peacock, J., Purvis, S., & Hazlett, R. L. (2011). Which broadcast medium better drives engagement? Measuring the powers of radio and television with electromyography and skin-conductance measurements. *Journal of Advertising Research*, *51*(4), 578–585.

Pearce, J., Strickland-Munro, J., & Moore, S. A. (2017). What fosters awe-inspiring experiences in nature-based tourism destinations? *Journal of Sustainable Tourism*, *25*(3), 362–378. https://doi.org/10.1080/09669582.2016.1213270

Pearce, M. T., Zaidel, D. W., Vartanian, O., Skov, M., Leder, H., Chatterjee, A., & Nadal, M. (2016). Neuroaesthetics: The cognitive neuroscience of aesthetic experience. *Perspectives on Psychological Science*, *11*(2), 265–279.

Pearce, P. (2020). Tourists' perception of time: Directions for design. *Annals of Tourism Research*, *83*, 102932.

Pearce, P. L. (2009). The relationship between positive psychology and tourist behavior studies. *Tourism Analysis*, *14*(1), 37–48.

Pearce, P. L. (2016). Australian contributions to tourist behaviour studies. *Journal of Hospitality and Tourism Management*, *26*, 84–90. https://doi.org/10.1016/j.jhtm.2016.01.004

Pearce, P. L., & Caltabiano, M. L. (1983). Inferring travel motivation from travelers' experiences. *Journal of Travel Research*, *22*(2), 16–20.

Pearce, P. L., & Lee, U. I. (2005). Developing the travel career approach to tourist motivation. *Journal of Travel Research*, *43*(3), 226–237.

Pearce, P. L., & Packer, J. (2013). Minds on the move: New links from psychology to tourism. *Annals of Tourism Research*, *40*, 386–411.

Pearce, P. L., & Wu, M.-Y. (2014). Tourists' evaluation of a romantic themed attraction: Expressive and instrumental issues. *Journal of Travel Research*, *55*(2), 220–232.

Pearce, P. L., & Zare, S. (2017). The orchestra model as the basis for teaching tourism experience design. *Journal of Hospitality and Tourism Management*, *30*, 55–64.

Pearson, P. H. (1970). Relationships between global and specified measures of novelty seeking. *Journal of Consulting and Clinical Psychology*, *34*(2), 199–204.

Peck, J., Barger, V., & Luangrath, A. (2013). In search of a surrogate for touch: The effect of haptic imagery on perceived ownership. *Journal of Consumer Psychology*, *23*(2), 189–196.

Peck, J., & Wiggins, J. (2006). It just feels good: Customers' affective response to touch and its influence on persuasion. *Journal of Marketing*, *70*(4), 56–69.

Pedro, R. M., Mendes, J., de Matos, N., & Ascenção, M. P. (2020). A conceptualization of the senses, emotions, and memories in memorable tourism experiences. In *Handbook of research on resident and tourist perspectives on travel destinations* (pp. 208–231). IGI Global.

Pee, L. G. (2020). Enhancing the learning effectiveness of ill-structured problem solving with online co-creation. *Studies in Higher Education*, *45*(11), 2341–2355.

Peeters, G., & Czapinski, J. (1990). Positive-negative asymmetry in evaluations: The distinction between affective and informational negativity effects. *European Review of Social Psychology*, *1*(1), 33–60.

Pera, R. (2017). Empowering the new traveller: Storytelling as a co-creative behaviour in tourism. *Current Issues in Tourism*, *20*(4), 331–338. https://doi.org/10.1080/13683500.2014.982520

Perez-Vega, R., Taheri, B., Farrington, T., & O'Gorman, K. (2018). On being attractive, social, and visually appealing in social media: The effects of anthropomorphic tourism brands on Facebook fan pages. *Tourism Management*, *66*, 339–347.

Perkins, H. C., & Thorns, D. C. (2001). Gazing or performing? Reflections on Urry's tourist gaze in the context of contemporary experience in the antipodes. *International Sociology*, *16*(2), 185–204.

Perugini, M., & Bagozzi, R. P. (2001). The role of desires and anticipated emotions in goal-directed behaviours: Broadening and deepening the theory of planned behaviour. *British Journal of Social Psychology*, *40*(1), 79–98. https://doi.org/10.1348/014466601164704

Pessoa, L. (2018). Understanding emotion with brain networks. *Current Opinion in Behavioral Sciences*, *19*, 19–25.

Peterson, C. K., Shackman, A. J., & Harmon-Jones, E. (2008). The role of asymmetrical frontal cortical activity in aggression. *Psychophysiology*, *45*(1), 86–92.

Petit, O., Velasco, C., & Spence, C. (2019). Digital sensory marketing: Integrating new technologies into multisensory online experience. *Journal of Interactive Marketing*, *45*, 42–61.

Peylo, C., Hilla, Y., & Sauseng, P. (2021). Cause or consequence? Alpha oscillations in visuospatial attention. *Trends in Neurosciences*, *44*(9), 705–713.

Pfister, H. R., & Böhm, G. (2008). The multiplicity of emotions: A framework of emotional functions in decision making. *Judgment and Decision Making*, *3*(1), 5–17.

Pfurtscheller, G., & Da Silva, F. L. (1999). Event-related EEG/MEG synchronization and desynchronization: Basic principles. *Clinical Neurophysiology*, *110*(11), 1842–1857.

Pham, M. T. (1998). Representativeness, relevance, and the use of feelings in decision making. *Journal of Consumer Research*, *25*(2), 144–159.

Pham, M. T., Cohen, J. B., Pracejus, J. W., & Hughes, G. D. (2001). Affect monitoring and the primacy of feelings in judgment. *Journal of Consumer Research*, *28*(2), 167–188.

Phan, K. L., Wager, T., Taylor, S. F., & Liberzon, I. (2002). Functional neuroanatomy of emotion: A meta-analysis of emotion activation studies in PET and fMRI. *NeuroImage*, *16*(2), 331–348.

Phi, G. T., & Dredge, D. (2019). Collaborative tourism-making: An interdisciplinary review of co-creation and a future research agenda. *Tourism Recreation Research*, *44*(3), 284–299.

Philiastides, M. G., Tu, T., & Sajda, P. (2021). Inferring macroscale brain dynamics via fusion of simultaneous EEG-fMRI. *Annual Review of Neuroscience*, *44*, 315–334.

Philips, B. (2017). Consumer imagination in marketing: A theoretical framework. *European Journal of Marketing*, *51*(11/12), 2138–2155.

Phillips, B. J., & McQuarrie, E. F. (2010). Narrative and persuasion in fashion advertising. *Journal of Consumer Research*, *37*(3), 368–392.

Piaget, J. (1926). *The thought and language of the child*. Harcourt, Brace, and Company.

Pieters, R., Hans, B., & Allen, D. (1995). A means-end chain approach to consumer goal structures. *International Journal of Research in Marketing*, *12*, 227–244.

Pieters, R., & Wedel, M. (2007). Goal control of attention to advertising: The Yarbus implication. *Journal of Consumer Research*, *34*(2), 224–233.

Piff, P. K., Dietze, P., Feinberg, M., Stancato, D. M., & Keltner, D. (2015). Awe, the small self, and prosocial behavior. *Journal of Personality and Social Psychology*, *108*(6), 883–899. https://doi.org/10.1037/pspi0 000018

Pillemer, D. B. (1984). Flashbulb memories of the assassination attempt on President Reagan. *Cognition*, *16*(1), 63–80.

Pine, B. J., & Gilmore, J. H. (1999). *The experience economy: Work is theatre and every business a stage*. Harvard Business School Press.

Pine, B. J., & Gilmore, J. H. (2013). The experience economy: Past, present, and future. In J. Sundbo & F. Sorensen (Eds.), *Handbook on the experience economy* (pp. 21–44). Edward Elgar Publishing.

Pinto, G., Tarchi, C., & Bigozzi, L. (2016). Development in narrative competences from oral to written stories in five-to seven-year-old children. *Early Childhood Research Quarterly*, *36*, 1–10. https://doi.org/10.1016/j.ecresq.2015.12.001

Pinto, Y., van der Leij, A. R., Sligte, I. G., Lamme, V. A., & Scholte, H. S. (2013). Bottom-up and top-down attention are independent. *Journal of Vision*, *13*(3), 1–14.

Piolino, P., Desgranges, B., Benali, K., & Eustache, F. (2002). Episodic and semantic remote autobiographical memory in ageing. *Memory*, *10*(4), 239–257.

Pizam, A., Neumann, Y., & Reichel, A. (1978). Dimensions of tourist satisfaction with a destination area. *Annals of Tourism Research*, *5*(3), 314–322. https://doi.org/10.1016/0160-7383(78)90115-9

Plass, J. L., & Kalyuga, S. (2019). Four ways of considering emotion in cognitive load theory. *Educational Psychology Review*, *31*(2), 339–359.

Plass, J. L., Moreno, R., & Brünken, R. (Eds.). (2010). *Cognitive load theory*. Cambridge University Press.

Plassmann, H., Venkatraman, V., Huettel, S., & Yoon, C. (2015). Consumer neuroscience: Applications, challenges, and possible solutions. *Journal of Marketing Research*, *52*(4), 427–435.

Platenkamp, V., & Botterill, D. (2013). Critical realism, rationality, and tourism knowledge. *Annals of Tourism Research*, *41*, 110–129.

Plato. (2003). In D. Lee (Trans.), *The Republic* (2nd ed.). Pinguin Books.

Plog, S. C. (1974). Why destination areas rise and fall in popularity. *Cornell Hotel and Restaurant Administration Quarterly*, *14*(4), 55–68.

Plutchik, R. (1980). *Emotion: A psycho evolutionary synthesis*. Harper & Row.

Plutchik, R. (2001). The nature of emotions: Human emotions have deep evolutionary roots, a fact that may explain their complexity and provide tools for clinical practice. *American Scientist*, *89*(4), 344–350.

Poels, K., & Dewitte, S. (2006). How to capture the heart? Reviewing 20 years of emotion measurement in advertising. *Journal of Advertising Research*, *46*(1), 18–37.

Pollai, M., Hoelzl, E., & Possas, F. (2010). Consumption-related emotions over time: Fit between prediction and experience. *Marketing Letters*, *21*(4), 397–411.

Ponsignon, F., Durrieu, F., & Bouzdine-Chameeva, T. (2017). Customer experience design: A case study in the cultural sector. *Journal of Service Management*, *28*(4), 763–787.

Popp-Baier, U. (2013). Narrative psychology. In A. L. C. Runehov & L. Oviedo (Eds.), *Encyclopedia of sciences and religions*. Springer. https://doi.org/10.1007/978-1-4020-8265-8_780

Posner, M. I., Snyder, C. R., & Solso, R. (2004). Attention and cognitive control. *Cognitive Psychology: Key Readings*, *205*, 55–85.

Postman, L. (1963). One-trial learning. In *Conference on verbal learning and verbal behavior 1961*, June 2, Ardsley-on-Hudson, McGraw-Hill Book Company.

Poulsson, S. H. G., & Kale, S. H. (2004). The experience economy and commercial experiences. *The Marketing Review*, *4*(3), 267–277.

Powell, R. B., Brownlee, M. T., Kellert, S. R., & Ham, S. H. (2012). From awe to satisfaction: Immediate affective responses to the Antarctic tour experience. *Polar Record*, *48*(245), 145–156. https://doi.org/10.1017/S0032247410000720

Prahalad, C. K., & Ramaswamy, V. M. (2004). Co-creation experiences: The next practice in value creation. *Journal of Interactive Marketing*, *18*(3), 5–14.

Pratt, M. A., & Sparks, B. (2014). Predicting wine tourism intention: Destination image and self-congruity. *Journal of Travel & Tourism Marketing*, *31*(4), 443–460.

Pratt, S., McCabe, S., Cortes-Jimenez, I., & Blake, A. (2010). Measuring the effectiveness of destination marketing campaigns: Comparative analysis of conversion studies. *Journal of Travel Research*, *49*(2), 179–190. https://doi.org/10.1177/0047287509336471

Prayag, G., Cohen, S., & Yan, H. (2015). Potential Chinese travellers to Western Europe: Segmenting motivations and service expectations. *Current Issues in Tourism*, *18*(8), 725–743. https://doi.org/10.1080/13683500.2013.868413

Prebensen, N. K., & Foss, L. (2011). Coping and co-creating in tourist experiences. *International Journal of Tourism Research*, *13*(1), 54–67.

Prebensen, N. K., Kim, H., & Uysal, M. (2016). Cocreation as moderator between the experience value and satisfaction relationship. *Journal of Travel Research*, *55*(7), 934–945.

Prebensen, N. K., Vittersø, J., & Dahl, T. I. (2013). Value co-creation significance of tourist resources. *Annals of Tourism Research*, *42*, 240 261.

Prebensen, N. K., & Xie, J. (2017). Efficacy of co-creation and mastering on perceived value and satisfaction in tourists' consumption. *Tourism Management*, *60*, 166–176.

Preston, A. R., & Eichenbaum, H. (2013). Interplay of hippocampus and prefrontal cortex in memory. *Current Biology*, *23*(17), 764–773. https://doi.org/10.1016/j.cub.2013.05.041

Pritchard, A., Richardson, M., Sheffield, D., & McEwan, K. (2020). The relationship between nature connectedness and Eudaimonia well-being: A meta-analysis. *Journal of Happiness Studies*, *21*(3), 1145–1167. https://doi.org/10.1007/s10902-019-00118-6

Próchniak, P. (2014). The wilderness novelty seeking scale. *Perceptual and Motor Skills*, *119*(2), 577–590.

Provis, J. M., Dubis, A. M., Maddess, T., & Carroll, J. (2013). Adaptation of the central retina for high acuity vision: Cones, the fovea, and the avascular zone. *Progress in Retinal and Eye Research*, *35*, 63–81.

Pucillo, F., & Cascini, G. (2014). A framework for user experience, needs and affordances. *Design Studies*, *35*(2), 160–179. http://doi.org/10.1016/j.destud.2013.10.001

Pullman, M. E., & Gross, M. A. (2004). Ability of experience design elements to elicit emotions and loyalty behaviors. *Decision Sciences*, *35*(3), 551–578.

Pung, J. M., Gnoth, J., & Del Chiappa, G. (2020). Tourist transformation: Towards a conceptual model. *Annals of Tourism Research*, *81*, 102885.

Purcell, A. T. (1986). Environmental perception and affect: A schema discrepancy model. *Environment and Behavior*, *18*(1), 3–30.

Pyke, G. H., Pullman, H. R., & Charnov, E. L. (1977). Optimal foraging: A selective review of theory and tests. *The Quarterly Review of Biology*, *52*, 137–154.

Pykett, J. (2018). Geography and neuroscience: Critical engagements with geography's "neural turn". *Transactions of the Institute of British Geographers*, *43*(2), 154–169.

Qiu, M., Sha, J., & Scott, N. (2021). Restoration of visitors through nature-based tourism: A systematic review, conceptual framework, and future research directions. *International Journal of Environmental Research and Public Health*, *18*(5), 2299. https://doi.org/10.3390/ijerph18052299

Rachão, S. A. S., Breda, Z., Fernandes, C., & Joukes, V. (2020). Food-and-wine experiences towards co-creation in tourism. *Tourism Review*, *76*(5), 1050–1066.

Rahman, N. H. A., Khalifah, Z., & Ismail, H. N. (2017). Addressing the importance of the sensory aspect in tourism studies—A literature review. *Advanced Science Letters*, *23*(4), 3167–3169. https://doi.org/10.1166/asl.2017.7690

Rangel-Gomez, M., & Meeter, M. (2016). Neurotransmitters and novelty: A systematic review. *Journal of Psychopharmacology*, *30*(1), 3–12.

Rasmussen, A. S., & Berntsen, D. (2013). The reality of the past versus the ideality of the future: Emotional valence and functional differences between past and future mental time travel. *Memory & Cognition*, *41*(2), 187–200.

Rather, R. A. (2020). Customer experience and engagement in tourism destinations: The experiential marketing perspective. *Journal of Travel & Tourism Marketing*, *37*(1), 15–32.

Ravaja, N. (2004). Contributions of psychophysiology to media research: Review and recommendations. *Media Psychology*, *6*(2), 193–235. https://doi.org/10.1207/s1532785xmep0602_4

Reeve, J., & Lee, W. (2014). Neuroscience and human motivation. In R. M. Ryan (Ed.), *The Oxford book on human motivation* (pp. 365–380). Oxford University Press.

Reger, R. K., Gustafson, L. T., Demarie, S. M., & Mullane, J. V. (1994). Reframing the organization: Why implementing total quality is easier said than done. *Academy of Management Review*, *19*(3), 565–584.

Reichardt, R., Polner, B., & Simor, P. (2020). Novelty manipulations, memory performance, and predictive coding: The role of unexpectedness. *Frontiers in Human Neuroscience*, *14*, 1–11.

Reisenzein, R. (2000). Exploring the strength of association between the components of emotion syndromes: The case of surprise. *Cognition & Emotion*, *14*(1), 1–38.

Reisinger, Y. (2013a). Transformation and transformational learning theory. *Transformational Tourism: Tourist Perspectives*, 17–26.

Reisinger, Y. (2013b). Connection between travel, tourism, and transformation. In Y. Reisinger (Ed.), *Transformational tourism: Tourist perspectives* (pp. 27–32). CABi.

Remmers, C., Topolinski, S., & Koole, S. L. (2016). Why being mindful may have more benefits than you realize: Mindfulness improves both explicit and implicit mood regulation. *Mindfulness*, *7*(4), 829–837.

Reyers, A., & Matusitz, J. (2012). Emotional regulation at Walt Disney World: An impression management view. *Journal of Workplace Behavioral Health*, *27*(3), 139–159.

Richards, G. (2011). Creativity and tourism: The state of the art. *Annals of Tourism Research*, *38*(4), 1225–1253.

Richards, G. (2018). Cultural tourism: A review of recent research and trends. *Journal of Hospitality and Tourism Management*, *36*, 12–21.

Richards, G. (2020). Designing creative places: The role of creative tourism. *Annals of Tourism Research*, *85*. https://doi.org/10.1016/j.annals.2020.102922

Richards, G., & Marques, L. (2012). Exploring creative tourism: Editors introduction. *Journal of Tourism Consumption and Practice*, *4*(2), 1–11.

Richardson, D. C., Griffin, N. K., Zaki, L., Stephenson, A., Yan, J., Curry, T., Noble, R., Hogan, J., Skipper, J. I., & Devlin, J. T. (2020). Engagement in video and audio narratives: Contrasting self-report and physiological measures. *Scientific Reports*, *10*, 11298. https://doi.org/10.1038/s41598-020-68253-2

Richins, M. L. (1997). Measuring emotions in the consumption experience. *Journal of Consumer Research*, *24*(2), 127–146.

Ricoeur, P. (1980). Narrative time. *Critical Inquiry*, *7*(1), 169–190.

Rihova, I., Buhalis, D., Gouthro, M. B., & Moital, M. (2018). Customer-to-customer co-creation practices in tourism: Lessons from customer-dominant logic. *Tourism Management*, *67*, 362–375.

Ritchie, B. J. R., & Hudson, S. (2009). Understanding and meeting the challenges of consumer/tourist experience research. *International Journal of Tourism Research*, *11*(2), 111–126.

Ritchie, B. J. R., Tung, V. W. S., & Ritchie, R. (2011). Tourism experience management research: Emergence, evolution, and future directions. *International Journal of Contemporary Hospitality Management*, *23*(4), 419–438.

Ro, H., Lee, S., & Mattila, A. S. (2013). An affective image positioning of Las Vegas hotels. *Journal of Quality Assurance in Hospitality & Tourism*, *14*(3), 201–217.

Ro, T., Friggel, A., & Lavie, N. (2007). Attentional biases for faces and body parts. *Visual Cognition*, *15*(3), 322–348.

Robertson, I. (1981). *Sociology*. Worth Publishers.

Robinson, J. (2005). *Deeper than reason, emotion and its role in literature, music and art*. Clarendon Press.

Robinson, M. J., Warlow, S. M., & Berridge, K. C. (2014). Optogenetic excitation of central amygdala amplifies and narrows incentive motivation to pursue one reward above another. *Journal of Neuroscience*, *34*(50), 16567–16580.

Roediger, H. L. (1990). Implicit memory: Retention without remembering. *American Psychologist*, *45*(9), 1043–1056.

Roediger, H. L., III, & Marsh, E. J. (2003). Episodic and autobiographical memory. In A. F. Healy & R. W. Proctor (Eds.), *Handbook of psychology: Experimental psychology* (Vol. 4, pp. 475–497). John Wiley & Sons Inc.

Roediger, H. L., & McDermott, K. B. (2000). Distortions of memory. In E. Tulving & F. Craik (Eds.), *The Oxford handbook of memory* (pp. 149–162). Oxford University Press.

Rohrer, T. (2005). Image schemata in the brain. In *From perception to meaning: Image schemas in cognitive linguistics* (pp. 165–196). Mouton De Guyter.

Roldan, S. M. (2017). Object recognition in mental representations: Directions for exploring diagnostic features through visual mental imagery. *Frontiers in Psychology*, *8*, 833. https://doi.org/10.3389/fpsyg.2017.00833

Roose, G., Vermeir, I., Geuens, M., & Van Kerckhove, A. (2019). A match made in heaven or down under? The effectiveness of matching visual and verbal horizons in advertising. *Journal of Consumer Psychology*, *29*(3), 411–427.

Rosbergen, E., Pieters, R., & Wedel, M. (1997). Visual attention to advertising: A segment-level analysis. *Journal of Consumer Research*, *24*(3), 305–314.

Roseman, I. J. (1984). Cognitive determinants of emotions: A structural theory. In P. Shaver (Ed.), *Emotions, relationships and health* (Vol. 5, pp. 11–36). SAGE.

Roseman, I. J. (1991). Appraisal determinants of discrete emotions. *Cognition & Emotion*, *5*(3), 161–200.

Roseman, I. J. (1996). Appraisal determinants of emotions: Constructing a more accurate and comprehensive theory. *Cognition & Emotion*, *10*(3), 241–278.

Roseman, I. J. (2001). A model of appraisal in the emotion system: Integrating theory, research, and application. In K. R. Scherer, A. Schorr, & T. Johnston (Eds.), *Appraisal processes in emotion: Theory, methods, research* (pp. 68–91). Oxford University Press.

Roseman, I. J. (2013). Appraisal in the emotion system: Coherence in strategies for coping. *Emotion Review*, *5*(2), 141–149.

Roseman, I. J., & Smith, C. A. (2001). Appraisal theory: Overview, assumptions, varieties, controversies. In K. R. Scherer, A. Schorr, & T. Johnstone (Eds.), *Appraisal processes in emotion: Theory, methods, research* (pp. 3–19). Oxford University Press.

Roseman, I. J., Spindel, M. S., & Jose, P. E. (1990). Appraisals of emotion-eliciting events: Testing a theory of discrete emotions. *Journal of Personality and Social Psychology*, *59*(5), 899–915. https://doi.org/10.1037/0022-3514.59.5.899

Rosenbaum, M. S., Fisk, R. P., & Massiah, C. (2011). An expanded servicescape perspective. *Journal of Service Management*, *22*(4), 471–490.

Rosenthal, D. (2004). Varieties of higher-order theory. In R. J. Gennaro (Ed.), *Higher-order theories of consciousness: An anthology* (pp. 17–440). John Benjamins Publishing Company.

Ross, D., & Saxena, G. (2019). Participative co-creation of archaeological heritage: Case insights on creative tourism in Alentejo, Portugal. *Annals of Tourism Research*, *79*, 102790.

Ross, D. E. L., & Iso-Ahola, S. E. (1991). Sightseeing tourists' motivation and satisfaction. *Annals of Tourism Research*, *18*(2), 226–237.

Roy, D. K., & Pentland, A. P. (2002). Learning words from sights and sounds: A computational model. *Cognitive Science*, *26*(1), 113–146.

Royo-Vela, M., & Black, M. (2020). Drone images versus terrain images in advertisements: Images' verticality effects and the mediating role of mental simulation on attitude towards the advertisement. *Journal of Marketing Communications*, *26*(1), 21–39.

Rubenstein, C. (1980). Report on how Americans view vacations. *Psychology Today*, *13*(12), 62–76.

Rubin, D. C. (2005). A basic-systems approach to autobiographical memory. *Current Directions in Psychological Science*, *14*(2), 79–83.

Ruhanen, L. M., McLennan, C.-L. J., & Moyle, B. D. (2012). Strategic issues in the Australian tourism industry: A 10-year analysis of national strategies and plans. *Asia Pacific Journal of Tourism Research*, *18*(3), 220–240. https://doi.org/10.1080/10941665.2011.640701

Rumelhart, D. E. (1984). Schemata and the cognitive system. In R. S. Wyer & T. K. Srull (Eds.), *Handbook of social cognition* (Vol. 1, pp. 161–188). Lawrence Erlbaum Associates Publishers.

Rumelhart, D. E., & Ortony, A. (1977). The representation of knowledge in memory. In R. C. Anderson, R. J. Spiro, & W. E. Montague (Eds.), *Schooling and the acquisition of knowledge*. Lawrence Erlbaum Associates, Inc.

Russell, J. A., & Pratt, G. (1980). A description of the affective quality attributed to environments. *Journal of Personality and Social Psychology*, *38*(2), 311–322.

Ruth, J. A., Brunel, F. F., & Otnes, C. C. (2002). Linking thoughts to feelings: Investigating cognitive appraisals and consumption emotions in a mixed-emotions context. *Journal of the Academy of Marketing Science*, *30*(1), 44–58.

Ryan, C. (2000). Tourist experiences, phenomeno graphic analysis, post-postivism and neural network software. *International Journal of Tourism Research*, *2*(2), 119–131.

Ryan, R. M., & Deci, E. L. (2001). On happiness and human potentials: A review of research on hedonic and eudaimonic well-being. *Annual Review of Psychology*, *52*(1), 141–166.

Ryan, R. M., Reeve, J. M., & Lee, W. (2019). Motivational neuroscience. In *The Oxford handbook of human motivation* (pp. 353–372). Oxford University Press.

Ryu, K., Lehto, X., Gordon, S., & Fu, X. (2019). Effect of a brand story structure on narrative transportation and perceived brand image of luxury hotels. *Tourism Management*, *71*, 348–363.

Salvaggio, M. (2018). The justification of reconstructive and reproductive memory beliefs. *Philosophical Studies*, *175*(3), 649–663.

Samarathunga, W. H. M. S., & Cheng, L. (2020). Tourist gaze and beyond: State of the art. *Tourism Review*, *76*(2), 344–357. https://doi.org/10.1108/tr-06-2020-0248

Sarbin, T. R. (Ed.). (1986). *Narrative psychology: The storied nature of human conduct*. Praeger Press.

Satpute, A. B., Ochsner, B., & Badre, D. (2011). The neuroscience of goal-directed behavior. In H. Aarts & A. Elliot (Eds.), *Goal-directed behavior*. NY Psychology Press.

Sayadaw, M. (1991). *Practical insight meditation: Basic and progressive stages*. Buddhist Publication Society.

Sæþórsdóttir, A. D. (2010). Planning nature tourism in Iceland based on tourist attitudes. *Tourism Geographies*, *12*(1), 25–52. https://doi.org/10.1080/14616680903493639

Schachtner, C. (2020). Storytelling as a cultural practice and life form. In *The narrative subject*. Palgrave Macmillan. https://doi.org/10.1007/978-3-030-51189-0_2

Schacter, D. L. (2013). Memory: Sins and virtues. *Annals of the New York Academy of Sciences*, *1303*(1), 56–60. https://doi.org/10.1111/nyas.12168

Schacter, D. L., & Addis, D. R. (2007). The cognitive neuroscience of constructive memory: Remembering the past and imagining the future. *Philosophical Transactions of the Royal Society B: Biological Sciences*, *362*(1481), 773–786.

Schacter, D. L., Addis, D. R., & Buckner, R. L. (2007). Remembering the past to imagine the future: The prospective brain. *Nature Reviews Neuroscience*, *8*(9), 657–661.

Schacter, D. L., Addis, D. R., & Buckner, R. L. (2008). Episodic simulation of future events. *Annals of the New York Academy of Sciences*, *1124*(1), 39–60.

Schacter, D. L., Addis, D. R., Hassabis, D., Martin, V. C., Spreng, R. N., & Szpunar, K. K. (2012). The future of memory: Remembering, imagining, and the brain. *Neuron*, *76*(4), 677–694.

Schacter, D. L., Gaesser, B., & Addis, D. R. (2013). Remembering the past and imagining the future in the elderly. *Gerontology*, *59*(2), 143–151.

Schaffer, V., & Kannis-Dymand, L. (2022). Awe. In D. Buhalis (Ed.), *Encyclopedia of tourism management and marketing* (pp. 233–236). Edward Elgar Publishing. https://doi.org/10.4337/9781800377486.awe

Schaffer, V., & Lawley, M. (2010). Sink it: But will they continue to come? Future of artificial dive tourism in Australia. In *Paper proceedings of the 2010 council for Australian University tourism and hospitality education conference. Tourism and hospitality: Challenge the limits, Hobart, Australia.*

Schank, R. C. (1990). *Tell me a story: A new look at real and artificial intelligence*. Scribner.

Schank, R. C. (1999). *Dynamic memory revisited*. Cambridge University Press.

Schank, R. C., & Abelson, R. (1977). *Scripts, plans, goals, and understanding*. Earlbaum Associates.

Schank, R. C., & Abelson, R. P. (1995). Knowledge and memory: The real story. In R. S. Wyer Jr. (Ed.), *Advances in social cognition* (Vol. 8, pp. 1–85). Lawrence Erlbaum Associates.

Scherer, K. R. (1984). On the nature and function of emotion: A component process approach. In K. R. Scherer & P. Ekman (Eds.), *Approaches to emotion* (pp. 293–317). Lawrence Erlbaum.

Scherer, K. R. (1988). Criteria for emotion-antecedent appraisal: A review. In V. Hamilton, G. Bower, & N. H. Frijda (Eds.), *Cognitive perspectives on emotion and motivation* (pp. 89–126). Kluwer Academic Publishers.

Scherer, K. R. (1993). Studying the emotion-antecedent appraisal process: An expert system approach. *Cognition & Emotion*, *7*(3–4), 325–355. https://doi.org/10.1080/02699939308409192

Scherer, K. R. (1997a). Profiles of emotion-antecedent appraisal: Testing theoretical predictions across cultures. *Cognition & Emotion*, *11*(2), 113–150.

Scherer, K. R. (1997b). The role of culture in emotion-antecedent appraisal. *Journal of Personality and Social Psychology*, *73*(5), 902–922.

Scherer, K. R. (1999). Appraisal theory. In T. Dalgleish & M. J. Power (Eds.), *Handbook of cognition and emotion* (pp. 637–663). John Wiley & Sons.

Scherer, K. R. (2001). Appraisal considered as a process of multilevel sequential checking. In K. R. Scherer, A. Schorr, & T. Johnstone (Eds.), *Appraisal processes in emotion: Theory, methods, research* (pp. 92–120). Oxford University Press.

Scherer, K. R. (2009). The dynamic architecture of emotion: Evidence for the component process model. *Cognition & Emotion*, *23*(7), 1307–1351. https://doi.org/10.1080/02699930902928969

Scherer, K. R. (2022). Theory convergence in emotion science is timely and realistic. *Cognition & Emotion*, *36*(2), 155–170. https://doi.org/10.1080/00000000.2020.000000

Scherer, K. R., & Ceschi, G. (1997). Lost luggage: A field study of emotion–antecedent appraisal. *Motivation and Emotion*, *21*(3), 211–235.

Scherer, K. R., & Moors, A. (2019). The emotion process: Event appraisal and component differentiation. *Annual Review of Psychology*, *70*, 719–745. https://doi.org/10.1146/annurev-psych-122216-011854

Scherer, K. R., Schorr, A., & Johnstone, T. (2001). *Appraisal processes in emotions: Theory, methods, research*. Oxford University Press.

Scheyvens, R., & Biddulph, R. (2018). Inclusive tourism development. *Tourism Geographies*, *20*(4), 589–609. https://doi.org/10.1080/14616688.2017.1381985

Schiff, B. (2012). The function of narrative: Toward a narrative psychology of meaning. *Narrative Works: Issues, Investigations, & Interventions*, *2*(1), 33–47.

Schimmack, U. (2021). The implicit association test: A method in search of a construct. *Perspectives on Psychological Science*, *16*(2), 396–414.

Schmidt, S. (2011). Mindfulness in east and west – Is it the same? In H. Walach, S. Schmidt, & W. B. Jonas (Eds.), *Neuroscience, consciousness and spirituality* (Vol. 1, pp. 23–38). Springer Netherlands.

Schmitt, B. (1999a). *Experiential marketing: How to get customers to sense, feel, think, act, and relate to your company and brands*. Free Press.

Schmitt, B. (1999b). Experiential marketing. *Journal of Marketing Management*, *15*(1–3), 53–67.

Schneider, D. J. (2004). *The psychology of stereotyping*. Guilford Press.

Schneider, I. K., Rutjens, B. T., Jostmann, N. B., & Lakens, D. (2011). Weighty matters: Importance literally feels heavy. *Social Psychological and Personality Science*, *2*(5), 474–478.

Schneider, K. J. (2009). *Awakening to awe: Personal stories of profound transformation*. Jason Aronson.

Schneider, K. J. (2011). Awakening to an awe-based psychology. *The Humanistic Psychologist*, *39*, 247–252. https://doi.org/10.1080/08873267.2011.592464

Schnitzer, M., & Barth, M. (2019). Does sport event satisfaction remain stable over time? *International Journal of Tourism Research*, *21*(6), 785–789.

Schomaker, J., & Meeter, M. (2018). Predicting the unknown: Novelty processing depends on expectations. *Brain Research*, *1694*, 140–148.

Schooler, J. W. (2002). Re-representing consciousness: Dissociations between experience and meta-consciousness. *Trends in Cognitive Sciences*, *6*(8), 339–344.

Schunk, D. H., & Usher, E. L. (2012). Social cognitive theory and motivation. In R. M. Ryan (Ed.), *The Oxford handbook of human motivation* (pp. 13–27). Oxford University Press.

Schwarz, N. (2011). Feelings-as-information theory. In P. Lange, A. Kruglanski, & T. Higgins (Eds.), *Handbook of theories of social psychology* (pp. 289–308). SAGE.

Schwarz, N., & Clore, G. L. (1996). Feelings and phenomenal experiences. In E. T. Higgins & A. W. Kruglanski (Eds.), *Social psychology: Handbook of basic principles* (pp. 433–465). The Guilford Press.

Schwarz, N., & Sudman, S. (Eds.). (2012). *Autobiographical memory and the validity of retrospective reports*. Springer Science & Business Media.

Scott, N. (2020). Cognitive psychology and tourism–Surfing the "cognitive wave": A perspective article. *Tourism Review*, *75*(1), 49–51. https://doi.org/10.1108/TR-06-2019-0217

Scott, N., Gao, J., & Ma, J. (2017). *Visitor experience design*. CABI.

Scott, N., Green, C., & Fairley, S. (2016). Investigation of the use of eye tracking to examine tourism advertising effectiveness. *Current Issues in Tourism*, *19*(7), 634–642. https://doi.org/10.1080/13683500.2014.1003797

Scott, N., Laws, E., & Boksberger, P. (2009). The marketing of hospitality and leisure experiences. *Journal of Hospitality Marketing & Management*, *18*(2–3), 1–12.

Scott, N., Le, D., Becken, S., & Connolly, R. M. (2020). Measuring perceived beauty of the Great Barrier Reef using eye-tracking technology. *Current Issues in Tourism*, *23*(20), 2492–2502.

Scott, N., Zhang, R., Le, D., & Moyle, B. (2019). A review of eye-tracking research in tourism. *Current Issues in Tourism*, *22*(10), 1244–1261.

Scuttari, A. (2021). Tourism experiences in motion. Mobile, visual and psychophysiological methods to capture tourists 'on the move'. *Tourism Management Perspectives*, *38*. https://doi.org/10.1016/j.tmp.2021.100825

Searle, J. R. (1990). Is the brain's mind a computer program? *Scientific American*, *262*(1), 25–31.

Searle, J. R. (1991). Cognitive science and the computer metaphor. In *Understanding the artificial: On the future shape of artificial intelligence* (pp. 127–138). Springer.

Sebastian, V. (2014). New directions in understanding the decision-making process: Neuroeconomics and neuromarketing. *Procedia-Social and Behavioral Sciences*, *127*, 758–762.

Seeler, S., Schänzel, H., & Lück, M. (2021). Sustainable travel through experienced tourists' desire for eudaemonia and immersion. *Scandinavian Journal of Hospitality and Tourism*, *21*(5), 494–513. https://doi.org/10.1080/15022250.2021.1974541

Seelos, C. (2010). Theorizing and strategizing with models: Generative models of business models (May 31, 2010). In *IESE business school working paper*. http://doi.org/10.2139/ssrn.1645200

Segal, Z., Williams, J., & Teasdale, J. D. (2002). *Mindfulness-based cognitive therapy for depression: A new approach to preventing relapse*. Guilford Press.

Seider, T. R., Fieo, R. A., O'Shea, A., Porges, E. C., Woods, A. J., & Cohen, R. A. (2016). Cognitively engaging activity is associated with greater cortical and subcortical volumes. *Frontiers in Aging Neuroscience*, *8*(94), 1–10. https://doi.org/10.3389/fnagi.2016.00094

Seligman, M. E., Railton, P., Baumeister, R. F., & Sripada, C. (2013). Navigating into the future or driven by the past. *Perspectives on Psychological Science*, *8*(2), 119–141.

Sentis, K., & Markus, H. (1986). Brand personality and self. *Advertising and Consumer Psychology*, *3*, 132–148.

Seo, Y., Li, X., Choi, Y. K., & Yoon, S. (2018). Narrative transportation and paratextual features of social media in viral advertising. *Journal of Advertising*, *47*(1), 83–95. https://doi.org/10.1080/00913367.2017.1405752

Sequeira, H., Hot, P., Silvert, L., & Delplanque, S. (2009). Electrical autonomic correlates of emotion. *International Journal of Psychophysiology*, *71*(1), 50–56.

Servidio, R., & Ruffolo, I. (2016). Exploring the relationship between emotions and memorable tourism experiences through narratives. *Tourism Management Perspectives*, *20*, 151–160. https://doi.org/10.1016/j.tmp.2016.07.010

Shahane, A. D., Lopez, R. B., & Denny, B. T. (2019). Implicit reappraisal as an emotional buffer: Reappraisal-related neural activity moderates the relationship between inattention and perceived stress during exposure to negative stimuli. *Cognitive, Affective, & Behavioral Neuroscience*, *19*(2), 355–365.

Shanahan, M. (2005). Global access, embodiment, and the conscious subject. *Journal of Consciousness Studies*, *12*(12), 46–66.

Sharma, A., Park, S., & Nicolau, J. L. (2020). Testing loss aversion and diminishing sensitivity in review sentiment. *Tourism Management*, *77*, 104020.

Sheldon, P. J. (2020). Designing tourism experiences for inner transformation. *Annals of Tourism Research*, *83*, 102935.

Shen, S., Sotiriadis, M., & Zhang, Y. (2020). The influence of smart technologies on customer journey in tourist attractions within the smart tourism management framework. *Sustainability*, *12*(10), 4157. https://doi.org/10.3390/su12104157

Sheng, M. L., Natalia, N., & Hsieh, C. Y. (2022). Reconceptualizing value creation: Exploring the role of goal congruence in the co-creation process. *Journal of Retailing and Consumer Services*, *66*, 102947.

Sherry, D. F., & Schacter, D. L. (1987). The evolution of multiple memory systems. *Psychological Review*, *94*(4), 439–454.

Shields, G. S., Sazma, M. A., McCullough, A. M., & Yonelinas, A. P. (2017). The effects of acute stress on episodic memory: A meta-analysis and integrative review. *Psychological Bulletin*, *143*(6), 636–675. https://doi.org/10.1037/bul0000100

Shin, J., & Mattila, A. S. (2018). When pizza doesn't sound as good as usual: Restrained versus unrestrained eaters' responses to gluten-free menu items. *Cornell Hospitality Quarterly*, *59*(4), 397–410.

Shin, S., Chung, N., Xiang, Z., & Koo, C. (2018). Assessing the impact of textual content concreteness on helpfulness in online travel reviews. *Journal of Travel Research*, *58*(4), 579–593. https://doi.org/10.1177/0047287518768456

Shiota, M. N., Keltner, D., & Mossman, A. (2007). The nature of awe: Elicitors, appraisals, and effects on self-concept. *Cognition & Emotion*, *21*(5), 944–963. https://doi.org/10.1080/02699930600923668

Shoval, N., Schvimer, Y., & Tamir, M. (2018). Real-time measurement of tourists' objective and subjective emotions in time and space. *Journal of Travel Research*, *57*(1), 3–16.

Shu, M., & Scott, N. (2014). Influence of social media on Chinese students' choice of an overseas study destination: An information adoption model perspective. *Journal of Travel & Tourism Marketing*, *31*(2), 286–302.

Sibicky, M., Klein, C. L., & Embrescia, E. (2021). Psychological misconceptions and their relation to students' lay beliefs of mind. *Teaching of Psychology*, *48*(2), 103–109.

Sica, E., Sisto, R., Bianchi, P., & Cappelletti, G. (2020). Inclusivity and responsible tourism: Designing a trademark for a national park area. *Sustainability*, *13*(1), 1–11. https://doi.org/10.3390/su13010013

Sie, L., Pegg, S., & Phelan, K. V. (2021). Senior tourists' self-determined motivations, tour preferences, memorable experiences, and subjective well-being: An integrative hierarchical model. *Journal of Hospitality and Tourism Management*, *47*, 237–251.

Siegel, D. J. (2009). Mindful awareness, mindsight, and neural integration. *The Humanistic Psychologist*, *37*(2), 137–158. https://doi.org/10.1080/08873260902892220

Siegel, D. J. (2010). Commentary on "integrating interpersonal neurobiology with group psychotherapy": Reflections on mind, brain, and relationships in group psychotherapy. *International Journal of Group Psychotherapy*, *60*(4), 483–485. https://doi.org/10.1521/ijgp.2010.60.4.483

Siegel, E. H., Sands, M. K., Van den Noortgate, W., Condon, P., Chang, Y., Dy, J., ... Barrett, L. F. (2018). Emotion fingerprints or emotion populations? A meta-analytic investigation of autonomic features of emotion categories. *Psychological Bulletin*, *144*(4), 343–393.

Siegel, R. D., Germer, C. K., & Olendzki, A. (2009). Mindfulness: What is it? Where did it come from? In F. Didonna (Ed.), *Clinical handbook of mindfulness* (pp. 17–35). Springer.

Siegel, W., & Ziff-Levine, W. (1990). Evaluating tourism advertising campaigns: Conversion vs. advertising tracking studies. *Journal of Travel Research*, *28*(3), 51–55. https://doi.org/10.1177/004728759002800312

Silvia, P. J. (2010). Confusion and interest: The role of knowledge emotions in aesthetic experience. *Psychology of Aesthetics, Creativity, and the Arts*, *4*(2), 75. https://doi.org/10.1037/a0017081

Simillidou, A., Christofi, M., Glyptis, L., Papatheodorou, A., & Vrontis, D. (2020). Engaging in emotional labour when facing customer mistreatment in hospitality. *Journal of Hospitality and Tourism Management*, *45*, 429–443.

Simmonds, L., Bogomolova, S., Kennedy, R., Nenycz-Thiel, M., & Bellman, S. (2020). A dual-process model of how incorporating audio-visual sensory cues in video advertising promotes active attention. *Psychology and Marketing*, *37*(8), 1057–1067.

Simon, H. A. (1956). Rational choice and the structure of the environment. *Psychological Review*, *63*, 129–138.

Simon, H. A. (1978). Rationality as process and as product of thought. *The American Economic Review*, *68*(2), 1–16.

Simon, H. A. (1988). *The sciences of the artificial.* MIT Press.

Simons, D. J., & Chabris, C. F. (1999). Gorillas in our midst: Sustained in attentional blindness for dynamic events. *Perception*, *28*, 1059–1074.

Simons, D. J., & Chabris, C. F. (2011). What people believe about how memory works: A representative survey of the U.S. population. *PLoS One*, *6*(8), e22757.

Sinay, L., Tham, A., Schaffer, V., Carter, R. W. B., & Scott, N. (2020). Does ecotourism influence environmental awareness? A methodological approach based on virtual reality and physiological responses. *Annals of Ecology and Environmental Science*, *4*(1), 1–10.

Singh, N., & Formica, S. (2007). Level of congruency in photographic representations of destination marketing organizations' websites and brochures. *Journal of Hospitality & Leisure Marketing*, *15*(3), 71–86.

Singh, N., & Hu, C. (2012). The role of incongruency and involvement in destination print advertising. *Asia Pacific Journal of Tourism Research*, *17*(4), 432–450.

Sirgy, M. J. (1991). Value-expressive versus utilitarian advertising appeals: When and why to use which appeal. *Journal of Advertising*, *20*(3), 23–33.

Sirgy, M. J., Lee, D.-J., & Grace, B. Y. (2016). Revisiting self-congruity theory in consumer behaviour: Making sense of the research so far. In C. V. Jansson-Boyd & M. J. Zawisza (Eds.), *Routledge international handbook of consumer psychology* (pp. 203–219). Routledge.

Sitzmann, T., & Bell, B. S. (2017). The dynamic effects of subconscious goal pursuit on resource allocation, task performance, and goal abandonment. *Organizational Behavior and Human Decision Processes*, *138*, 1–14. https://doi.org/10.1016/j.obhdp.2016.11.001

Siu, O. L., Lu, C. Q., & Spector, P. E. (2007). Employees' well-being in Greater China: The direct and moderating effects of general self-efficacy. *Applied Psychology*, *56*(2), 288–301.

Skandalis, A., Byrom, J., & Banister, E. (2019). Experiential marketing and the changing nature of extraordinary experiences in post-postmodern consumer culture. *Journal of Business Research*, *97*, 43–50.

Skavronskaya, L., Moyle, B., & Scott, N. (2020). The experience of novelty and the novelty of experience. *Frontiers in Psychology*, *11*, 322.

Skavronskaya, L., Moyle, B., Scott, N., & Schaffer, V. (2020). Novelty, unexpectedness, and surprise: A conceptual clarification. *Tourism Recreation Research*, *36*(4), 548–552.

Skavronskaya, L., Scott, N., Moyle, B., & Kralj, A. (2020). The psychology of novelty in memorable tourism experiences. *Current Issues in Tourism*, *23*(21), 2683–2698.

Skavronskaya, L., Scott, N., Moyle, B., Le, D., Hadinejad, A., Zhang, R., Gardiner, S., Coghlan, A., & Shakeela, A. (2017). Cognitive psychology and tourism research: State of the art. *Tourism Review*, *72*(2), 221–237.

Skinner, B. F. (1985). Cognitive science and behaviourism. *British Journal of Psychology*, *76*(3), 291–301.

Skinner, E. A., Edge, K., Altman, J., & Sherwood, H. (2003). Searching for the structure of coping: A review and critique of category systems for classifying ways of coping. *Psychological Bulletin*, *129*(2), 216–269.

Śliwa, M., & Riach, K. (2012). Making scents of transition: Smellscapes and the everyday in 'old' and 'new' urban Poland. *Urban Studies*, *49*(1), 23–41.

Small, J., Darcy, S., & Packer, T. (2012). The embodied tourist experiences of people with vision impairment: Management implications beyond the visual gaze. *Tourism Management*, *33*(4), 941–950. https://doi.org/10.1016/j.tourman.2011.09.015

Smallwood, J., & Schooler, J. W. (2006). The restless mind. *Psychological Bulletin*, *132*(6), 946–958.

Smith, C. A., & Ellsworth, P. C. (1985). Patterns of cognitive appraisal in emotion. *Journal of Personality and Social Psychology*, *48*(4), 813–838.

Smith, E. R., & DeCoster, J. (2000). Dual-process models in social and cognitive psychology: Conceptual integration and links to underlying memory systems. *Personality and Social Psychology Review*, *4*(2), 108–131.

Smith, M., & Puczkó, L. (2014). *Health, tourism and hospitality: Spas, wellness and medical travel.* Routledge.

Smith, M. K., & Diekmann, A. (2017). Tourism and wellbeing. *Annals of Tourism Research*, *66*, 1–13. https://doi.org/10.1016/j.annals.2017.05.006

Smith, S., & Costello, C. (2009). Segmenting visitors to a culinary event: Motivations, travel behavior, and expenditures. *Journal of Hospitality Marketing & Management*, *18*(1), 44–67. https://doi.org/10.1080/19368620801989022

Snepenger, D., King, J., Marshall, E., & Uysal, M. (2006). Modeling Iso-Ahola's motivation theory in the tourism context. *Journal of Travel Research*, *45*(2), 140–149.

Sniehotta, F. F., Presseau, J., & Araújo-Soares, V. (2014). Time to retire the theory of planned behaviour. *Health Psychology Review*, *8*(1), 1–7. https://doi.org/10.1080/17437199.2013.869710

Sole, D., & Wilson, D. G. (2002). Storytelling in organizations: The power and traps of using stories to share knowledge in organizations. *LILA, Harvard, Graduate School of Education*, *9*(1), 1–12.

Somerfield, M. R., & McCrae, R. R. (2000). Stress and coping research: Methodological challenges, theoretical advances, and clinical applications. *American Psychologist*, *55*(6), 620.

Somervuori, O., & Ravaja, N. (2013). Purchase behavior and psychophysiological responses to different price levels. *Psychology and Marketing*, *30*(6), 479–489.

Song, H., Dwyer, L., Li, G., & Cao, Z. (2012). Tourism economics research: A review and assessment. *Annals of Tourism Research*, *39*(3), 1653–1682.

Song, H., Finn, E. S., & Rosenberg, M. D. (2021). Neural signatures of attentional engagement during narratives and its consequences for event memory. *Proceedings of the National Academy of Sciences*, *118*(33), e2021905118. https://doi.org/10.1073/pnas.2021905118

Song, H., You, G.-J., Reisinger, Y., Lee, C.-K., & Lee, S.-K. (2014). Behavioral intention of visitors to an oriental medicine festival: An extended model of goal directed behavior. *Tourism Management*, *42*(0), 101–113.

Sonnemans, J., & Frijda, N. H. (1995). The determinants of subjective emotional intensity. *Cognition & Emotion*, *9*(5), 483–506.

Soscia, I. (2007). Gratitude, delight, or guilt: The role of consumers' emotions in predicting post consumption behaviors. *Psychology and Marketing*, *24*(10), 871–894.

Spector, F., & Maurer, D. (2009). Synesthesia: A new approach to understanding the development of perception. *Developmental Psychology*, *45*(1), 175–189.

Spielmann, N. (2016). Is it all or nothing? Testing schema congruity and typicality for products with country origin. *Journal of Business Research*, *69*(3), 1130–1137.

Spielmann, N., & Mantonakis, A. (2018). In virtuo: How user-driven interactivity in virtual tours leads to attitude change. *Journal of Business Research*, *88*, 255–264.

Sporns, O., & Betzel, R. F. (2016). Modular brain networks. *Annual Review of Psychology*, *67*, 613–640. https://doi.org/10.1146/annurev-psych-122414-033634

Spreng, R. N., Mar, R. A., & Kim, A. S. N. (2008). The common neural basis of autobiographical memory, prospection, navigation, theory of mind, and the default mode: A quantitative meta-analysis. *Journal of Cognitive Neuroscience*, *21*(3), 489–510.

Squire, L. R., & Zola, S. M. (1996). Structure and function of declarative and nondeclarative memory systems. *Proceedings of the National Academy of Sciences*, *93*(24), 13515–13522.

Stancato, D. M., & Keltner, D. (2021). Awe, ideological conviction, and perceptions of ideological opponents. *Emotion*, *21*(1), 61.

Stankov, U., Filimonau, V., Gretzel, U., & Vujičić, M. D. (2020). E-mindfulness–The growing importance of facilitating tourists' connections to the present moment. *Journal of Tourism Futures*, *6*(3), 239–245.

Stankov, U., Filimonau, V., & Vujičić, M. D. (2020). A mindful shift: An opportunity for mindfulness-driven tourism in a post-pandemic world. *Tourism Geographies*, *22*(3), 703–712. https://doi.org/10.1080/14616688.2020.1768432

Stasiak, A. (2019). Creating tourism experiences—Theoretical foundations. In D. Jelincic & Y. Mansfeld (Eds.), *Creating and managing experiences in cultural tourism* (pp. 17–41). World Scientific.

Statler, I. C. (2015). How do we have feelings? *Journal of Behavioral and Brain Science*, *5*(11), 471–489.

Steffen, P. R., Hedges, D., & Matheson, R. (2022). The brain is adaptive not triune: How the brain responds to threat, challenge, and change. *Front Psychiatry*, *13*, 802606. https://doi.org/10.3389/fpsyt.2022.802606

Steffensen, M. S., & Joag-Dev, C. (1984). Cultural knowledge and reading. *Reading in a Foreign Language*, 48–61.

Sthapit, E., & Coudounaris, D. N. (2018). Memorable tourism experiences: Antecedents and outcomes. *Scandinavian Journal of Hospitality and Tourism*, *18*(1), 72–94. https://doi.org/10.1080/15022250.2017.1287003

Stein, N. L., & Albro, E. R. (1997). Building complexity and coherence: Children's use of goal-structured knowledge in telling storie. In M. Bamberg (Ed.), *Narrative development: Six approaches* (Vol. 1, pp. 5–44). Lawrence Erlbaum.

Steinvorth, S., Corkin, S., & Halgren, E. (2006). Ecphory of autobiographical memories: An fMRI study of recent and remote memory retrieval. *NeuroImage*, *30*(1), 285–298.

Stephens, G. J., Silbert, L. J., & Hasson, U. (2010). Speaker-listener neural coupling underlies successful communication. *Proceedings of the National Academy of Sciences*, *107*(32), 14425–14430. https://doi.org/10.1073/pnas.1008662107

Stephenson, E., & DeLongis, A. (2020). Coping strategies. In *The Wiley encyclopaedia of health psychology* (pp. 55–60). John Wiley & Sons, Incorporated.

Stern, L. D., Marrs, S., Millar, M. G., & Cole, E. (1984). Processing time and the recall of inconsistent and consistent behaviors of individuals and groups. *Journal of Personality and Social Psychology*, *47*(2), 253–262.

Sternberg, R. J., & Sternberg, K. (2011). *Cognitive psychology* (6th ed.). Cengage Learning.

Sthapit, E. (2019). Memories of gastronomic experiences, savoured positive emotions and savouring processes. *Scandinavian Journal of Hospitality and Tourism*, *19*(2), 115–139.

Sthapit, E., Björk, P., & Coudounaris, D. N. (2017). Emotions elicited by local food consumption, memories, place attachment and behavioural intentions. *Anatolia*, *28*(3), 363–380.

Sthapit, E., Björk, P., Jiménez-Barreto, J., & Stone, M. J. (2021). Spillover effect, positive emotions, and savouring processes: Airbnb guests' perspective. *Anatolia*, *32*(1), 33–45.

Stienmetz, J. L., Maxcy, J. G., & Fesenmaier, D. R. (2015). Evaluating destination advertising. *Journal of Travel Research*, *54*(1), 22–35. https://doi.org/10.1177/0047287513514295

Still, A., & Costall, A. (Eds.). (1991). *Against cognitivism: Alternative foundations for cognitive psychology*. Harvester Wheatsheaf.

Stokols, D. (1978). Environmental psychology. *Annual Review of Psychology*, *29*, 253–295.

Strahan, E. J., Spencer, S. J., & Zanna, M. P. (2002). Subliminal priming and persuasion: Striking while the iron is hot. *Journal of Experimental Social Psychology*, *38*(6), 556–568.

Strauss, A., & Corbin, J. (1997). *Grounded theory in practice*. SAGE Publications.

Strijbosch, W., Mitas, O., van Blaricum, T., Vugts, O., Govers, C., Hover, M., Gelissen, J., & Bastiaansen, M. (2021). When the parts of the sum are greater than the whole: Assessing the peak-and-end-theory for a heterogeneous, multi-episodic tourism experience. *Journal of Destination Marketing & Management*, *20*, 100607.

Strijbosch, W., Vessel, E. A., Welke, D., Mitas, O., Gelissen, J., & Bastiaansen, M. (2022). On the neuronal dynamics of aesthetic experience: Evidence from electroencephalographic oscillatory dynamics. *Journal of Cognitive Neuroscience*, *34*(3), 461–479.

Strizhakova, Y., & Tsarenko, Y. (2010). Consumer response to service failures: The role of emotional intelligence and coping. *Advances in Consumer Research*, *37*, 304–311.

Strizhakova, Y., Tsarenko, Y., & Ruth, J. A. (2012). "I'm mad and I can't get that service failure off my mind" coping and rumination as mediators of anger effects on customer intentions. *Journal of Service Research*, *15*(4), 414–429.

Stronza, A. (2001). Anthropology of tourism: Forging new ground for ecotourism and other alternatives. *Annual Review of Anthropology*, *30*(1), 261–283.

Stroop, J. R. (1935). Studies of interference in serial verbal reactions. *Journal of Experimental Psychology*, *18*(6), 643–662.

Stumpf, C., & Baum, M. (2016). Customer referral reward–brand–fit: A schema congruity perspective. *Psychology and Marketing*, *33*(7), 542–558.

Su, L., Cheng, J., & Swanson, S. R. (2020). The impact of tourism activity type on emotion and storytelling: The moderating roles of travel companion presence and relative ability. *Tourism Management*, *81*, 104138.

Su, L., & Hsu, M. K. (2013). Service fairness, consumption emotions, satisfaction, and behavioral intentions: The experience of Chinese heritage tourists. *Journal of Travel & Tourism Marketing*, *30*(8), 786–805.

Su, L., Hsu, M. K., & Boostrom, R. E., Jr. (2020). From recreation to responsibility: Increasing environmentally responsible behavior in tourism. *Journal of Business Research*, *109*, 557–573.

Su, L., Yang, X., & Huang, Y. (2022). Tourists' goal-directed behaviors: The influences of goal disclosure, goal commitment, and temporal distance. *Journal of Travel Research*, *61*(4), 940–960.

Su, X., & Zhang, H. (2022). Tea drinking and the tastescapes of wellbeing in tourism. *Tourism Geographies*, *24*(6–7), 1061–1081.

Suddendorf, T., & Corballis, M. C. (2007). The evolution of foresight: What is mental time travel, and is it unique to humans? *Behavioral and Brain Sciences*, *30*(3), 299–313.

Sujan, M., Bettman, J. R., & Sujan, H. (1986). Effects of consumer expectations on information processing in selling encounters. *Journal of Marketing Research*, *23*(4), 346–353.

Sun, J., & Lv, X. (2021). Feeling dark, seeing dark: Mind–body in dark tourism. *Annals of Tourism Research*, *86*, 103087.

Suntikul, W., & Jachna, T. (2016). The co-creation/place attachment nexus. *Tourism Management*, *52*, 276–286.

Sutter, R., & Kaplan, P. W. (2017). The history of continuous EEG monitoring. In A. Husain & S. Sinha (Eds.), *Continuous EEG monitoring* (pp. 3–12). Springer.

Swarbrooke, J., & Horner, S. (2004). *Consumer behaviour in tourism*. Butterworth, Heinemann.

Swartz, J. J., Braxton, D., & Viera, A. J. (2011). Calorie menu labeling on quick-service restaurant menus: An updated systematic review of the literature. *International Journal of Behavioral Nutrition and Physical Activity*, *8*, 1–8.

Sweller, J., Van Merrienboer, J. J. G., & Paas, F. (2019). Cognitive architecture and instructional design: 20 years later. *Educational Psychology Review*, *31*(2), 261–292.

Szpunar, K., Addis, D., & Schacter, D. (2012). Memory for emotional simulations: Remembering a rosy future. *Psychological Science*, *23*(1), 24–29.

Szpunar, K., & McDermott, K. (2008). Episodic future thought and its relation to remembering: Evidence from ratings of subjective experience. *Consciousness and Cognition*, *17*(1), 330–334.

Szpunar, K. K., & Schacter, D. L. (2013). Get real: Effects of repeated simulation and emotion on the perceived plausibility of future experiences. *Journal of Experimental Psychology: General*, *142*(2), 323–327. https://doi.org/10.1037/a0028877

Szpunar, K. K., Spreng, R. N., & Schacter, D. L. (2014). A taxonomy of prospection: Introducing an organizational framework for future-oriented cognition. *Proceedings of the National Academy of Sciences*, *111*(52), 18414–18421.

Tajzadeh-Namin, A. (2012). A review on value creation in tourism industry. *Management Science Letters*, *2*(1), 203–212.

Tan, S. K., Luh, D. B., & Kung, S. F. (2014). A taxonomy of creative tourists in creative tourism. *Tourism Management*, *42*, 248–259.

Tanaka, K., Dam, H. C., Kobayashi, S., Hashimoto, T., & Ikeda, M. (2016). Learning how to learn through experiential learning promoting metacognitive skills to improve knowledge co-creation ability. *Procedia Computer Science*, *99*, 146–156.

Tang, L. (2014). The application of social psychology theories and concepts in hospitality and tourism studies: A review and research agenda. *International Journal of Hospitality Management*, *36*, 188–196.

Tapper, A. R., & Molas, S. (2020). Midbrain circuits of novelty processing. *Neurobiology of Learning and Memory*, *176*, 107323. https://doi.org/10.1016/j.nlm.2020.107323

Tasci, A. D., & Godovykh, M. (2021). An empirical modeling of transformation process through trip experiences. *Tourism Management*, *86*, 104332.

Tasci, A. D., & Ko, Y. J. (2016). A fun-scale for understanding the hedonic value of a product: The destination context. *Journal of Travel & Tourism Marketing*, *33*(2), 162–183.

Taylor, E. W. (2001). Transformative learning theory: A neurobiological perspective of the role of emotions and unconscious ways of knowing. *International Journal of Lifelong Education*, *20*(3), 218–236.

Taylor, P. M., & Uchida, Y. (2019). Awe or horror: Differentiating two emotional responses to schema incongruence. *Cognition & Emotion*, *33*(8), 1548–1561. https://doi.org/10.1080/02699931.2019.1578194

Taylor, S. E., & Crocker, J. (1981). Schematic bases of social information processing. In *Social cognition* (pp. 89–134). Routledge.

Taylor, S. E., & Stanton, A. L. (2007). Coping resources, coping processes, and mental health. *Annual Review of Clinical Psychology*, *3*, 377–401.

Teichert, T. A., & Schöntag, K. (2010). Exploring consumer knowledge structures using associative network analysis. *Psychology and Marketing*, *27*(4), 369–398.

Teixeira, T., & Wedel, M. (2012). Emotion-induced engagement in Internet video advertisements. *Journal of Marketing Research*, *49*(2), 144–159.

Teoh, M. W., Wang, Y., & Kwek, A. (2019). Coping with emotional labor in high stress hospitality work environments. *Journal of Hospitality Marketing & Management*, *28*(8), 883–904.

Tercia, C., Teichert, T., Sirad, D., & Soehadi, A. (2020). Conveying pre-visit experiences through travel advertisements and their effects on destination decisions. *Journal of Destination Marketing & Management*, *16*, 100331.

Thaler, R. H., & Sunstein, C. R. (2021). *Nudge*. Yale University Press.

Thích, N. H. (1976). *The miracle of mindfulness*. Beacon.

Thích, N. H. (1988). *The sun my heart: From mindfulness to insight contemplation*. Parallax Press.

Thiel, C. E., Connelly, S., & Griffith, J. A. (2011). The influence of anger on ethical decision making: Comparison of a primary and secondary appraisal. *Ethics & Behavior*, *21*(5), 380–403.

Thomas, M. S., Ansari, D., & Knowland, V. C. (2019). Annual research review: Educational neuroscience: Progress and prospects. *Journal of Child Psychology and Psychiatry*, *60*(4), 477–492.

Thorndyke, P. W. (1977). Cognitive structures in comprehension and memory of narrative discourse. *Cognitive Psychology*, *9*, 77–110.

Thorndyke, P. W., & Hayes-Roth, B. (1979). The use of schemata in the acquisition and transfer of knowledge. *Cognitive Psychology*, *11*(1), 82–106.

Tkaczynski, A., & Rundle-Thiele, S. R. (2011). Event segmentation: A review and research agenda. *Tourism Management*, *32*(2), 426–434.

Tomej, K., & Xiang, Z. (2020). Affordances for tourism service design. *Annals of Tourism Research*, *85*, 103029. https://doi.org/10.1016/j.annals.2020.103029

Tomkins, S. S. (1962). Affect, imagery, consciousness. In *The positive affects* (Vol. 1). Springer.

Tomkins, S. S. (1978). *Script theory: Differential magnification of affects*. University of Nebraska Press.

Tong, E. M. (2015). Differentiation of 13 positive emotions by appraisals. *Cognition & Emotion*, *29*(3), 484–503.

Tooby, J., & Cosmides, L. (1990). The past explains the present: Emotional adaptations and the structure of ancestral environments. *Ethology and Sociobiology*, *11*(4–5), 375–424.

Tosun, C., Ozdemir, S., & Cubuk, F. (2016). Usage of neuro-tourism methods in hotel preferences of the consumers. In *The 2016 WEI international academic conference proceedings* (pp. 19–26). The West East Institute, Boston.

Tran, X., & Ralston, L. (2006). Tourist preferences: Influence of unconscious needs. *Annals of Tourism Research*, *33*(2), 424–441. https://doi.org/10.1016/j.annals.2005.10.014

Trauer, B. B. (2007). *Special interest tourism: The nature of adventure tourism experiences*. Ph.D. Dissertation/Thesis, The University of Queensland, School of Tourism Brisbane.

Tribe, J. (1997). The indiscipline of tourism. *Annals of Tourism Research*, *24*(3), 638–657.

Tribe, J. (2010). Tribes, territories, and networks in the tourism academy. *Annals of Tourism Research*, *37*(1), 7–33.

Tribe, J., & Liburd, J. J. (2016). The tourism knowledge system. *Annals of Tourism Research*, *57*, 44–61.

Trope, Y., & Liberman, N. (2010). Construal-level theory of psychological distance. *Psychological Review*, *117*(2), 440–463.

Troy, A. S., Wilhelm, F. H., Shallcross, A. J., & Mauss, I. B. (2010). Seeing the silver lining: Cognitive reappraisal ability moderates the relationship between stress and depressive symptoms. *Emotion*, *10*(6), 783–795.

Tryon, W. W. (2014). *Cognitive neuroscience and psychotherapy: Network principles for a unified theory*. Academic Press.

Tsaur, S.-H., & Lo, P.-C. (2020). Measuring memorable dining experiences and related emotions in fine dining restaurants. *Journal of Hospitality Marketing & Management*, *29*(8), 887–910.

Tse, W. T. S., & Tung, V. W. S. (2020). Assessing explicit and implicit stereotypes in tourism: Self-reports and implicit association test. *Journal of Sustainable Tourism*, 1–24.

Tubillejas-Andrés, B., Cervera-Taulet, A., & Calderón García, H. (2020). How emotional response mediates servicescape impact on post consumption outcomes: An application to opera events. *Tourism Management Perspectives*, *34*, 100660.

Tuerlan, T., Li, S., & Scott, N. (2021). Customer emotion research in hospitality and tourism: Conceptualization, measurements, antecedents, and consequences. *International Journal of Contemporary Hospitality Management*, *33*(8), 2741–2772.

Tulving, E. (1972). Episodic and semantic memory. In E. Tulving & W. Donaldson (Eds.), *Organization of memory* (pp. 381–403). Academic Press.

Tulving, E. (1974). Cue-dependent forgetting: When we forget something we once knew, it does not necessarily mean that the memory trace has been lost; it may only be inaccessible. *American Scientist*, *62*(1), 74–82.

Tulving, E. (1985). *Elements of episodic memory*. Oxford University Press.

Tulving, E., & Kroll, N. (1995). Novelty assessment in the brain and long-term memory encoding. *Psychonomic Bulletin & Review*, *2*(3), 387–390.

Tummers, L. L., Bekkers, V., Vink, E., & Musheno, M. (2015). Coping during public service delivery: A conceptualization and systematic review of the literature. *Journal of Public Administration Research and Theory*, *25*(4), 1099–1126.

Tung, V. W. S., Lin, P., Qiu Zhang, H., & Zhao, A. (2017). A framework of memory management and tourism experiences. *Journal of Travel & Tourism Marketing*, *34*(7), 853–866.

Tung, V. W. S., & Ritchie, J. B. (2011). Exploring the essence of memorable tourism experiences. *Annals of Tourism Research*, *38*(4), 1367–1386.

Tussyadiah, I. (2020). A review of research into automation in tourism: Launching the annals of tourism research curated collection on artificial intelligence and robotics in tourism. *Annals of Tourism Research*, *81*, 102883.

Tussyadiah, I., Park, S., & Fesenmaier, D. (2011). Assessing the effectiveness of consumer narratives for destination marketing. *Journal of Hospitality & Tourism Research*, *35*(1), 64–78.

Tussyadiah, I., Wang, D., Jung, T., & Tom Dieck, M. C. (2018). Virtual reality, presence, and attitude change: Empirical evidence from tourism. *Tourism Management*, *66*, 140–154.

Tussyadiah, I. P. (2014). Toward a theoretical foundation for experience design in tourism. *Journal of Travel Research*, *53*(5), 543–564.

Tussyadiah, I. P., & Fesenmaier, D. R. (2008). Marketing places through first-person stories—An analysis of Pennsylvania road tripper blog. *Journal of Travel & Tourism Marketing*, *25*(3–4), 299–311.

Tussyadiah, I. P., Fesenmaier, D. R., & Yoo, Y. (2008). Designing interactions in tourism mediascape–Identification of patterns for mobile 2.0 platform. In P. O'Connor, W. Hopken, & U. Gretzel (Eds.), *Information and communication technologies in tourism* (pp. 395–406). Springer.

Tussyadiah, I. P., Jung, T. H., & Dieck, M. C. T. (2018). Embodiment of wearable augmented reality technology in tourism experiences. *Journal of Travel Research*, *57*(5), 597–611. https://doi.org/10.1177/0047287517709090

Tversky, B., & Zacks, J. M. (2013). Event perception. In D. Reisberg (Ed.), *The Oxford handbook of cognitive psychology* (pp. 83–94). Oxford University Press.

Uhrich, S., Koenigstorfer, J., & Groeppel-Klein, A. (2014). Leveraging sponsorship with corporate social responsibility. *Journal of Business Research*, *67*(9), 2023–2029.

Ulker-Demirel, E., & Ciftci, G. (2020). A systematic literature review of the theory of planned behavior in tourism, leisure, and hospitality management research. *Journal of Hospitality and Tourism Management*, *43*, 209–219.

Ulrich, R. S. (1983). Aesthetic and affective response to natural environment. In I. Altman & J. F. Wohlwill (Eds.), *Behavior and the natural environment. Human behavior and environment* (Vol. 6, pp. 85–125). Springer.

UNWTO. (2020). *Join us on the 2030 journey: Tourism 4 SDGS*. https://www.unwto.org/tourism4sdgs

UNWTO. (2022). *Tourism 4 SDGS*. https://www.unwto.org/tourism4sdgs

Urbany, J. E., Dickson, P. R., & Wilkie, W. L. (1989). Buyer uncertainty and information search. *Journal of Consumer Research*, *16*(2), 208–215.

Uriely, N. (2005). The tourist experience: Conceptual developments. *Annals of Tourism Research*, *32*(1), 199–216.

Urry, J. (1990). *The tourist Gaze: Leisure and travel in contemporary societies*. SAGE Publications.

Urry, J. (1992). The tourist gaze "revisited". *American Behavioral Scientist*, *36*(2), 172–186.

Urry, J., & Larsen, J. (2011). *The tourist gaze 3.0*. SAGE.

Uusberg, A., Taxer, J. L., Yih, J., Uusberg, H., & Gross, J. J. (2019). Reappraising reappraisal. *Emotion Review*, *11*(4), 267–282. https://doi.org/10.1177/1754073919862617

Vaccaro, A. G., Scott, B., Gimbel, S. I., & Kaplan, J. T. (2021). Functional brain connectivity during narrative processing relates to transportation and story influence. *Frontiers in Human Neuroscience*, *351*, 665319. https://doi.org/10.3389/fnhum.2021.665319

Vada, S., Prentice, C., & Hsiao, A. (2019). The role of positive psychology in tourists' behavioural intentions. *Journal of Retailing and Consumer Services*, *51*, 293–303.

Vakratsas, D., & Ambler, T. (1999). How advertising works: What do we really know? *Journal of Marketing*, *63*(1), 26–43.

Valentine, P. S., Birtles, A., Curnock, M., Arnold, P., & Dunstan, A. (2004). Getting closer to whales – Passenger expectations and experiences, and the management of swim with dwarf minke whale interactions in the Great Barrier Reef. *Tourism Management*, *25*(6), 647–655. https://doi.org/10.1016/j.tourman.2003.09.001

Van Laer, T., De Ruyter, K., Visconti, L., & Wetzels, M. (2014). The extended transportation-imagery model: A meta-analysis of the antecedents and consequences of consumers' narrative transportation. *Journal of Consumer Research*, *40*(5), 797–817. https://doi.org/10.1086/673383

Van Laer, T., Feiereisen, S., & Visconti, L. (2019). Storytelling in the digital era: A meta-analysis of relevant moderators of the narrative transportation effect. *Journal of Business Research*, *96*, 135–146.

Van Yperen, N. W., & Hagedoorn, M. (2003). Do high job demands increase intrinsic motivation or fatigue or both? The role of job control and job social support. *Academy of Management Journal*, *46*(3), 339–348.

Varela, F., Lachaux, J. P., Rodriguez, E., & Martinerie, J. (2001). The brainweb: Phase synchronization and large-scale integration. *Nature Reviews Neuroscience*, *2*(4), 229–239.

Vargo, S. L., & Lusch, R. F. (2004). Evolving to a new dominant logic for marketing. *Journal of Marketing*, *68*, 1–17.

Vargo, S. L., & Lusch, R. F. (2008). Service-dominant logic: Continuing the evolution. *Journal of the Academy of Marketing Science*, *36*(1), 1–10.

Vecchio, A., & De Pascalis, V. (2020). EEG resting asymmetries and frequency oscillations in approach/avoidance personality traits: A systematic review. *Symmetry*, *12*(10), 1712.

Verhoef, P. C., Lemon, K. N., Parasuraman, A., Roggeveen, A., Tsiros, M., & Schlesinger, L. A. (2009). Customer experience creation: Determinants, dynamics, and management strategies. *Journal of Retailing*, *85*(1), 31–41. https://doi.org/10.1016/j.jretai.2008.11.001

Verleye, K. (2015). The co-creation experience from the customer perspective: Its measurement and determinants. *Journal of Service Management*, *26*(2), 321–342.

Verschoor, S., Weidema, M., Biro, S., & Hommel, B. (2010). Where do action goals come from? Evidence for spontaneous action-effect binding in infants. *Frontiers in Psychology*, *1*, 201. https://doi.org/10.3389/fpsyg.2010.00201

Vione, K. C., & Kotera, Y. (2021). Mindfulness-based approaches for COVID-19 mental health in working from home. *International Journal of Mental Health and Addiction*, 1–7. https://doi.org/10.1007/s11469-021-00647-3

Völlm, B. A., Taylor, A. N., Richardson, P., Corcoran, R., Stirling, J., McKie, S., … Elliott, R. (2006). Neuronal correlates of theory of mind and empathy: A functional magnetic resonance imaging study in a nonverbal task. *NeuroImage*, *29*(1), 90–98. https://doi.org/10.1016/j.neuroimage.2005.07.022

Volo, S. (2009). Conceptualizing experience: A tourist based approach. *Journal of Hospitality Marketing & Management*, *18*(2–3), 111–126.

Volo, S. (2017). Emotions in tourism: From exploration to design. In D. R. Fesenmaier & Z. Xiang (Eds.), *Design science in tourism: Foundations of destination management* (pp. 29–40). Springer International Publishing.

Volo, S. (2021). The experience of emotion: Directions for tourism design. *Annals of Tourism Research*, *86*, 103097.

Vom Lehn, D. (2006). Embodying experience. *European Journal of Marketing*, *40*(11/12), 1340–1359.

Walker, B., & Olson, J. (1997). The activated self in consumer behavior: A cognitive structure perspective. *Research in Consumer Behavior*, *8*, 135–172.

Walla, P., Koller, M., Brenner, G., & Bosshard, S. (2017). Evaluative conditioning of established brands: Implicit measures reveal other effects than explicit measures. *Journal of Neuroscience, Psychology, and Economics*, *10*(1), 24–41.

Wallace, B. A. (1999). The Buddhist tradition of Samatha: Methods for refining and examining consciousness. *Journal of Consciousness Studies*, *6*(2–3), 175–187.

Wallace, B. A. (2006). *The attention revolution: Unlocking the power of the focused mind.* Wisdom Publications.

Wallace, B. A. (2007). *Contemplative science: Where Buddhism and neuroscience converge*. Columbia University Press.

Wallace, B. A., & Shapiro, S. L. (2006). Mental balance and well-being: Building bridges between Buddhism and Western psychology. *American Psychologist*, *61*(7), 690–701. https://doi.org/10.1037/0003-066x.61.7.690

Walls, A. R., Okumus, F., & Wang, Y. (2011). Cognition and affect interplay: A framework for the tourist vacation decision-making process. *Journal of Travel & Tourism Marketing*, *28*(5), 567–582.

Walsh, G., Shiu, E., Hassan, L. M., Michaelidou, N., & Beatty, S. E. (2011). Emotions, store-environmental cues, store-choice criteria, and marketing outcomes. *Journal of Business Research*, *64*(7), 737–744.

Walter, W. G. (1968). The contingent negative variation: An electro-cortical sign of sensori-motor reflex association in man. *Progress in Brain Research*, *22*, 364–377.

Walters, G., & Sparks, B. (2012). The impact of consumption vision and emotion on the tourism consumer's decision behavior. *Journal of Hospitality & Tourism Research*, *36*(3), 366–389. https://doi.org/10.1177/1096348010390815

Walters, G., Sparks, B., & Herington, C. (2007). The effectiveness of print advertising stimuli in evoking elaborate consumption visions for potential travelers. *Journal of Travel Research*, *46*(1), 24–34.

Walters, G., Sparks, B., & Herington, C. (2012). The impact of consumption vision and emotion on the tourism consumer's decision behavior. *Journal of Hospitality & Tourism Research*, *36*(3), 366–389.

Wan, C., Cai, P., Wang, M., Qian, Y., Huang, W., & Chen, X. (2020). Artificial sensory memory. *Advanced Materials*, *32*(15), 1902434.

Wan, L. C., & Wyer, R. S., Jr (2015). Consumer reactions to attractive service providers: Approach or avoid? *Journal of Consumer Research*, *42*(4), 578–595.

Wang, K. Y., Chih, W. H., Hsu, L. C., & Lin, W. C. (2020). Investigating apology, perceived firm remorse and consumers' coping behaviors in the digital media service recovery context. *Journal of Service Management*, *31*(3), 421–439.

Wang, L., & Lyu, J. (2019). Inspiring awe through tourism and its consequence. *Annals of Tourism Research*, *77*, 106–116. https://doi.org/10.1016/j.annals.2019.05.005

Wang, Y., & Sparks, B. A. (2016). An eye-tracking study of tourism photo stimuli: Image characteristics and ethnicity. *Journal of Travel Research*, *55*(5), 588–602.

Wang, T. C., Tsai, C. L., & Tang, T. W. (2019). Restorative quality in tourist hotel marketing pictures: Natural and built characteristics. *Current Issues in Tourism*, *22*(14), 1679–1685.

Wang, Y.-C., Chen, P.-J., Shi, H., & Shi, W. (2021). Travel for mindfulness through Zen retreat experience: A case study at Donghua Zen temple. *Tourism Management*, *83*, 104211. https://doi.org/10.1016/j.tourman.2020.104211

Wästlund, E., Otterbring, T., Gustafsson, A., & Shams, P. (2015). Heuristics and resource depletion: Eye-tracking customers' in situ gaze behavior in the field. *Journal of Business Research*, *68*(1), 95–101.

Watson, D., Clark, L. A., & Tellegen, A. (1988). Development and validation of brief measures of positive and negative affect: The PANAS scales. *Journal of Personality and Social Psychology*, *54*(6), 1063–1070.

Watson, D., & Tellegen, A. (1985). Toward a consensual structure of mood. *Psychological Bulletin*, *98*(2), 219–235.

Watson, L., & Spence, M., T. (2007). Causes and consequences of emotions on consumer behaviour: A review and integrative cognitive appraisal theory. *European Journal of Marketing*, *41*(5/6), 487–511.

Watson, P., van Wingen, G., & de Wit, S. (2018). Conflicted between goal-directed and habitual control, an fMRI Investigation. *eNeuro*, *5*(4), 1–11. https://doi.org/10.1523/ENEURO.0240-18.2018

Wattanacharoensil, W., & La-ornual, D. (2019). A systematic review of cognitive biases in tourist decisions. *Tourism Management*, *75*, 353–369.

Weaver, D. B. (2005). Comprehensive and minimalist dimensions of ecotourism. *Annals of Tourism Research*, *32*(2), 439–455.

Wedel, M., & Pieters, R. (2017). A review of eye-tracking research in marketing. In *Review of marketing research*. Routledge.

Wei, C., Zhao, W., Zhang, C., & Huang, K. (2019). Psychological factors affecting memorable tourism experiences. *Asia Pacific Journal of Tourism Research*, *24*(7), 619–632.

Wei, W., Qi, R., & Zhang, L. (2019). Effects of virtual reality on theme park visitors' experience and behaviors: A presence perspective. *Tourism Management*, *71*, 282–293.

Weick, K. E., & Putnam, T. (2006). Organizing for mindfulness: Eastern wisdom and Western knowledge. *Journal of Management Inquiry*, *15*(3), 275–287. https://doi.org/10.1177/1056492606291202

Weick, K. E., & Sutcliffe, K. M. (2006). Mindfulness and the quality of organizational attention. *Organization Science*, *17*(4), 514–524. https://doi.org/10.1287/orsc.1060.0196

Weierich, M. R., Wright, C. I., Negreira, A., Dickerson, B. C., & Barrett, L. F. (2010). Novelty as a dimension in the affective brain. *NeuroImage*, *49*(3), 2871–2878.

Weiler, B., Moyle, B., & McLennan, C. (2012). Disciplines that influence tourism doctoral research: The United States, Canada, Australia, and New Zealand. *Annals of Tourism Research*, *39*(3), 1425–1445.

Weiler, B., Moyle, B., Wolf, I., de Bie, K., & Torland, M. (2016). Assessing the efficacy of communication interventions for shifting public perceptions of park benefits. *Journal of Travel Research*, *56*(4), 468–481.

Weiler, B., Torland, M., Moyle, B. D., & Hadinejad, A. (2018). Psychology-informed doctoral research in tourism. *Tourism Recreation Research*, *43*(3), 277–288.

Weiner, B. (1985). An attributional theory of achievement motivation and emotion. *Psychological Review*, *92*(4), 548–573.

Weinmann, M., Schneider, C., & Brocke, J. V. (2016). Digital nudging. *Business & Information Systems Engineering*, *58*(6), 433–436.

Weinstein, N., Przybylski, A. K., & Ryan, R. M. (2009). Can nature make us more caring? Effects of immersion in nature on intrinsic aspirations and generosity. *Personality and Social Psychology Bulletin*, *35*(10), 1315–1329. https://doi.org/10.1177/0146167209341649

Wen, J., Zheng, S., Huang, D., Mai, X., & Liu, X. (2022). Co-creation experience in travel photography service: Exploring the formative mechanism of repurchase intention from the perspective of script theory. *Current Issues in Tourism*, 1–19.

Werner, K., & Gross, J. J. (2010). *Emotion regulation and psychopathology: A conceptual framework* (pp. 13–37). The Guilford Press.

White, C. J., & Thompson, M. (2009). Self determination theory and the wine club attribute formation process. *Annals of Tourism Research*, *36*(4), 561–586. https://doi.org/10.1016/j.annals.2009.04.001

White, J. D., & Carlston, D. E. (1983). Consequences of schemata for attention, impressions, and recall in complex social interactions. *Journal of Personality and Social Psychology*, *45*(3), 538–549.

Widha, L., Rahmat, H. K., & Basri, A. S. H. (2021). A review of mindfulness therapy to improve psychological well-being during the COVID-19 pandemic. *The Proceeding International Conference on Science and Engineering*, *4*(February), 383–386.

Willems, R. M., Nastase, S. A., & Milivojevic, B. (2020). Narratives for neuroscience. *Trends in Neurosciences*, *43*(5), 271–273.

Williams, D. R., & Vaske, J. J. (2003). The measurement of place attachment: Validity and generalizability of a psychometric approach. *Forest Science*, *49*(6), 830–840.

Williams, L. E., Huang, J. Y., & Bargh, J. A. (2009). The scaffolded mind: Higher mental processes are grounded in early experience of the physical world. *European Journal of Social Psychology*, *39*(7), 1257–1267.

Williams, R. N. (1987). The human and the cognitive models: Criticism and reply. *The Journal of Mind and Behavior*, *8*(2), 317–323.

Williams, A. (2006). Tourism and hospitality marketing: Fantasy, feeling and fun. *International Journal of Contemporary Hospitality Management*, *18*(6), 482–495.

Williams, L. E., & Poehlman, T. A. (2017). Conceptualizing consciousness in consumer research. *Journal of Consumer Research*, *44*(2), 231–251. https://doi.org/10.1093/jcr/ucw043

Wilson, E. O. (1984). *Biophilia*. Harvard University Press.

Wilson, K., & Desha, C. (2016). Engaging in design activism and communicating cultural significance through contemporary heritage storytelling. *Journal of Cultural Heritage Management and Sustainable Development*, *6*(3), 271–286. https://doi.org/10.1108/jchmsd-10-2015-0039

Wilson, T., & Gilbert, D. (2005). Affective forecasting: Knowing what to want. *Current Directions in Psychological Science*, *14*(3), 131–134.

Wilson, T. D., Wheatley, T., Meyers, J. M., Gilbert, D. T., & Axsom, D. (2000). Focalism: A source of durability bias in affective forecasting. *Journal of Personality and Social Psychology*, *78*(5), 821–836.

Winkielman, P., & Berridge, K. C. (2004). Unconscious emotion. *Current Directions in Psychological Science*, *13*(3), 120–123.

Winstanley, J. (2006). *Key concepts in psychology*. Palgrave Macmillian.

Wirtz, D., Kruger, J., Scollon, C., & Diener, E. (2003). What to do on spring break? The role of predicted, on-line, and remembered experience in future choice. *Psychological Science*, *14*(5), 520–524.

Wirtz, J., & Bateson, J. E. G. (1999). Consumer satisfaction with services: Integrating the environment perspective in services marketing into the traditional disconfirmation paradigm. *Journal of Business Research*, *44*(1), 55–66.

Wirtz, J., Mattila, A. S., & Tan, R. (2000). The moderating role of target-arousal on the impact of affect on satisfaction—An examination in the context of service experiences. *Journal of Retailing*, *76*(3), 347–365.

Wise, R. A. (2002). Brain reward circuitry: Insights from unsensed incentives. *Neuron*, *36*(2), 229–240.

Witt, C. A., & Wright, P. L. (1992). Tourist motivation: Life after Maslow. In P. Johnson & B. Thomas (Eds.), *Tourist motivation: Life after Maslow* (pp. 33–35). Mansell Publishing.

Wolf, I. D., Stricker, H. K., & Hagenloh, G. (2015). Outcome-focused national park experience management: Transforming participants, promoting social well-being, and fostering place attachment. *Journal of Sustainable Tourism*, *23*(3), 358–381. https://doi.org/10.1080/09669582.2014.959968

Wong, I. A., Wan, Y. K. P., Huang, G. I., & Qi, S. (2021). Green event directed pro-environmental behavior: An application of goal systems theory. *Journal of Sustainable Tourism*, *29*(11–12), 1948–1969. https://doi.org/10.1080/09669582.2020.1770770

Wong, I. A., & Wu, J. S. (2013). Understanding casino experiential attributes: An application to market positioning. *International Journal of Hospitality Management*, *35*, 214–224.

Woo, E., Kim, H., & Uysal, M. (2015). Life satisfaction and support for tourism development. *Annals of Tourism Research*, *50*, 84–97.

Woodside, A. G. (1990). Measuring advertising effectiveness in destination marketing strategies. *Journal of Travel Research*, *29*(2), 3–8. https://doi.org/10.1177/004728759002900201

Woodside, A. G. (2010). Brand-consumer storytelling theory and research: Introduction to a psychology & marketing special issue. *Psychology and Marketing*, *27*(6), 531–540. https://doi.org/10.1002/mar.20342

Woodside, A. G., & Reid, D. M. (1974). Tourism profiles versus audience profiles: Are upscale magazines really upscale? *Journal of Travel Research*, *12*(4), 17–23. https://doi.org/10.1177/004728757401200404

Woodside, A. G., Sood, S., & Miller, K. E. (2008). When consumers and brands talk: Storytelling theory and research in psychology and marketing. *Psychology and Marketing*, *25*(2), 97–145. https://doi.org/10.1002/mar.20203

Wörfel, P. (2021). Unravelling the intellectual discourse of implicit consumer cognition: A bibliometric review. *Journal of Retailing and Consumer Services*, *61*, 101960.

Worthington, E. L., & Scherer, M. (2004). Forgiveness is an emotion-focused coping strategy that can reduce health risks and promote health resilience: Theory, review, and hypotheses. *Psychology and Health*, *19*(3), 385–405.

Wu, Z., Lai, I. K. W., & Tang, H. (2021). Evaluating the sustainability issues in tourism development: An adverse-impact and serious-level analysis. *Sage Open*, *11*(4), 21582440211050384. https://doi.org/10.1177/21582440211050384

Wyer, R. S. (1974). Some implications of the "Socratic effect" for alternative models of cognitive consistency. *Journal of Personality*, *42*(3), 399–419.

Xie, Y., & Sun, J. (2018). How does embodiment work in dark tourism "field"? Based on visitors' experience in memorial hall of the victims in Nanjing Massacre. *International Journal of Tourism Cities*, *4*(1), 110–122. https://doi.org/10.1108/ijtc-12-2016-0055

Xie, Z., Zhang, M., & Ma, Z. (2022). The impact of mental simulation on subsequent tourist experience–Dual evidence from eye tracking and self-reported measurement. *Current Issues in Tourism*, 1–16. https://doi.org/10.1080/13683500.2022.2106194

Xu, X., Huang, D., & Chen, Q. (2021). Stress and coping among micro-entrepreneurs of peer-to-peer accommodation. *International Journal of Hospitality Management*, *97*, 103009.

Xu, X., Zhang, J., & Xie, P. (2014). Eye-tracking technology and effectiveness of tourism advertising. *Business and Economic Review*, *15*(6), 27–32.

Yabe, M., Oshima, S., Eifuku, S., Taira, M., Kobayashi, K., Yabe, H., & Niwa, S. I. (2018). Effects of storytelling on the childhood brain: Near-infrared spectroscopic comparison with the effects of picture-book reading. *Fukushima Journal of Medical Science*, *64*(3), 125–132. https://doi.org/10.5387/fms.2018-11

Yaden, D. B., Kaufman, S. B., Hyde, E., Chirico, A., Gaggioli, A., Zhang, J. W., & Keltner, D. (2019). The development of the Awe Experience Scale (AWE-S): A multifactorial measure for a complex emotion. *The Journal of Positive Psychology*, *14*(4), 474–488. https://doi.org/10.1080/17439760.2018.1484940

Yan, G., Xiong, J., Zang, C., Yu, L., Cui, L., & Bai, X. (2013). Review of eye-movement measures in reading research. *Advances in Psychological Science*, *21*(4), 589–605.

Yang, S. (2012). Eye movements on restaurant menus: A revisitation on gaze motion and consumer scanpaths. *International Journal of Hospitality Management*, *31*(3), 1021–1029. https://doi.org/10.1016/j.ijhm.2011.12.008

Yang, Z., Izuma, K., & Cai, H. (2023). Nostalgia in the brain. *Current Opinion in Psychology*, *49*, 101523.

Yantis, S., & Johnston, J. C. (1990). On the locus of visual selection: Evidence from focused attention tasks. *Journal of Experimental Psychology: Human Perception and Performance*, *16*(1), 135–149.

Ye, H., & Tussyadiah, I. (2011). Destination visual image and expectation of experiences. *Journal of Travel & Tourism Marketing*, *28*(2), 129–144. https://doi.org/10.1080/10548408.2011.545743

Yi, S., & Baumgartner, H. (2004). Coping with negative emotions in purchase-related situations. *Journal of Consumer Psychology*, *14*(3), 303–317.

Yi, Y., & Gong, T. (2013). Customer value co-creation behavior: Scale development and validation. *Journal of Business Research*, *66*(9), 1279–1284.

Yim, M. Y. C., Baek, T. H., & Sauer, P. L. (2018). I see myself in service and product consumptions: Measuring self-transformative consumption vision (SCV) evoked by static and rich media. *Journal of Interactive Marketing*, *44*(1), 122–139.

Yim, M. Y. C., Kim, Y. K., & Lee, J. (2021). How to easily facilitate consumers' mental simulation through advertising: The effectiveness of self-referencing image dynamics on purchase intention. *International Journal of Advertising*, *40*(5), 810–834.

Yim, M. Y.-C., & Yoo, C. Y. (2020). Are digital menus really better than traditional menus? The mediating role of consumption visions and menu enjoyment. *Journal of Interactive Marketing*, *50*, 65–80.

Yin, C.-Y., Poon, P., & Su, J.-L. (2017). Yesterday once more? Autobiographical memory evocation effects on tourists' post-travel purchase intentions toward destination products. *Tourism Management*, *61*, 263–274. https://doi.org/10.1016/j.tourman.2017.02.014

Ying, T., Tan, X., Wei, W., Zheng, Y., Ye, S., & Wu, M. (2021). "I have to watch my back": Exploring Chinese hotel guests' generalized distrust and coping behavior. *Tourism Management*, *86*, 104355.

Yoo, J., & Kim, M. (2014). The effects of online product presentation on consumer responses: A mental imagery perspective. *Journal of Business Research*, *67*(11), 2464–2472.

Yoon, Y., & Uysal, M. (2005). An examination of the effects of motivation and satisfaction on destination loyalty: A structural model. *Tourism Management*, *26*(1), 45–56. https://doi.org/10.1016/j.tourman.2003.08.016

Young, J. E., Klosko, J. S., & Weishaar, M. E. (2003). *Schema therapy*. Guilford Press.

Yu, H., Lee, L., Popa, I., & Madera, J. M. (2021). Should I leave this industry? The role of stress and negative emotions in response to an industry negative work event. *International Journal of Hospitality Management*, *94*, 102843.

Yuan, X., Pei, W., Wu, W., Lin, X., & Du, Y. (2019). Tourism consumer's demand on online hotel decision-making—Based on eye-tracking study. *Resource Development & Market*, *35*(11), 1438–1443.

Yung, R., & Khoo-Lattimore, C. (2019). New realities: A systematic literature review on virtual reality and augmented reality in tourism research. *Current Issues in Tourism*, *22*(17), 2056–2081.

Yung, R., Khoo-Lattimore, C., & Potter, L. E. (2021). VR the world: Experimenting with emotion and presence for tourism marketing. *Journal of Hospitality and Tourism Management*, *46*, 160–171.

Zacks, J. M., Speer, N. K., Swallow, K. M., Braver, T. S., & Reynolds, J. R. (2007). Event perception: A mind-brain perspective. *Psychological Bulletin*, *133*(2), 273–293. https://doi.org/10.1037/0033-2909.133.2.273

Zaichkowsky, J. L. (2012). Consumer involvement: Review, update, and links to decision neuroscience. In V. Wells & G. R. Foxall (Eds.), *Handbook of developments in consumer behaviour* (pp. 523–546). Edward Elgar Publishing Ltd.

Zak, P. J. (2013, December 17). *How stories change the brain. Mind & body*. https://greatergood.berkeley.edu/article/item/how_stories_change_brain

Zak, P. J. (2014). Why your brain loves good storytelling. *Harvard Business Review*, *28*, 1–5. https://hbr.org/2014/10/why-your-brain-loves-good-storytelling

Zak, P. J. (2015, January). Why inspiring stories make us react: The neuroscience of narrative. In *Cerebrum: The Dana forum on brain science* (Vol. 2015). Dana Foundation.

Zhang, R. (2018). *Changing beach destination image schemas through storytelling: A study of Hainan Island, China*. PhD, Griffith University.

Zhang, W., & Fesenmaier, D. R. (2018). Assessing emotions in online stories: Comparing self-report and text-based approaches. *Information Technology & Tourism*, *20*(1), 83–95.

Zhang, T. C., Jahromi, M. F., & Kizildag, M. (2018). Value co-creation in a sharing economy: The end of price wars? *International Journal of Hospitality Management*, *71*, 51–58.

Zhang, M., Liversedge, S. P., Bai, X., Yan, G., & Zang, C. (2019). The influence of foveal lexical processing load on parafoveal preview and saccadic targeting during Chinese reading. *Journal of Experimental Psychology: Human Perception and Performance*, *45*(6), 812–825.

Zhang, K., Qiu, H., Wang, J., Li, C., Zhang, J., & Chen, D. D. (2022). Tourist gaze through computer vision: Where, what, how and why? *Tourism Review*, *77*(4), 1186–1202.

Zhang, R., Scott, N., & Gardiner, S. (2018). The role of image schemas in developing the beach destination image of Hainan Island. In *Paper presented at the CAUTHE 2018: Get smart: Paradoxes and possibilities in tourism, hospitality and events education and research Newcastle, NSW*.

Zhang, P., & Soergel, D. (2016). Process patterns and conceptual changes in knowledge representations during information seeking and sensemaking: A qualitative user study. *Journal of Information Science*, *42*(1), 59–78.

Zhang, X., Chen, Z., & Jin, H. (2021). The effect of tourists' autobiographical memory on revisit intention: Does nostalgia promote revisiting? *Asia Pacific Journal of Tourism Research*, *26*(2), 147–166.

Zhao, H., Zhang, H., Xu, Y., He, W., & Lu, J. (2019). Why are people high in dispositional awe happier? The roles of meaning in life and materialism. *Frontiers in Psychology*, *10*, 1208. https://doi.org/10.3389/fpsyg.2019.01208

Zhao, M., Hoeffler, S., & Dahl, D. (2009). The role of imagination-focused visualization on new product evaluation. *Journal of Marketing Research*, *46*(1), 46–55.

Zheng, D., Luo, Q., & Ritchie, B. W. (2021). Afraid to travel after COVID-19? Self-protection, coping and resilience against pandemic 'travel fear'. *Tourism Management*, *83*. https://doi.org/10.1016/j.tourman.2020.104261

Zheng, C., Chen, Z., Zhang, Y., & Guo, Y. (2022). Does vivid imagination deter visitation? The role of mental imagery processing in virtual tourism on tourists' behavior. *Journal of Travel Research*, *61*(7), 1528–1541.

Zheng, D., Ritchie, B. W., & Benckendorff, P. (2020). Tourism performing arts development in China. In *Handbook on tourism and China*. Edward Elgar Publishing.

Zheng, D., Ritchie, B. W., Benckendorff, P. J., & Bao, J. (2019). Emotional responses toward tourism performing arts development: A comparison of urban and rural residents in China. *Tourism Management*, *70*, 238–249.

Zhong, Y. Y. S., Busser, J., & Baloglu, S. (2017). A model of memorable tourism experience: The effects on satisfaction, affective commitment, and storytelling. *Tourism Analysis*, *22*(2), 201–217. https://doi.org/10.3727/108354217x14888192562366

Zhou, L., Zhang, Y. Y., Wang, Z. J., Rao, L. L., Wang, W., Li, S., & Liang, Z. Y. (2016). A Scanpath analysis of the risky decision-making process. *Journal of Behavioral Decision Making*, *29*(2–3), 169–182.

Zhou, X., Tang, C., Lv, X., & Xing, B. (2020). Visitor engagement, relationship quality, and environmentally responsible behavior. *International Journal of Environmental Research and Public Health*, *17*(4), 1151. https://doi.org/10.3390/ijerph17041151

Ziegler, D. J. (2001). The possible place of cognitive appraisal in the ABC model underlying rational emotive behavior therapy. *Journal of Rational-Emotive and Cognitive-Behavior Therapy*, *19*(3), 137–152.

Zouni, G., & Kouremenos, A. (2008). Do tourism providers know their visitors? An investigation of tourism experience at a destination. *Tourism and Hospitality Research*, *8*(4), 282–297.

Zuckerman, M. (1979). *Sensation seeking: Beyond the optimum level of arousal*. Lawrence Erlbaum.

Zuckerman, M. (2014). *Sensation seeking (psychology revivals): Beyond the optimal level of Arousal* (1st ed.). Psychology Press.

INDEX

Printed in the USA
CPSIA information can be obtained
at www.ICGtesting.com
JSHW062226030624
64255JS00002B/5

9 781802 625806